The Perfection of Nature

The Perfection of Nature

ANIMALS, BREEDING, AND RACE IN THE RENAISSANCE

Mackenzie Cooley

The University of Chicago Press   CHICAGO AND LONDON

The University of Chicago Press, Chicago 60637
The University of Chicago Press, Ltd., London
© 2022 by The University of Chicago
Published 2022
Printed in the United States of America

31 30 29 28 27 26 25 24 23 22   1 2 3 4 5

ISBN-13: 978-0-226-82226-6 (cloth)
ISBN-13: 978-0-226-82228-0 (paper)
ISBN-13: 978-0-226-82227-3 (e-book)
DOI: https://doi.org/10.7208/chicago/9780226822273.001.0001

Library of Congress Cataloging-in-Publication Data

Names: Cooley, Mackenzie, author.
Title: The perfection of nature : animals, breeding, and race in the
    Renaissance / Mackenzie Cooley.
Description: Chicago : University of Chicago Press, 2022. | Includes
    bibliographical references and index.
Identifiers: LCCN 2022006312 | ISBN 9780226822266 (cloth) |
    ISBN 9780226822280 (paperback) | ISBN 9780226822273 (ebook)
Subjects: LCSH: Animal breeding—History. | Eugenics—History. |
    Renaissance—Europe.
Classification: LCC SJ105 .C67 2022 | DDC 636.08/2—dc23/eng/20220223
LC record available at https://lccn.loc.gov/2022006312

♾ This paper meets the requirements of ANSI/NISO Z39.48-1992
(Permanence of Paper).

For B. A. B.,
K. A. C.,
A. O. C.,
A. K. B.

SOCRATES: Tell me this, Glaucon: I see that you have hunting dogs and quite a flock of noble fighting birds at home. Have you noticed anything about their mating and breeding?

GLAUCON: Like what?

SOCRATES: In the first place, although they're all noble, aren't there some that are the best and prove themselves to be so?

GLAUCON: There are.

SOCRATES: Do you breed them all alike, or do you try to breed from the best as much as possible?

GLAUCON: I try to breed from the best . . .

SOCRATES: And do you think that if they weren't bred in this way, your stock of birds would get much worse?

GLAUCON: I do.

SOCRATES: What about horses and other animals? Are things any different with them?

GLAUCON: It would be strange if they were.

SOCRATES: Dear me! If this also holds true of human beings, our need for excellent rulers is indeed extreme. . . . It follows from our previous agreements, first, that the best men must have sex with the best women as frequently as possible, while the opposite is true of the most inferior men and women, and, second, that if our herd is to be of the highest possible quality, the former's offspring must be reared but not the latter's.

Plato, *Republic*, Book V, 459–460

Contents

Illustrations

A Note on Terms and Orthography

This book is a work of history that attempts to be attentive to the categories that historical actors developed to make sense of their surroundings. Different fields have various orthographic traditions and I have sought to offer a logical middle ground when bringing them together. An argument of the book is that terms relating to race and selective breeding have changed their principal valence, sometimes repeatedly, over the course of the early modern, modern, and postmodern periods. For this reason, some key terms, such as *razza* (plural *razze*), have been left in the original language. What was meant by *razza* in Italian in the sixteenth century is not the same sense of fixed, often human, difference that comes with the modern term and its English translation. To preserve the subtle differences of these meanings, I have generally tried to retain the original language of the texts interpreted. Therefore, the reader will find *razza* in Italian, *raza* in Spanish, *race* in French, *raça* in Portuguese, and *race* in English mentioned throughout the book. I understand these terms to have overlapping meanings and highlight these in the text itself. As is often the case with late medieval and early modern sources, the spelling of these terms varies in documents from the period. To maintain a focus on actors' categories, I have often left the orthography unstandardized such that, for example, if an Italian document spells *razza* with one *z*, as was sometimes the case, I have preserved that spelling despite the risk of confusing these terms with Castilian variants. Likewise, sometimes camelids are vicuñas and other times they are vigunas. This means that some spellings differ from those in the modern languages. By contrast, I have generally standardized the classical Nahuatl according to modern orthographic conventions unless the section explicitly focuses on Europeans' changes to and misinterpretation of those terms. This approach seeks to increase readers' facility in making comparisons between this analysis and others engaged in modern Nahuatl.

Finally, a comment on translation, as history is nothing if not an act of translation between past and present. When possible, I have compared original sources with critical editions of those texts. When, as is often the case with archival and rare materials, such sources are not available, I have translated the texts myself. Unless otherwise noted, the translations are my own. So, too, are any errors that doubtless remain.

The Perfection of Nature

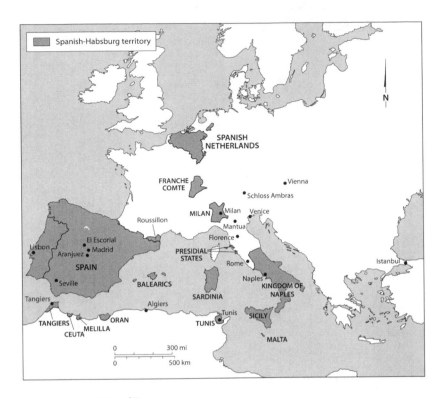

FIGURE 0.1. Map of Europe, 1550.

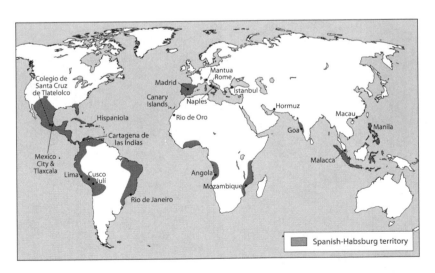

FIGURE 0.2. Map of Europe and the American territories under Spanish control, 1600.

Introduction

Like many before him and many after, Tommaso Campanella (1568–1639) imagined a utopia. The Dominican friar envisioned a more perfect world at the turn of the seventeenth century, just as Spanish imperial power had transformed his homeland into a wayward province. In his thirtieth year, Campanella had returned home to Calabria, in Southern Italy, after a brush with the Inquisition. Increasingly convinced that astrological signs and prophetic texts foretold great upheaval, he was denounced for fomenting a rebellion in order to transform Calabria into a republic—the only chance, he believed, to save it from the tyrannical rule of the Spanish crown. Two of his fellow conspirators cracked and revealed his plot to the Spanish authorities. Campanella was arrested. Despite the pressures of interrogation under torture sanctioned by Pope Clement VIII, he refused to accept the accusation of rebellion, continually insisting that he had simply been following the prophesies from ancient texts and an unusual number of eclipses. The year 1600, he said, foretold great, turbulent changes. He was not rebelling, but simply acting upon the signs of nature. The authorities thought this insane, which worked out well for Campanella, as it made him unable to repent and thus not an appropriate victim for the death penalty.[1]

Following this conflict with the Church and viceregal authorities, Campanella penned *La città del sole* (The City of the Sun). The book transformed parts of Plato's *Republic* into an imagined new world located somewhere near equatorial Taprobane in the Indian Ocean, an island that had long floated on the margins of European knowledge and wonder.[2] There lived the Solarians, whose society enacted key Renaissance ideals. With them, whether he meant their predilections to be read sincerely or in jest, Campanella's present-day reality bled inconspicuously into early modern science fiction. Just like elite Europeans, the Solarians invested in their animals; more to the point, "highly esteemed among them is the art of breeding horses, bulls, sheep, dogs, and every sort of domestic animals, just as it was in the time of Abraham." The Solarians would have found similar breeding practices in Renaissance European stables, such as those throughout Campanella's native Southern Italy. In both places, experts monitored animal breeding. Stallions and mares were not "set loose in the meadows" but instead brought "together outside their farm stables at the opportune time." These breeders even orchestrated animals' pairings to match the constellations. Horses required Sagittarius ascendant, in conjunction with Mars and Jupiter, while Taurus yielded the best oxen, and Aries improved sheep. In Europe, beautiful images were hung upon barn walls during mating season as a means of spurring the imaginations of the livestock, which was thought to produce ever lovelier offspring. Likewise, the Solarians employed "magic to induce these creatures to breed in the presence of paintings of horses, bulls, or sheep."[3]

However, while Europeans and Solarians shared practices for systematically developing animals, they diverged in their treatment of humans. Campanella's central narrator, "the Genoese Sea-Captain," had encountered Solarian society and returned to Italy to explain its wonders. In one of many such explanatory passages, the captain described how Solarians looked down on what they believed to be a contradiction among Europeans: "Indeed, they laugh at us who exhibit a studious care for our breeding of horses and dogs [ch'attendemo alla razza delli cani e cavalli], but neglect the breeding of human beings." By contrast, the Solarians had created a superior society because they were willing to apply the rational principles of good animal husbandry to crafting generations of humans.[4] They understood breeding to be broadly defined—encompassing education and reproduction alike, and pertaining to the fruits of human wombs just as to the seeds of the earth. The Solarians had mastered a Renaissance version of eugenics—a feat that many Europeans would have envied, others loathed, and still others doubted.

Campanella's vision of controlling the breeding of men and women so that "they bring forth the best offspring," just like other domestic animals, was built around the language of *razza*, a term that, as Dániel Margócsy has put it, "meant race, breed, and stud all at once."[5] In Campanella's time, breeders across the European countryside and Europe's American colonies applied terminology with its origins in the stable to describe the populations of livestock and other domesticated animals that they had bred.[6] To Campanella, his Solarians, and other Renaissance thinkers, the word *razza* was associated with a specific population that could share qualities, and was often employed in breeding projects aimed at creating the "perfect" animal.[7] A *razza* did not have inflexibly fixed characteristics, though;

these were instead evanescent and easily lost, and their persistence resulted from reproductive work. Most of all—although, unlike widespread
animal breeding, it was rarely realized—Campanella's fiction encouraged
readers to use this attention to *razza* to reshape human populations. Just
as animal breeding required careful staging and rational decision-making,
Campanella's narrative suggested that similarly, human breeding could—
and should—be carefully controlled.

Beyond that, Campanella's vision of improved nature reflected widespread beliefs and more than a century of real investment in animal breeding projects. *Razze* of horses joined collections of books and exotica from
around the globe, displayed in their stables like objects in *kunstkammer*
in order to evoke wonder and princely power.[8] Numerous princely families, from the Spanish Habsburgs to the Gonzaga of Mantua, created their
own "races" of horses, dogs, and other domesticated animals, their experts' efforts recorded in a mass of bureaucratic texts.[9] A paper trail bal-
looned around such projects, complete with the brands stamped into the
animals' flesh, their diets, lifespans, coloring, and other details. Through
the ceaseless writing down, labeling, and categorizing of animal life, the
language of *razza* was cemented and increasingly but unsystematically tied
to specific traits. Each time these categories were written down and used
in a sale, or read at the palace, or referred to in court proceedings, these
documents helped to consolidate the idea of *razza* as a nameable, visible,
and legible reality. Race, that complex concept, emerged through efforts
both conscious and unconscious, but animal records represent one of the
many spokes of the wheel. For Campanella, humans were the telos of
breeding's powers, animals the epiphenomenon. This book partially in-
verts that anthropocentric emphasis.

Campanella wrote almost a century after the cascade of American
encounters—from Christopher Columbus's voyages (1492–1504) to
Hernán Cortés's seizure of Mexico-Tenochtitlan (1519–21) and Francisco
Pizarro's thuggish domination of Cusco (1533)—when the glow of discovery had started to fade, leaving in its wake questions about the feasibility
of long-term domination and permanent conversion. As Campanella sat
in prison, King Philip III's Spain was zealously guarding a vast swath of
the known world. The domains of his father, Philip II (1527–1598), had
encompassed the Iberian Peninsula, stretched across the Italian peninsula, extended up to the rebellious Low Countries, spread over large sections of North and South America, and extended to a smattering of islands across the Atlantic and Pacific and to strongholds up and down the
coasts of Africa and Asia.

As a Habsburg subject in the non-Spanish possessions, Campanella wrote extensively about the Spanish monarchy, especially the Habsburg family, and the implications their imperial actions had for their dynastic fate.[10] Freed from his Neapolitan prison and writing from Paris at the outset of a long war between France and Spain (1635–59), he prophesized that Spain's failure to change its tactics to improve integration meant that it would lose its power to the French, who would reunify Christians.[11] His reflections on the monarchy took the Catholic mandate of universal conversion seriously, but he increasingly believed that the Spaniards were squandering their position as a superpower through their pride.[12] Consistent with the ideas he articulated in *City of the Sun,* marriage and the problem of population emerged as central to Campanella's critique of the Spanish monarchy, as he saw Spain's population as declining, with young men dying in war and women growing infertile. For Campanella, demography was a yardstick of power, and what is demography but the creation of a human population through choices made by generation upon generation? Although he overestimated its demographic collapse, Campanella had a point about Spain's demographic travails and the role of war and empire. The Habsburgs, however, did not follow Campanella, either in terms of imperial population or in applying the lessons of animal husbandry to elite marriages, deferring instead to the pressures of honor, family loyalty, and dynastic strategy.[13]

Renaissance Worlds in Counterpoint

Each chapter in this book juxtaposes humans and nonhuman animals. This structure aims to suggest enduring parallels, offer space for irreducible differences, and foreground the distinctiveness of each case study. I turn from horse *razze* to collections that included humans, from livestock branding to human branding, from the language of *xinachtli* in maize to the inheritance of human nobility, from American dogs to discussions of human *mestizaje,* from camelids to changing approaches to human blood purity, from ancient animal migration to human migration, and from physiognomic techniques deployed by husbandmen to their potential to predict human action. Across these narratives, several thematic lines emerge—animal breeding, animal-human categorization, utopian theorization, physiognomic vision, *limpieza de sangre* (purity of blood) discrimination, Christian nature, and the history of natural history—that, as in musical counterpoint, speak in parallel with their own rhythms but are harmonically interdependent. A dominant theme is early modern breeders' self-conscious struggle to produce and maintain *razza.* Another countermelody is natural philosophers' preoccupation with the origins of animal kinds. A third line is how the goal of bringing the Old World to the New intersected with utopian trends in Renaissance thought. This counterpoint method is meant to preserve space for the multiplicity of approaches to understanding the likenesses and differences of humans and nonhuman animals, and the imperfect overlaps between the history of the discourses of *razza*, human race, and taxonomization, as well as divergences in European and American understandings of animal difference. Some thought breeders to be practical experts in physiognomy, the art of reading internal character from external features; others emphasized environmental and dietary requirements rather than heredity in the making of *razze*.[14] Selection for control over animals did not equate to the search for purity. Animal *razza* did not unidirectionally become human race, although it did contribute to the application of that category. Likewise, at both the beginning and the end of the sixteenth century, the roles of nature and nurture remained contested, with animals mobilized on both sides of the debate.[15] This approach is meant to emphasize complexity and multiplicity. To return to the idea of counterpoint, these melodic lines often resonate, but they are not meant to dominate one another.

Although the term *eugenics* was not coined until the nineteenth century, its essential principles were already practiced centuries prior; thus, I use the term to suggest the human-led, biotechnical means of shaping the results of inheritance across generations. This is largely uncontroversial when enacted through traditional methods in plants and animals. Who would not want a larger tomato or a faster horse?[16] These practices have long been standard in breeding animals. The term *eugenics* was proposed by Francis Galton (1822–1911), who defined it as "the science of improving the inherited stock not only by judicious matings, but by all other influences."[17] Competing with his half-cousin Charles Darwin's theory of evolution, in his law of ancestral heredity, Galton suggested that humans ought to be bred according to a "golden book" to produce the best human offspring, just like basset hounds and horses.[18] That vision was soon put into practice by American and European eugenicists who traced family trees with an emphasis on pedigrees that led to intelligence or debauchery. Indeed, it is the perpetuation of Galton's own false sense of ownership over the idea that has preserved this nineteenth-century origin story for eugenics.[19] By turning to the modern term *eugenics* for the early modern period with a focus on the sixteenth century, I aim to redefine a longer history of a practice of which nineteenth- and twentieth-century scientific eugenics was only one small part. This breeding philosophy has been seized upon cyclically across the ages as a means of shaping the future to meet the interests of the present. This project traces the oscillation between interest in what have been termed "eugenic principles" and concerns about their potential to distort human life; in the process, it engages with fundamental questions about nature versus nurture, difference versus shared characteristics, determinism versus free will, and individual autonomy versus society's claims.

Renaissance attempts to perfect nature shared characteristic methods and utopian aspirations with other periods. They featured aesthetic ideals matched to a clear system of hierarchy, sustained cross-generational engagement, and detailed record keeping of traits and genealogies. Despite these likenesses, as a rule, Renaissance patrons were more interested in growing than in weeding. While efforts to perfect society have often been accompanied by the eviction of unwanted types (including the infamous purge of Muslims and Jews in Iberia), sixteenth-century documents imagining and recording animal breeding projects do not dwell on elimination.[20] Rather, Renaissance concerns often focused largely on the porosity of categories, nature's overwhelming fecundity, and the risk of producing aberrations on the fault lines between living things.[21] John Florio's 1611 Italian-English dictionary shows the importance of this breadth in its tripartite definition of the Italian verb *generáre*: "to ingender or beget as the male. Also to conceive, to beare or bring forth as the female. Also to create, to breed, to make, to invent."[22] The last definition particularly stands out. Sex and generation offered one of humanity's ultimate tools of invention, which could be used to manipulate present and future populations of humans and animals. In bringing together traditionally isolated narratives—about the Renaissance, the history of science and the Scientific Revolution, the Columbian Exchange, and the Spanish Empire—this book centers Catholic Southern Europeans for a revised understanding of how early moderns thought through animal difference.[23]

Speaking of animal difference relies on the language of species, breed, genus, and taxon, which in turn are challenging to pin down in both a biological and historical sense. Given that this is a book about early modern categorical ambiguities more than modern ones, whenever possible I have included the early modern sources' language for these terms. I hope that this will avoid unnecessary anachronism, highlight the tensions between different historical actors' employment of these terms, and allow the reader to build a chronology for the deployment of these ideas across different regions and topics in the sixteenth and early seventeenth centuries.

Nonetheless, a preliminary guide to the inter-mapping of these concepts is in order. *Species* (pl. *speciei*) emerged in Latin as a translation of the classical Greek *eidos*, meaning "idea" or "form." *Genos* meant kind, as derived from the word for generation. Today, *species* refers to a unit of classification more specific than the genus. While species can be defined by DNA sequence, morphology, or karyotype, the most common definition of species is that it represents the largest group of organisms within which any two individuals can produce fertile offspring. Species concepts are fluid, heuristic distinctions, for the closer one looks at the boundaries between beings, the more hybrids or crosses between species emerge. Some organisms are reproductively isolated, with no interbreeding, while others are characterized by *panmixia*, or the tendency of all individuals in a population to represent potential partners. Species, in turn, break down into subspecies (often defined by range and morphology), varieties, and forms. Other issues pervade definitions of *genus*. Today, scientists in macroevolution and biological systematics are hotly debating the treatment of taxonomic ranks above the species level. Some suggest that higher-ranked groupings are arbitrary inventions; others contend that species are real and formed through evolution; yet other fields accept Linnean ranked taxa as biologically "real" without further debate.[24] A modern genus can contain a single species, or multiple species; it is the lowest commonly used supraspecific classificatory rank.[25]

Breed is more specific but likewise perplexing. A breed is a group of domesticated animals that look similar (having a homogenous appearance or phenotype) and behave similarly in ways that can be differentiated from other members of their species. As Donna Landry has suggested in respect to modern equine breeds, "What counts as a breed is therefore ideologically loaded: genotype, phenotype, and character or attitude all participate in the definitional protocols."[26] Examples of modern breeds include the Lusitano horse and the purebred dogs featured in the Westminster Kennel Club dog show. In modern science, crosses between breeds are called crossbreeds or mixed breeds, while crosses at higher taxonomic levels, such as species, are called hybrids. Despite these designations, there is no single accepted definition of the term *breed*, which is more a result of art than science.[27]

Thinkers and practitioners cited throughout this book balanced between two broad approaches to kinds and the differences that separated them, described in shorthand as "fixing" and "mixing." The conceptual history of species is defined by these two competing lines of thought. One was committed to essentialism and fixism: in this view, "deriving from Plato and Aristotle," as John Wilkins has put it, "all members of a type were defined by their possession of a set of necessary and sufficient properties or traits, which were fixed, and between which there was no transformation." In the logical tradition, species is a class separated out of a larger class of genus: "There is a clear distinction between the formal definitions of logical species and the material characters and powers of the biological organisms of a biological species pretty well from the beginning of modern natural history." Another view centered on population thinking ran concurrently, although Wilkins contends it was "developed in full by Charles Darwin, in which taxa are populations of organisms with variable traits, which are polytypic . . . and which can transform over time from one to another taxon, as the species that comprise them, or the populations that comprise a species, evolve. There are no necessary and sufficient traits."[28] In sixteenth-century documents, early adoption of elements of this second definition are present in thinkers committed to mixing, multiplicity, and the potential for slipping between kinds, whether at the level of *razze* or broader animal types.[29]

In the sixteenth century, as Fernand Braudel put it, the "Mediterranean impressed its own image on the Spanish New World"; reciprocally, more recent scholarship by historians, such as Marcy Norton, has revealed how New World products and ideas became part of European thought and life.[30] Italian and Spanish histories, once separated by nationalist narratives, were likewise deeply intertwined, as Renaissance thought and politics had major consequences for the conquest of the Americas and Spanish governance. Spanish courts emulated Italian models of etiquette and art. Spanish forces fought simultaneously in the New World and Renaissance Italy, providing critical examples for Machiavelli's political analysis and benefiting from deep European traditions of strategic thought in making choices during early colonization. Naples, as the largest city in the Spanish empire and a key political and military center on which Spain's Mediterranean position relied, had, by 1550, become deeply embedded in the networks of circulation and power that linked the Spanish Habsburg monarchy to its possessions throughout the world. As the foremost Catholic power, in control of the Kingdom of Naples and the Duchy of Milan, the Spanish Habsburgs wielded formidable influence in Rome through Catholic religious orders, the College of Cardinals, the Inquisition, and the presence of major military forces nearby. Inhabitants of this Spanish-inflected world often saw themselves as connected, and negotiated a shared knowledge base.[31] This interconnectedness of Europe and the colonial world is essential to understanding the practice of animal husbandry and its implications for the negotiation of human and natural difference.

Spanish and Italian worlds overlapped but were by no means identical, as is particularly clear in the case of the *limpieza de sangre* conceptual and legal system, which developed in mid-fifteenth-century Castile as so-called *convivencia* collapsed, constituting a conservative reaction to

the Spanish crown's and Catholic Church's emphasis on universal conversion. A belief in nurture over nature fell victim to a radical vision of
natural determinism. This embodied vision of human potential itself allowed characteristics of behavior, such as heretical acts, to be passed on
to subsequent generations through a peculiar, unscholarly theory of the
body. María Elena Martínez has traced different outcomes of the *limpieza
de sangre* system in Castile and in the Americas to study how religious
connotations constructed and promoted classification over the sixteenth
and seventeenth centuries.[32] *Limpieza de sangre*, literally meaning purity
of blood, developed alongside growing interest in blood and circulation
by natural philosophers, doctors, and experimentalists.[33] As I will show,
limpieza de sangre discrimination indirectly shaped the reception of camelids and the welcome José de Acosta received among the Jesuits. The Roman Inquisition was normally hostile to determinative claims like those
advanced by physiognomists and astrologers, but *limpieza de sangre* determinism entered global Catholic practice. For all its importance, *limpieza
de sangre*, like any of the lines of counterpoint, did not always feature in
thinking about animal breeding.

This history moves across the Atlantic Ocean and back again to consider approaches to breeding on either side and the deleterious effects
of their collision. In so doing, I establish both Europe and the Americas
as having had their own local ideas about breeding. Adopting the earlier
meaning of *indigenous* as born or produced naturally in a land or region, I
suggest that peoples on either side of the ocean developed their own specific indigenous knowledges of breeding and generation that explained
their husbandry practices. In the Americas, that knowledge tended to emphasize cyclicity. In the European context, by contrast, the idea of *razze*
emerged from attempts to intentionally improve nature through breeding.
Europeans' inclination to keep an abundance of records is therefore both
useful and analytically significant to this story, as it reveals the scale of resources invested in creating and maintaining these animal populations.
By contrast, while there are an abundance of written European sources
before and after 1492, pre-Columbian New World sources are fundamentally different, often not recorded on paper, and sometimes hard to parse.
It remains challenging to access comparable details of how Mesoamericans and Andeans quantified their animal charges, why they preferred specific colors in their domesticates, or whether they thought nature could
be perfected in any real sense. To navigate this gap, I turn toward Camilla
Townsend's methods, using Nahuatl-language annals and natural histories to access Mexican history before the Europeans' arrival, despite the
fact that they were not written down in Latin characters until after Euro-

peans had established control over the region.[34] Through these sources, particularly those about the growing of corn and raising of turkeys, dogs, and camelids, we see hints of animal cultures, and how terms of breeding were employed differently before, during, and after European intervention in the Americas.[35]

Animality and the Slipperiness of Metaphor

A tide of analogies and recognition of their power runs through the heart of this book. As ancient practices, agriculture and husbandry have long provided fodder for metaphors in human health, sexual reproduction, and social relations; as Frances Dolan has pointed out using early modern English sources, plowing and planting, cultivation and barrenness, growth and degeneration all shift between the literal and metaphorical.[36] It was through analogy that Socrates linked the breeding of horses and dogs to the breeding of humans, and likewise through analogy that Campanella elaborated the comparison. William Shakespeare gave voice to the profane equation of Othello, a soldier of Moorish background in the Venetian Republic, to an imported stud when he had Iago say to Brabantio, "You'll have your daughter covered with a Barbary horse; you'll have your nephews neigh to you; you'll have coursers for cousins and gennets for germans."[37] Brabantio counters by calling Iago a "profane wretch"; Iago presses on to tell him that his "daughter and the Moor are now making the beast with two backs."[38] The double inversion is not only that Othello is equated with a horse, but that many of the horse *razze* mentioned had their origins in human ethnic categories. The word *jennets*, for example, was transferred to the horse breed from the name of a Berber tribe, and in turn became the name for a type of light cavalryman.[39] More broadly, how literally was the insult meant, and what is required of a history that aims to include both the figurative slander and the real horses to which it referred?

In the Renaissance, as now, people wrote at multiple levels of literalism. Their world was infused by emblems and religious tradition that demanded allegorical readings. Take, for example, Martín de Villaverde's comparison of turkeys (*pavos de la yndia*, today *Meleagris gallopavo*) with peacocks (*pavones reales*, today *Pavo cristatus*) in his 1570–71 *Bestiario de Don Juan de Austria* (fig. 0.3). In both illustrations, the male, complete with large tail, beckons the female, raising his left leg as a sign of courtship. Villaverde wrote of the peahen and peacock that "their *honestidad* [virtue] is so notable that they have a thousand advantages over other birds, that is to see the serenity of the peacock and the calm of the peahen. It is the example of a married couple. Between them passes none of the effrontery

FIGURE 0.3. Peacocks (*left*) and peacocks of the Indies (*right*). *Bestiario de Don Juan de Austria*, 1570–71, 115v–116r. Reproduced with permission of Monastery of Santa María de la Vid. Image by Juan José García Gil in *Bestiario de Don Juan de Austria* [Texto impreso]: S. XVI: original conservado en la Biblioteca del Monasterio de Santa Maria de la Vid (Burgos), Monasterio de Santa María de la Vid Biblioteca (Burgos: Gil de Siloé, 1998).

of the chicken, the goose, or the sparrow that before our eyes use their dishonesty; that is not what the peacock does."[40] To be dishonest meant one was lascivious, scandalous, a bad example; it was a euphemism for carnal vice.[41] In contrast to the peacocks' restrained behavior, Villaverde described turkeys as jealous and easily upset. The female turkey laid an egg every day and sat on her eggs for eight days more than the chicken. "She is a good breeder," but not self-controlled, as "this turkey gets angry very easily," wrote Villaverde. "As honest" as that Old World couple was, the Americans "were dishonest." Villaverde considered even their singing to be inferior; "gobble gobble" was a "sound like retching with sadness."[42] In this case, tropes of human honor mapped onto animals. A turkey is not simply a turkey in this text, but the emblem of two unsophisticated continents. Throughout this book, I have sought to flag the extent to which passages ought to be read literately and indicate when my own prose has employed an agricultural metaphor, as in the case of grafting in chapter 4.

Breeding and working with animals more generally encouraged thinking about life cycles and generations, and the likening of animal bodies to human ones.[43] Still, the analogy meets its limits in tautology for those who saw human bodies as animal bodies. From a series of specific metaphorical linkages—between animal heredity and human lineage—race eventually emerged as a predominant feature of discourse to describe human populations, born of their chosen unions, in contrast to the carefully fashioned progeny of livestock. Following George Lakoff and Mark Johnson, I understand metaphors to construct human understanding. It is through metaphor that we render abstract concepts comprehensible and maintain abstract reasoning. When William Harvey, for example, argued *ex ovo omnia*, that all animals come from eggs, he had successfully expanded the category of "egg" metaphorically to build a new model for generation across the plant and animal kingdoms. Linguist Gilles Fauconnier and literary critic Mark Turner have argued that human thought progresses through acts of construction of new conceptual blends from already existing ones. These combinations can be intentionally or unconsciously made. To blend ideas requires combining packets of meaning, through selection and constraints, to create a new packet with a new, emergent meaning. These packets or blends need not be consistent, but rather evolve and change through use and citation, leaving behind trickles of evidence for the genealogy of the idea.[44]

Dictionaries provide a window into this process of transference. Renaissance vernacular dictionaries collected meanings, juxtaposing terms with different connotations in their original contexts in one entry, sometimes to purify the language. Their many definitions, as we saw with Florio's *generáre*, at once preserve multiplicity and lump diverse contexts together. For example, in Sebastián de Covarrubias's 1611 dictionary, *raza* refers to a "*casta* of chaste horses, those that are marked with iron such that they are known," indicating that they are branded.[45] But *raza* was also an imperfection in a cloth, as Javier Irigoyen-García has suggested, which Covarrubias then links to an Italian term.[46] The definition also includes a meaning associated with human lineages, in which *raza* "is taken as a bad term, as having some *raza* of Moor or Jew." Overlapping semantic fields also juxtaposed animals and humans in Covarrubias's discussion of *mulato*, which meant at once "the son of a black woman and a white man or the reverse, and for being an out of the ordinary mixture, they compared it to the nature of the mule." This is not to say that someone talking about horse breeding was always thinking about *raza* as linked to what Covarrubias saw as a stain on human lineage, or about *mulato* as akin to the sterile nature of a horse-donkey hybrid. But it is to say that these meanings were semantically parallel and sometimes linked. As readers increasingly turned to dictionaries, utopian treatises, or theories of population that connected the two, they blended such that it is challenging to distinguish where animal *razze* ended and human race began.

Archival materials that include rare manuscripts with small readerships can be revelatory. However, when trying to access a pan-European discourse about breeding, it is also important to consider the most widely read texts and the influential authors who produced them. Writers like Baldassare Castiglione, Federico Grisone, Ulisse Aldrovandi, Francisco Hernández, José de Acosta, and Giovanni Battista della Porta were read throughout Europe and the wider world in numerous editions and translations, not least by the elite audiences and expert advisors who operated at the interface between ideas of animals and real animals. Their print histories serve as rough-and-ready guides to the scale of readership they commanded, which was considerable for the period.

The adoption of selective breeding practices in Renaissance Europe, which followed on a long tradition in many other regions, made possible formidable surpluses in agriculture. Control over earthly cycles of fertility, whether in the fields or the stables, meant more bountiful harvests and larger, healthier flocks of livestock. Nonetheless, appreciating the violence of conceptual blending as agriculture mapped onto human culture is central to understanding this process and its discontents. While animal *razza* had roots in European nobility and attempts to control nature, it came to adhere to some human bodies more than others. As Joshua Bennet has shown, blackness has been a caesura between human and nonhuman, forced into this role through the "all-too-fraught proximity between the black person and the nonhuman animal."[47] In the wake of the early modern Atlantic slave trade, ideas of race became particularly attached to the black human body, stuck to individuals who were traded like chattel and stripped of their right to personhood. Bennett identifies the analytical resources embedded in the extended animal comparison, but also its fundamental ambivalence. The animal valences of race were sometimes used as a means of enforcing narrow hierarchies and keeping people in their place, but they were also capable of being turned back on themselves and used as a resource by those same people. This potential reappropriation does not negate the violence and horror of dehumanization, but it does highlight that the historical individuals who were described in these terms sometimes intentionally inverted them as a tool of their own empowerment. Similar narratives were doubtless invoked by early moderns but are challenging to show in this source base. Relatedly, the most tragic sections of this book emerge in the consequences of ubiquitous thinking about difference through breeding and the messiness of Renaissance metaphors across human and nonhuman animal worlds. The impact of an interest in controlling nature through breeding is, in the end, ambivalent.

Structure

The book begins with two thematic chapters that elaborate common issues developed across the rest of the narrative: philosophy of breeding and the role of branding and commodification. Chapter 1 builds on a strain of scholarship in early modern history of science that sees the epistemological work accomplished by artisans as central to the Scientific Revolution. I employ a capacious definition of philosophy to delve into the world of breeders as thinkers and understand the questions evoked by thinking through animal breeding on a daily basis. Some breeders were Aristotelians, who emphasized the importance of male seed. Some saw themselves as practitioners of natural magic. Most, however, created their own commonsensical vision of the natural world that emphasized surprising slippage, but also the overarching tendency of like to produce like. Animal breeding is as ancient and widespread as agriculture itself, and by the early modern period, many groups, such as indigenous American breeders, emphasized the presence of likenesses across various types of animal being. European breeders, meanwhile, became increasingly eager to collect particular *razze* of animals, which were sometimes described as the most "noble" or "perfect" versions of those animals. While breeders' daily work was relatively uncontroversial, Italian breeding practices risked entering into debates about the nature of nobility, the role of education, and the extent of predictive powers. Some breeders who embraced the importance of animals' imaginations in deciding the color and quality of their offspring covered their stables in lavish tapestries to influence their stock. These practices of making "designer horses" encountered staunch opposition from trainers like Federico Grisone, who emphasized animal education and nurture over an overt focus on breeding and nature. More generally, like any determinative science, creating designed bodies through sustained breeding raised problematic theological questions about free will.

Italian breeding practices and the horses they crafted were prized throughout Europe. Grisone's students, along with those of fellow Neapolitan Giovanni Battista Pignatelli, trained influential equestrians who would go on to shape the court riding traditions in France, Spain, and England; another Neapolitan, Prospero d'Osma, was asked to report on the

English Royal Studs in 1576.[48] Italian, specifically Neapolitan and Mantuan, and Spanish horses were collected alongside Arabian, Turkish, and
barb horses imported to "grade up" other European stock, a practice intended to transform local breeds into something more akin to foreign ones
with the repeated use of an imported sire.[49] Leon Battista Alberti had theorized the qualities of "perfect" horses, and others correspondingly followed in his footsteps in the hopes of breeding their own.

Chapter 2 explores the tools by which early modern breeders and patrons rendered these *razze* more permanent. This chapter centers on the
changing role of sixteenth-century property ownership in the history of
razza-making and the slippage that would mark human slaves with marks
similar to those used for animal property. Through a close reading of the
records of Haniballo Musulino, a Neapolitan horse master in Spanish Italy, I explore how massive programs of breeding aimed at creating "house
races" of animals that exemplified the excellence of the noble family who
owned and bred them; each house-bred variety of horse, dog, cat, or other
animal was called a *razza*. However, the hubris of breeding far outran the
ability to write differences into an animal population that could be sustained across several generations. Differences had a way of trickling away
over time, hastened by new environments, feed, and what breeders called
degeneration, or the tendency to revert to a more common type. Branding offered a solution. With iron and fire, breeders marked their animal
charges with the symbol of the house that had made and owned them,
emblazoning their status as property into their flanks. Through brands
and the bureaucracy surrounding such marks, breeders identified individual horses, mules, cattle, and other animals along with their pedigrees,
rendering difference legible even as the effects of breeding alone waned
generationally.

Part II begins the chronological arc of the book circa 1492, at a model
Renaissance European court that epitomized practices echoed across the
continent. Chapter 3 turns to the Gonzaga court in Mantua, Italy, where,
according to Jacob Burckhardt, the founder of Renaissance studies, "the
foundations of scientific zoology and botany were laid."[50] Driven by a
competitive passion for collecting, the ruling Gonzaga dynasty experimented with selective breeding to develop their own animal stock, a process they termed "seeding." Burckhardt slid between animal breeding and
human collections, noting that "even human menageries were not lacking"
before describing how Cardinal Ippolito de' Medici "kept at his strange
court a troop of barbarians who spoke no less than twenty different languages and who were all perfect specimens of their races," which included
North African Moors, Tartars, Africans, Indians, and Turks.[51] This elision

seems to have been consistent with documents from this period. Using practices similar to those employed by her husband's horsemen, the patroness Isabella d'Este (1474–1539) developed a "race" of courtly people, sending the products as gifts to other courts. Courtiers accepted that everyone, from horses to princes, possessed a race because their unions were deliberately selected to achieve a preferred social and biological outcome. Still, as the court became a living laboratory for breeding, intellectuals worried that race—unlike sex or gender—was temporary. Without constant labor, it would slip away after only a few generations.

Chapter 4 turns from European courts to parallel cultural centers in the Americas. It begins by tracing how members of Mexica emperor Moteuczoma's court sorted curiosities among palatial rooms. The emperor's collection was described as housing dwarves and hunchbacks, who were prohibited from intermingling with other inhabitants, in their own halls adjacent to his animal-filled menagerie. Evolving in parallel, courts in both Renaissance Europe and early sixteenth-century America had become laboratories for shaping nature, enculturating a fascination with the wondrous and a desire to collect it. Nahuas employed their own ideas of heredity; corn provided the intellectual framework. Just as the Mantuans discussed "seeding," Mexican notions of inheritance centered on seeds through *xinachtli*—a linguistic concept linking maize cultivation, semen, and brotherhood. The chapter follows definitions of *xinachtli* in European-indigenous dictionaries and its meaning in the recollections of Nahua chronicler Don Domingo Francisco de San Antón Muñón Chimalpahin Quauhtlehuanitzin (1579–1660). Through the metaphor of grafting—an Old World agricultural technique requiring one to cut a branch from an existing plant and splice it onto a living plant—this chapter suggests that indigenous American approaches to animal husbandry were partially displaced by European ones.

Part III, "A Brave New Natural World," returns to living animals in a dynamic relationship with Renaissance natural history. Chapter 5 examines the emergence of new mestizo dog populations in sixteenth-century New Spain. Most pre-contact American and European mammals differed so greatly that they could not effectively interbreed post-contact. However, dogs, like humans, were exceptions to this rule. By the fifteenth century, both Mesoamericans and Europeans had developed a mosaic of distinct domesticated dog breeds, including the hairless Mexican *xoloitzcuintli* and the European shepherd. Following contact, Old World dogs brought over by the Spanish interbred with New World dogs. Suddenly mixed with shepherds, by the sixteenth century the *xoloitzcuintli* had become a *mestizo* (mixed) dog, just as its human companions increasingly became of mixed ancestry. *Mestizaje* was itself an early modern zoological term referring to hybrid offspring; this eventually transformed into a racial category indicating mixing.[52] For humans and animals alike, colonial societies not only produced new races, but also aimed to lock them in stable categories.

Chapter 6 focuses on camelids, broadly defined to include llamas, alpacas, dromedaries, and the sheep early moderns confused them with. Through a suggestive juxtaposition, this chapter deploys the counterpoint method to interpret attempts to integrate camelids into Spanish New World nature. The first "voice" is Spanish blood purity regulations. In the late fifteenth century, Castilians had embraced a millenarian optimism that non-Christians could be successfully converted. However, just over a century later, with the inquisitorial persecution of New Christians and the final expulsion of the Moriscos, Spaniards abandoned the hope that Jews and Muslims could escape their blood. Having just expelled the Moors, Castilians did not want to see animals associated with Islam in the Andes, instead envisioning llamas as the Andean variety of safe, Christian sheep. Other Europeans, however, saw camels. As such, attempts to understand and categorize Andean camelids offer the second "voice." In concert together, these voices showcase tensions between New World mixing and Iberians' desire to render difference permanent by establishing and enforcing a new taxonomy that conflated ethnicity, blood, and religion. Shaped by a combination of environmental and social factors, both the Spaniards' limits on camelid use and their tacit and intentional classification of camelids reflected a program of cultural differentiation from Islamic North Africa and the Ottoman Empire.

Part IV, "Difference in European Thought," shifts from American animals themselves to their role in European thought, with a focus on the stage of Rome and Southern Italy, and a particular attention to parallel intellectual developments around 1592. Chapter 7 returns across the Atlantic with José de Acosta, the prominent Jesuit naturalist whose popular and nuanced descriptions of the New World in his 1590 *Historia natural y moral de las Indias* (Natural and Moral History of the Indies) became a mainstay of European libraries. Indeed, it is Acosta's framing that contributes to this book's title, as he considered the theological and practical problem of how "perfect animals" (*animales perfectos*), such as lions, tigers, wolves, and, most perfect of all, humans, populated the Americas after the Noahic flood. The "perfection of nature" references the hierarchy of the Great Chain of Being, with God at the top, then an unbridgeable gap, followed by humans, perfect animals, and imperfect animals.[53] Acosta's definition of perfection comes from the concept of "clean beasts," defined by their means of multiplying. He cites Genesis: "Of all clean beasts take seven and seven, the male and the female in order that their seed [*generación*] may be saved upon the earth." These beasts are contrasted with the "imperfect" animals, such as spiders, rats, and insects, which are "engendered from the earth without generation."[54] The phrase "perfection of nature," then, is meant to echo this focus on generation as a theologically important category. Acosta puzzled about differences among kinds of animals, asking how European-type dogs could be missing from the Americas and unique Andean camelids be so plentiful there and absent elsewhere, just as he inquired about how human behavior ranged so dramatically across societies. Through population thinking, Acosta contended that minor differences could be explained as Aristotelian accidents and required a unified natural world; as such, any human could be successfully converted to Christianity. Meanwhile, however, the European religious establishment had become even more embroiled in Spain's poisonous politics of *limpieza de sangre*. The chapter juxtaposes Acosta's theories about species and kinds with his arguments in favor of convertibility, accusations that Acosta himself was a New Christian, and the Society of Jesus's decision to prohibit New Christians from joining their order.

Just like the preceding chapter, chapter 8 begins in 1592 in Rome, although it explores a parallel approach to animal difference, one focused on predicting the qualities of individuals rather than understanding the origins of populations. In the latter half of his career, Giovanni Battista della Porta (1535?–1615) argued that he had discovered the key to nature's secrets: that human physiognomy could be explained through the careful study of animals. With its base in Southern Italy in the 1580s and 1590s, this chapter considers how physiognomy—the art of reading embodied signs to predict characteristics and thus behavior—not only offered tools to explain patterns in proclivities of humans and nonhumans, but also raised questions about the place of free will. Wielding his personal mantra, *aspicit et inspicit* ("He that sees the outside sees the inside"), Porta analytically divided animals into their various physical features, understood those features in relation to the animal's essential character, and used the animal's parts to explain variation in temperament as a function of external appearance. As Porta joined the physiognomic fad sweeping Europe with the publication of *On Human Physiognomy*, he again risked raising the hackles of the Roman Inquisition, which had placed his previous work, *Natural Magic* (1558), on the Index of Prohibited Books and given him a sentence of *purgatio canonica*. Despite these dangers, Porta forged on with books on physiognomic links across humans, animals, plants, and celestial bodies, moving beyond his laboratory to root his observations in morphological patterns in nature and their potential consequences for

behavior and medicine. Through his collaborator Federico Cesi, founder
of the Accademia dei Lincei, Porta's ideas shaped the practice of taxonomy. The Lincei embraced his motto of *aspicit et inspicit*, through which
Porta connected human characteristics to bestial features, actively undermining the animal-human divide in the service of understanding human
types. The Lincei's effort to wrestle with New World natural diversity in
its Mexican Treasury (the most systematic account of American nature
to date) led its members to combine ideas of classification and fixity of
difference from multiple sources, including Porta's approach to scientific
observation and Francisco Hernández's writings about plants, minerals,
and animals in Mexico.

This Renaissance history has many afterlives. The projects it traces
made real animals; many of these breeds stuck, ossified by generations of
stud registries that recorded details of their brands, characteristics, owners, and careers. Today, breeders and enthusiasts lay claim to the (albeit
often dubious) historical legacies of animals from the early modern period. As for horses, the Royal Stables of Córdoba still breed and train Andalusians and the *Pura Raza Española*, which some claim originate in stock
preserved by Carthusian monks or King Philip II's initiative to improve
Spanish horses.[55] The Habsburgs' Lipizzaners are bred for the Spanish
Riding School in Vienna.[56] The English tradition of breeding Thoroughbred horses, which emerged as a distinct breed in the eighteenth century,
set the standard for genealogical paperwork that registered a breed's lineage and corresponding purity, and these recorded pedigrees were subsequently used for other animals.[57] For dogs, look no further than the
spaniel, literally "Spanish dog," or the Mexican *xoloitzcuintli*.[58] As modern nations consolidated, *razza*-makers sought to make animal capital that
represented their regional priorities. Systems, both social and scientific,
consolidated taxonomy and racial hierarchy throughout the seventeenth
and eighteenth centuries. In the eighteenth century, Latin American *casta*
paintings sought to define a taxonomy of human racial types. Simultaneously, the Comte de Buffon, an oft-cited founder of zoology, sought to
breed peasants in the French countryside, and eighteenth-century writers envisioned how they might "manufacture" a human population of
male mulatto soldiers to address social, political, and military threats in
Saint Domingue.[59] From one century to the next, efforts to shape living
nature, both human and nonhuman animals, waxed and waned with human hubris.

PART I

Knowing and Controlling Animal Generation

CHAPTER 1

Breeders as Philosophers

Breeders in the New Philosophy

In his *New Atlantis*, Campanella's contemporary Francis Bacon (1561–1626) presented a society focused on the pursuit of scientific advancement, a manifestation of his vision of the "New Philosophy" of "active science" that built knowledge on collection, experimentation, and analysis. In it, Bacon developed the implications of the power of breeding along lines parallel to those laid out by Campanella. In his unfinished treatise, published posthumously in 1627, Bacon described animal bodies as crucial to the experimental process—poisons and medicines were tested on them in his Atlantis—and argued that through breeding their utility could be increased.[1] Animal bellies provided furnaces; sound-houses captured the voices of birds and animals and trained the human ear; and enclosures contained curious beasts and birds for dissection, trial, and observation. With great care, the Atlanteans created new creatures and controlled the fertility and size of beasts. One of Bacon's Atlanteans described the process:

> By art likewise, we make them greater or taller than their kind is; and contrariwise dwarf them, and stay their growth: we make them more fruitful and bearing than their kind is; and contrariwise barren and not generative. Also, we make them differ in color, shape, activity, in many ways. We find means to make commixtures and copulations of different kinds; which have produced many new kinds, and them not barren, as the general opinion is. We make a number of kinds of serpents, worms, flies, fishes, of putrefaction; whereof some are advanced (in effect) to be perfect creatures, like beasts or birds; and have sexes, and do propagate. Neither do we this by chance, but we know beforehand, of what matter and commixture what kind of those creatures will arise. We have also particular pools, where we make trials upon fishes, as we have said before of beasts and birds.[2]

Behind Bacon's vision of the ideal research center in the court play *Gesta Grayorum* (1594) and Salomon's house in *New Atlantis* rested spaces that were already key to European knowledge-making. These included libraries; gardens, menageries, aviaries, and lakes brimming with living nature; cabinets filled with art and natural wonders; and stillroom distilleries and instrument-laden laboratories in which to test nature.[3] Real breeding efforts focused on "perfect animals" were more limited than those of the Atlanteans. Even so, early modern breeders shaped a wide array of animals and plants; developed variations of size, fertility, color, and shape in these populations; explored the idea of "commixture," or mixing across seemingly fixed types; and sometimes relied on systematic records.

In biological terms, *breeding* is the controlled mating and production of offspring in animals; that term is bound up with ideas of "good breeding," or the manners characteristic of human aristocracy. I invoke *breeders* in a capacious sense to refer to a group of people who, in practice and in theory, concerned themselves with generation and the fruits of inheritance. Breeding was enacted on both an informal basis and, in many cultural contexts, a formal one. Courtly breeders acquired professional standing as stable managers, keepers of the dogs, husbandmen, and experts in animal health. In some European courts, elites developed a systematic breeding practice with clear goals (such as fostering speed in racehorses), a defined population under consideration (such as Turkish racehorses), an organizational apparatus capable of handling and observing the population (stud farms), a correspondence tradition (letters among traders, shoers, managers, health experts, trainers, and competitive owners), and extensive bureaucratic record keeping (reporting details of the stock to patrons, rulers, or administrators). Breeding meant acting as a matchmaker for specific creatures, first imagining the yields of their combination and then going through the practice of pairing them, whether through direct intervention, as in a controlled covering of a stud over a mare, or through organized pasturage to make space for such an encounter to take place. It meant raising young offspring and finding them appropriate homes. Attentive breeding efforts required constant worry about whether stock represented the desired characteristics, came from a particular heredity, benefited from their environment, and developed well.

Breeders are some of the oldest experimentalists. Their ability to domesticate and mold animal generations through trial, testing, rejection, acceptance, and the opportunity to create future offspring contributed importantly to the first agricultural revolution, as well as to the early modern scientific revolutions with which this book is concerned.[4] Shaping qualities of herds by selecting for strong animals and manipulating color through animals' vision during copulation pre-dated the writing down of Genesis 30:32–43, wherein Jacob experimented with breeding animals as he begot his own human family. Jacob struck a deal with Laban, whose herds he was keeping, wherein he would build his own flock from Laban's animals. Jacob agreed to remove all spotted and speckled cattle, brown cattle among the sheep, and speckled among the goats, such that every goat or brown sheep that was not spotted and speckled in his possession "shall be counted stollen by me." To breed speckles into his herd, Jacob "pilled white strakes" in poplar, hazel, and chestnut rods and set the rods in front of the watering trough, as the King James Bible translates this passage. When the livestock would come to conceive, they would see the white rods, that "brought forth cattle that were ringstraked, speckled, and spotted." As he separated his flock from Laban's, he laid the rods before the stronger cattle as they conceived, and through that process Jacob bred stronger, speckled animals for himself.[5] Although many basic breeding strategies have remained the same in the eras of Jacob, Bacon, and modern breeders, sixteenth-century Europeans increased the scale and recording of breeding efforts. Pastoral societies like Jacob's understood livestock to represent wealth and avoided overstocking, aware that undernutrition would result in unproductive animals. Irigoyen-García has shown how "the motif of Laban's livestock was used as a metaphor for miscegenation in early modern Europe, and was commonly associated with the belief that the imagination can leave an impression on women during conception, as the biblical passage was read in terms of phenotypical difference."[6]

Breeding strategies—as Nicholas Russell defines them through horse, cow, and sheep breeding in England from 1500 to 1800—are techniques breeders use to influence their stock. These strategies presume an excess of animals. As Russell argues, one might choose, for example, whether to buy into a particular breed, acquire a herd or a flock, or completely replace a herd or a flock. A breeder might allow random mating (wherein natural selection takes its course without human intervention) or use a negative breeding strategy (wherein one maximizes profit by using the worst animals as breeding stock and selling the best ones). A breeder might alternatively adopt a positive breeding strategy through selection based on phenotype (appearance), either culling the worst phenotypes or selecting the better phenotypes and allowing only that subset of the stock to breed. Alternatively, a breeder could use selection based on geography, on the premise that the environment influences character as much as heredity does. This strategy entails moving animals of one type into a new climate with the presumption that they will become more like the new climate; through selection by geography, Philip II's breeders hoped that the "King's

Race" of horses would become more Neapolitan when they were bred in Southern Italy. A breeder might hybridize through sire or group rotation (importing a sire into a local herd for a time, or selling off female stock) or create hybrid stock (selecting parent stock from different breeds and varieties to combine characteristics in hybrid offspring, an approach that is effective only under limited conditions). The breeder might try to "grade up" in a crossbreeding practice in which an imported sire mates with a local breed, on the premise that, through the use of foreign sires over multiple generations, the local population will become genetically more like the imported type with each generation. In all these cases, breeders would need to avoid incestuous pairings, although they often did not. More modern breeding practices focused on pedigree value, replacing interest in the animal's individual character with its ability to pass on desirable traits to the next generation. Through pedigree analysis, the breeder would analyze an animal's parentage, noting if it had other desirable animals in its family tree. This strategy required careful bureaucratic record keeping, but it could help in a confused livestock trade, as Russell notes in the case of the seventeenth-century English horses that had such varied external appearance that it was hard to know a horse's qualities. An early modern buyer could purchase pedigree with confidence derived from the parallel process of record keeping of human nobility and a popular belief that nobility conferred superiority. In more recent times, modern breeders employed the highly fallible strategy of progeny testing (monitoring parents' offspring for performance to see if the parental phenotypes were reflected in their offspring). Since the advent of modern quantitative analyses and population genetic theory, modern breeding practices have become more effective, but the strategies themselves remained similar from antiquity through the early modern era.[7]

From Generational Theory to Artisanal Knowledge

Breeders' knowledge of the implications of their actions represents a category of artisanal knowledge: the know-how of people who worked with their hands. Recent scholarship has sought to restore artisans' bodily knowledge to a central role in the scientific innovations of the early modern period.[8] Aristotle had denounced artisans, who, unlike scholars, could not be full citizens because a life of labor lacked room for virtue and morality. Practical craftsmen—like painters, actors, musicians, sculptors, and metalworkers—who labored with their bodies, not their minds, were not thinkers so much as doers in the Aristotelian scheme of knowledge, which continued to hold weight throughout the medieval period. Theory was elite knowledge, and its *episteme* or *scientia* was explored in universities devoted to Aristotelian logic and textual knowledge. *Praxis* or *experientia*, being a collection of experiences, was the means by which things were done and made, filled with details of experience that could not be transformed into a deductive system. Finally, *technē* was the practice of making things through bodily labor, which, as Pamela Smith has characterized it, "had nothing to do with certainty but instead was the lowly knowledge of how to make things or produce effects practiced by animals, slaves, and craftspeople."[9] Artisans might have collected recipes or followed rules, but they did not often record the abstract implications of their work.

In a complication of this schema, despite having artisanal origins, animal knowledge was often valued highly by elites, for whom the archetypal category of noble in Romance languages—*cavaliere* in Italian, *caballero* in Spanish, *chevalier* in French—literally means horseman. A robust heritage of classical thought on generation was available for breeders to mobilize.[10] The study of animal generation and fetal development had a central place in the edifice of Greek philosophy.[11] In *Categories*, Aristotle argued that both horses and humans had the potential to come into being, and thereby participate in *ousia*, the fullest level of substance or essence.[12] He turned to the details of conception and fetal development to understand that status of being. The idea of seed or semen (*seme*-related words were used interchangeably) helped explain the first cause by which a living thing came into being. Aristotle held that seeds came from a specific organ in the male while the mother contributed matter for the body of the fetus. Subsequent Aristotelians debated whether the seed originated in the brain, blood, or bone marrow. The male semen would then carry an immaterial soul to the female fluid. From there, like must beget like as perishable substances took on a new life in a new form.[13] In turn, for the third-century physician Galen, both parents were capable of contributing to form, even though the seed provided the formal principle of generation. He understood females to be naturally colder than males and argued that female animals would generate their imperfect seed in their ovaries, which awaited the male semen.[14] Avicenna, the eleventh-century Persian philosopher, sought to synthesize the Aristotelian and Galenic arguments, transforming the female seed into menstrual blood drawn from Aristotle's model.[15] Acceptance of these ideas was contingent on one's disciplinary camp in the early modern period: Aristotle became the authoritative voice of natural philosophy, while Galen was the authoritative voice for medicine.

In the sixteenth and seventeenth centuries, new attention to direct observation and the epistemological implications of experience eventually led scholars to question categories of Aristotelian knowledge, leading to a shift in widespread beliefs about nature and the material world. To know about nature increasingly became to sense, experience, observe, and record nature, although Renaissance libraries were still filled with ancient texts of agriculture and husbandry, including Xenophon, Columella, and Varro, alongside the writings of Aristotle, Avicenna, and Galen.

One strand of scholarly thought on generation developed into embryology. At the University of Padua, Hieronymus Fabricius d'Acquapendente sought to revive the Aristotelian program for animal embryology and the parts of creatures shared across animal kingdoms. William Harvey, famous for discovering the circulation of blood, studied with Fabricius, focusing on an expansive definition of eggs in nature, from trees to oviparous animals to viviparous animals.[16] Using the microscope, he zoomed in on animal anatomy, and subsequent seventeenth-century research built on his approach.[17] By the early eighteenth century, a fierce debate raged in the field between preformationists (who believed that all beings were already formed in the egg) and epigenesists (who argued for the creation of new beings with every insemination).

While a growing field of embryological theorists wrestled over the philosophical implications of animal generation, another community of naturalists and medical experts debated the role of monstrosity in defining and deviating from the normal body. Aristotle's system had focused on normal inheritance; birth defects clearly showed how progeny's traits could differ from those of their parents. Publications by Zurich surgeon Jakob Rüff and French surgeon Ambroise Paré featured bodies deemed monstrous, such as conjoined twins and babies born with severe deformities. Paré argued that too much seed produced a monstrous child. Monsters took on multiple meanings as wondrous, as portents of religious intervention, as entertainers, and as a living philosophical challenge, as Katharine Park and Lorraine Daston have shown.[18] Deviations from the norm were also thought to be caused by an irregular position in conception or the dalliance of the maternal imagination.[19]

Despite this scholarly apparatus, documents left by Renaissance animal experts are generally unconcerned about fitting their work into Aristotelian, Galenic, or Avicennan philosophy or embryological theory. Most breeders had seen irregular births at least once in their career. Given that experience, they were more likely to adopt a popular understanding of monstrosity over a refined attention to developmental embryological patterns that would have been more apparent had they cut up their animal charges for dissection. Popular understandings saw the mysterious process of generation as part of *Naturae arcana*, or the secrets of nature, hid-

den in the womb and influenced by processes that seemed to walk a fine
line between the earthly, the divine, and potentially the demonic.[20] Some
theorists and practitioners of natural magic, like Giovanni Battista della
Porta, saw breeders as potential practitioners of such arts, as explored as
a case study further along in this chapter.

Were breeders philosophers? Although they commanded a varied level
of scholarly knowledge about the maintenance of life and questions of
generation, environmental influence, and inheritance, breeders occupied
a central place in fostering the discourse of differences and likenesses
through which to understand both animal and human life. Rather than
remaining wedded to a definition of philosophy as an a priori discipline
focused on conceptual analysis—theory over praxis anew—a new generation of scholars has sought a more expansive vision of what it meant to be
a philosopher. Justin Smith has argued for six types of philosophers: the
Curiosus, who experimented and collected wonders; the Sage, who mediated between earthly and transcendent realms; the Gadfly, who sought
to correct nearsightedness in his or her own society; the Ascetic, who rejected opulence in favor of clean thought; the pejoratively phrased Mandarin, who joined a class of bureaucrats and sought to produce what Thomas
Kuhn called "normal science"; and the Courtier, who strove for worldly
prowess and the social glory of knowing.[21] Amid this cast of characters,
European breeders were often variations on the Mandarin, committed to
upholding a bureaucratic process of data gathering and analysis that stayed
carefully within the lines assigned to them at court. In the preserved remnants of Mesoamerican records, some breeders appear to have taken on
a ritual role akin to the Sage, mediating between current and future life,
between ancestry and present potential. One might argue that José de
Acosta, who observed American nature and theorized about animal and
population migration, was a mix of a Sage and a Gadfly. In the exceptional
case of Porta, with whom this book ends, a breeder-theorist took on the
role of a Courtier, delighted by the extent of his own power to shape nature through manipulation and prediction. One unified breeding epistemology would be an oversimplification of a diverse group of individuals
committed to a wide array of projects.

While biological knowledge and theories benefited from artisanal
knowledge from breeders, breeders often did not seem particularly focused on understanding generation through theoretical principles.[22] As
Russell has put it, "neither pangenesis nor Aristotle's generative theory
could guide the practical breeder. Columella's consideration of three-
generation character transmission in asses and sheep would not have benefited from Aristotle's theoretical discussion of how characteristics could

pass to filial generations from grandparental or other ancestral individu-als."[23] Breeders working directly with animals were more concerned about the practical impediments to their craft than creating a unified system through which to understand inheritance.

If artisanal metalworkers created enduring objects, the natural things breeders made were ephemeral.[24] A long-lived dog would be fortunate to live fifteen to twenty years, while a long-lived horse or llama might reach its athletic peak for five to ten and might live until twenty. Artis-tic representations often included animals and the work of breeders, but the traditions of representation often preferred idealized steeds and dogs of war over naturalistic representations of such animals, making these an imperfect window onto past animal bodies. Branding tools have proved difficult to date. Leather dog collars from the Renaissance are few and far between and reveal little of the animals that wore them. Saddles and armor give clearer material indications of the size and use to which the animals adorned in them were put.[25] The high pommel and horn of a Renaissance sidesaddle adorned in velvet reveals a bit about the size of the rider and the ways she might have left the control of her horse to the hands of an expert groom, while the width of the triangular tree that dis-persed her weight atop the horse's back hints at the size of the animal's body and the height of its withers.[26] Bits that rested in horses' mouths, linked to reins to control the position of their necks and the direction of their movement, likewise offer trace evidence about riders' priorities. The length of the long curb bit worked as a lever to arch the neck and drive down the head of a horse, and its width needed to be fitted to the horse's mouth.[27] With padding, cloth, and adjustable leather additions, these physical remnants of animal husbandry are an imperfect guide to the animals that wore them. Zooarchaeology offers a potential window into past populations. Archaeological evidence has suggested that be-tween 1500 and 1700, sheep and draft horses increased in overall size.[28] However, genetic analyses carry their own challenges, from accessing an-imal remains for sampling to isolating past variations without modern corollaries among animal genomes, which are less well mapped than those of humans.[29] For cattle, for example, despite the many samples scattered across early colonial archaeological sites, no aDNA study has yet conclusively identified the animals' breed or traced their origins back to specific domesticated populations in Europe.[30] Modern plants and animals—whether corn or spaniels—carry influences of breeders' knowledge and work, although their connections to the historical lega-cies to which they lay claim are murky at best. Centuries of Mesoamer-ican agricultural knowledge is folded into maize, the domesticated crop that differed so profoundly from the wispy, wild teosinte grass from which it was first domesticated. However, living populations were shaped by both previous and subsequent centuries of natural and human selection, making them an imperfect window to the past. To catch early modern breeders in the act of thinking about generation and difference, I turn to archival records, as well as the published writings of naturalists concerned with breeders' efforts to shape nature through seeding, grafting, and mating.[31] The remainder of this chapter explores breeders' work and theories with a focus on horses. I turn first to normal breeding practices, then to the potential for natural magic in breeding, the continued importance of animal education, and finally, the tendency of breeders' work to slip away.

Breeding Practice in Renaissance Italy

The Renaissance Italian tradition included an ideal of perfection that could extend to animals, particularly horses.[32] Some breeders linked levels of perfection in the animal to its internal organization of heat according to a Hippocratic tradition. Throughout Europe, as Russell has noted, many breeders imagined themselves "breeding towards the recovery of past God-given perfection" rather than some artificial version of the animal designed by humans.[33]

Leon Battista Alberti—architect, essayist, and quintessential Renaissance man—encouraged breeders of horses to start by finding the stallion, whose excellence could be ascertained through distinct conformational qualities including body, beauty, head shape, ear shape, brow shape, black eyes, wavy mane, flowing tail, flared nostrils, straight neck, bony shoulders, prominent backbone, broad chest, ample stomach, balanced testicles, perfectly proportioned thighs, well-built knees, light cannon bones, and well-formed hooves. Then one could turn to the mare, with attention to selecting one who looked and behaved like the stallion, had an open chest, well-muscled and sound legs, and a large stomach.[34] The need to define perfection reflects an increasing desire to create a perfect horse, Sarah Duncan has suggested. Anthony Grafton has shown how Alberti saw likenesses in family and equine management, noting that Alberti's treatises *On the Family* and *On the Horse* both explored "eugenic selection."[35] In Grafton's words, "Alberti's aesthetic and pedagogical ideals could be realized more fully, it seems, in the world of horses than in the world of men. The painter who emulated Zeuxis could produce, at best, an ideal image of a woman more beautiful than any real human. The breeder of horses, by contrast, could produce an ideal mare and stallion."[36] Alberti had been asked to judge the equine statue of Niccolò III of Ferrara. Aware that the Este, one of the most important families of horse breeders in Italy, were interested, Alberti wrote the thirteen-folio *De equo animante* on the breeding, stabling, and trading of horses in 1443. As part of a growing tradition of treatises "not written for the men who look after horses but for a prince who is a great savant," Alberti presumed that the noble horse owner took on the role of breeder-in-chief.[37] Like the Solarians, Alberti emphasized balance. Emotions that were neither weak nor strong, balanced humors, regular habits, and training and exercise were central to keeping a horse well. On breeding itself, *De equo animante* offered a great deal of advice. Horses were not equal: some were good for farm work and civilian duties, others were able to compete for Olympic crowns. "See to it that those that you choose for parents and creators of the herd are of attractive appearance," he wrote, "and of proper age, and appropriate to the task."[38]

Breeders balanced environmental influence with confidence in family resemblances rather than distinguishing between heredity and environment.[39] A sixteenth-century treatise on "the ways and order of breeding *razze* of horses" in the Biblioteca Apostolica Vaticana emphasized the relationship between landscape, humoral balance, and the qualities of horses of different colors. It urged breeders to select pleasant countryside, neither too dry nor mountainous, and certainly not too cold in the winter.[40] The markings and color of horses in the herd determined their reaction to different elements: a sorrel horse might be cheeky due to her hot humors; dark red horses might behave coldly and require humidity instead; dark and light bays required yet other balances to accommodate their lack of humidity.[41] Herds were to be moved according to the season. Some fodder was appropriate for June with newborn foals, but not for fall. Special attention was given to the availability of certain herbs, which in turn influenced the quality and quantity of the mare's milk.[42] Colts were to be separated from their mothers during the spring of their second year of age so that the young male horses would not feel the heat of mounting their relatives.[43] Seasonal temperature brought risks. If a mare was too warm in the summer her milk might become too warm, leading to the abortion of her foal. Some environments caused worms, for which the latter pages of the treatise suggested remedies.[44] This emphasis on environmental control was exported to Elizabeth I's England with Prospero D'Osma's 1576 recommendations for good English stock, which would also require appropriate soil, grass, and water.[45]

"Good air, good water, good pasturage, the good *razza*, for producing good horses" was the popular refrain in Giovanni Bernardino Papa's manuscript "Treatise on the *Razze* of Horses, Directed to the Grand Duke of

Tuscany on March 25, 1607." The province of Calabria had "most fertile and abundant pasturage, good water, and good air" and correspondingly fostered healthy stock, although Papa made sure to compliment the grand duke of Tuscany for his good horses, too. Where the *razze* pastured, the air needed to be temperate, the water clear, on dry land, forested with oaks, with local herbs, and open but between mountains to protect from the wind, as well as the rain and snow; and mares of the *razze* should not share their feeding ground with "any other kind of animal and in particular buffalos, pigs, geldings, sheep, and cows." Water was paramount to animal health, and supplying healthful water—as Papa attributed to the rivers in the Maremma, and around Pistoia and Pisa—was a key part of the breeder's work.[46]

Breeding certainly required an intimate knowledge of one's charges. Alberti had noted that the breeder should encourage mating by directly stimulating the stallion, especially if "they are somewhat frigid." He continued, "The place where the body explodes with pleasure must be touched with rings, nettles, chopped pepper, and stimulants of this kind; it is also necessary to massage with the hand and smear it with such aphrodisiacs, as well as applying them to the nostrils of the subjects who are preparing to mate."[47] In his treatise for the Tuscan grand duke, Papa emphasized Galenic humoral theory, arguing that "stallions with large testicles should be bathed with cold water in order to warm the humors of their loins" and that the breeder should think twice about having the stallion mount in order to further the balance of his humors. If things remained imbalanced, he should be treated by a *marescalco* (a combined farrier, blacksmith, and veterinarian).[48] Papa explained that stallions should not be older than twelve, and that there should be no more than fourteen mares allocated to a stallion.[49] Mares were to be given to the stallion when they were between the years of three and twelve or fourteen; they should "not be covered immediately after they have calved, but should wait at least six days," otherwise the breeder risked making a mistake. Papa recommended the Neapolitan practice of bringing the foals to the stables from March to October, and advised breeding in hand; if the breeder simply left the stallion in the field day in and day out, he might not "content a large part of the mares." Rather, Papa suggested that the stallion be bred "to the number of mares four or five times; that is, each stallion a mare in the morning and another in the evening until the number is complete." Other experts opposed covering in the stable in hand, as they argued it would create sluggish and sterile horses.[50] Regardless, a breeder should keep the mares desirous of the stallion by manipulating when he joined the herd and should "observe when the mare does not run from the stallion." Good breeding

required careful timing, an awareness of the environment, and a willingness to manipulate the intimate lives of animals.[51] Papa, like other animal
experts writing to princes, included a sharp class critique: Poor people and
ignorant people might not know what animals were mixed. Poor people
might want foals very badly, and they would have "bad ones" that were
"born of that *razza*, but they are not noble horses," and only fit to serve
those poor people.[52] He did not complain of lack of purity—when he did
mention blood, it was only in the humoral sense—but of lack of maintenance. Contemporary English breeders understood the Italian school to
expound a belief in the "purity of blood" with a special emphasis on heredity beyond the need to control environment alone, but in most Italian
sources, blood was not thought to determine the whole of equine potential, and authors of equine treatises often emphasized the importance of
the horses' ability to learn.[53]

A good breeder thought across years to understand the dynamic cycles
of animals' lifespan and gestation. A "Book of the *Razza dei Cavalli* of his
Most Serene Highness" recorded the breeding efforts of Francesco Maria II della Rovere, the duke of Urbino, between 1614 and 1618. This project might be read as an attempt to hybridize through sire or group rotation. The book's three-column structure featured the mare's identity in the
present alongside her reproduction in the last year and subsequent year.
The central column of each folio indicated the "animals covered by Belladonna, courser" in a particular year. In 1615, the mares bred included Baia
(a seven-year-old chestnut bay of the marquis of Pescara's *razza*), Zingana
(a three-year-old dark bay of Pescara), Balzanella (a five-year-old golden
bay with star and socks on two feet, of the prince of Santo Buono), Turca
(a six-year-old bay from Santo Buono), Favorita (a four-year-old bay of the
duke of Celenza), Montagnola (a three-year-old chestnut), Pavoneina (a
three-year-old dapple gray of Pescara), Stornella (a nine-year-old dapple
gray of Pescara), Bellafronte (a four-year-old gray of the prince of Stigliano), Mordecchia (a four-year-old gray of Stigliano), Serpentina (a three-
year-old dapple gray of the marquis of San Eramo), and Learda (an eight-
year-old dapple gray of the duke of Termini). Of these mares, Baia had a
foal already—a bay without marking that "came from the Regno in the
pregnant Mother." One other mare, Innamorata, was struck from the list.[54]
The following year, a similar list was repeated with one year added to the
age of each mare, meaning that each mare was bred to Belladonna again.
New mares like Galeotta (a bay of the Archduke with a chestnut filly) and
Morella (with her morel colt) were added to the list. By 1615, Stornella and
Pavoneina had colts; by 1616, Galeotta had a new foal, as did Baia, Favor-

ita, Montagnola, Turca, and Serpentina. As the years passed, the writing grew messier.[55]

While some breeders specialized, breeders often managed multiple *razze* at once, including livestock such as sheep, goats, cattle, and pigs. Southern Italian equine stable records sometimes also include information on other domesticated animals, such as the 126 large cattle noted alongside the horses of Puglia in 1541.[56] As with horses, the cattle's heritage was monitored, as male calves were named as the "son" or "daughter" of their parents.[57] Elite breeding enterprises generally kept these animals separate, as Papa encouraged; less well-financed operations likely pastured them together. Still other breeders focused on dogs and cats, as chapter 3 traces, often using the same language of maintaining *razze* to characterize their work.

Breeding and Natural Magic

Renaissance writers entwined the practical realities of husbandry with enchanted stories of natural magic preserved from the classical tradition. For example, a wind mare appears alongside the thirty European horses representing their regions depicted in the *Equile of Don Juan of Austria*, a series of prints by Joannes Stradanus (also known as Jan Van der Straet, 1523–1605). Most of the plates in the series employ a "place-based racialized interpretation" of the stable of Philip II's illegitimate son, Don Juan d'Austria, to use Dániel Margócsy's phrasing. Margócsy has argued that this successful print series "directly influenced seventeenth-century natural history" such that encyclopedias of zoology "devoted much of their attention to the racial diversity of horses." He reads this series as representing the commonly held belief that "the qualities of a horse depended on its place of origin and the local temperature, humidity, and quality of pasture there."[58] However, unlike the other horses—all stallions—that fill the *Equile*, this equine protagonist is female; the artists gendered her with a focus on sex, depicting her as a horse so lusty that she impregnates herself through imagination and the west wind alone (fig. 1.1). Aristotle had likewise taken up the claim, locating these mares in Crete; Varro, Aelian, and Columella adopted a similar account of the unpersonified wind's fertilizing power.[59] In this story lingers a tacit recognition of unintended pregnancy. Breeding was subject to natural forces beyond human agency. Philosophical texts likewise followed Aristotle in holding that hens touched by a finger could produce smaller, sterile "wind eggs."[60] These unplanned births fused with verses memorized from Virgil's *Georgics*—an agricultural

FIGURE 1.1. A mare impregnated by her lust and the power of the west wind. Joannes Stradanus (Jan van der Straet), *Equile Ioannis Austriaci* (Antwerp: Phillipe Galle, ca. 1578). The image's caption quotes Virgil's *Georgics* 3.273, which was quoted in Columella's *Res Rustica* 6.27.6. Image taken from the holdings of the Biblioteca Nacional de España.

poem based in part on the ancient Mantuan poet's experiences in Calabria and Naples—which suffused the writing of Porta and his contemporaries. Virgil contended that the "frenzy of mares" was a conspicuous madness of nature. Wind took on the form of Venus, who "endowed them with passion" such that the mares traversed the wilds of mountains and rivers and faced the west wind as "spring revived the heat in their bones." Wind and their feverish passion alone was enough to impregnate them: "catching the light air, and often without union, made pregnant by the breeze."[61] Breeders might have loathed the wind mare's image or seen her as an excuse for the limitations of their craft, as doubtless there were some unexpected pregnancies that must have happened in the fields along the Bay of Naples.

While most animal management was quotidian, some experts offered additional secrets to their patrons. The Gonzaga family's *marescalco* Zanino de Ottolengo sought to improve animal health using astrological charts, magic, and prayer; a later note by Ludovico Gonzaga, duke of Nevers, accused him of using "spells, heresy, and superstition."[62] In his later treatise, Papa described these "secret experiments," which may or may not have been known to the other *marescalci*.[63] At the same moment that breeders fashioned horses to fit noble *razze*, experimentalists took up the practice of natural magic; this drove the Church to clarify its position on magic to an unprecedented degree.[64] Saint Thomas Aquinas (1225–1274), born in Sicily and first educated in Naples, had taught that "natural" powers, wielded by someone wise to their strength, could manipulate nature through mastery over occult forces that controlled vision, smell, touch, and hearing.[65] There was nothing unlawful about understanding cause and effect in nature, but a magical art that moved beyond nature could be more dangerous and less useful. As experimentation and wonder became a central part of fledgling scientific practice, a growing class of educated men tested their hands at combining metals through alchemy and causing images to grow and shrink through lenses.[66] However, Thomas Aquinas had also warned that preternatural secret pacts with the devil, mediated by his demons, would claim to provide the same powers, although he deemed such "unnatural" magic both unlawful and futile. Geomancers might animate earth, hydromancers shape water, pyromancers fire, and necromancers summon the dead to their bidding. Those who followed in the intellectual legacy of Thomas Aquinas differed as to whether those who claimed such powers were simply deluded, or misled into working with demons, or both. While the literate poured ink into debating these issues, the *populo* turned to magical solutions to find love and avenge crimes.[67]

In the 1550s, the young magus-philosopher Giovanni Battista della Porta argued that occult forces and animal imagination alike could shape future generations of domesticated creatures through active manipulation of nature.[68] Precocious, bold, and a consummate performer, Porta published his first book, *Magia naturalis* (Natural Magic) in 1558, at only twenty-three years old.[69] He saw magic as "of two sorts." He practiced magic that "is natural which everyone reveres," that required nothing "save the consummate knowledge of natural things and a perfect philosophy." Uneducated observers would call its results miracles because they surpassed human intellect. A magician could, through skill and talent, harness the secrets of nature, which Porta dutifully recorded in the multiple volumes of *Natural Magic*. He purported to avoid that other magic "which is wicked, full of superstitions and incantations, and proceeds through daemons' revelations." No demons were necessary for shaping nature; on the contrary, all that was needed was to know its secrets.[70]

Porta saw staging animal sex as natural magic. Like Pliny, who contended that "it is a well-known fact also, that marks, moles, and even scars, are reproduced in members of the same family in successive generations," Porta believed that parents' wounds had immediate implications for the next generation.[71] For example, he believed that given proper conditions, dogs could be made to degenerate from four-footed creatures to two-footed ones. By amputating dogs across successive generations and compelling the amputees to reproduce, one could compel "nature to produce bipedal canines." As Porta explained in *Natural Magic*,

> Thus, if we desire to generate two-footed dogs we consider how it will be carried out. When still quite young, truncate the feet, and diligently mend them so that they can be made whole. Mate those whose feet have been cut. And if from these are born those who are not bipedal, deprive these of feet. And so on. Through this operation, for however long it takes, nature herself will bring forth bipedal dogs.[72]

Like the Atlanteans, Porta argued that crossbreeding between different kinds of animals was possible; he observed it both in nature and through the orchestration of the magus. His vision of the natural world was inherently fecund and social. Living beings had a propensity to unite, returning to the universal natural fabric from which they all came. Porta explained these inherent connections by likening species to tributaries of a river, eventually mixing with one another and thereby unifying despite their differences.[73] To him, all animals were a mix of different animals. Just as the magus could cross fruits with one another, one could cross animals to bear a mixed litter. So long as animals' reproductive anatomy was of comparable size and their reproductive cycles were similar, a "monster of variegated nature" could be conceived. Wolves and dogs could mate and create the wild animals of Ethiopia. Together, lions and panthers produced "ignoble lions," whose spotted hides and lack of mane made them look rather different from either parent. Wolves, too, could mate with leopards and produce offspring, as reported in ancient texts. Such creatures were by no means only the result of human interventions. Lured by the prospect of fresh water, a wide array of African animals crossed the parched earth to the watering hole, where Porta supposed they did a bit more than hydrate. In his vision, Africa brimmed with life and provided an environment in which animals readily mixed. If animals were naturally inclined to be promiscuous and ready to mingle outside of their species, all bets were off, even for humans. For instance, citing Aelian, Porta wrote that a shepherd named Crathin, who lived near Sybarus, "became very much enamored with the most beautiful she-goat and he was tortured severely by his fancy." He lured her close with "lush, beautiful grass" and embraced his cloven-hoofed lover.[74]

Distressing as the shepherd's bestiality is, it reveals that Porta believed humans and goats shared certain reproductive elements. Likewise, animals and humans were both agents who processed the world around them through their senses, chiefly their vision, and could impact future generations through the scenes they internalized in the process of reproduction. Breeders, like the magus or the animals themselves, could intentionally erode the boundary separating different types of creatures. Aristotle had reported monstrous fusions of lions and hyenas, and the mule was a famous example of an infertile offspring from two different mammals: horses and donkeys. As they are almost always infertile, every generation of mules must be bred anew, from horses and asses, with human intervention.[75] Common knowledge held that mares often did not want to be bred to asses, and needed their pride struck down with a brutal cut to their long manes before they could be prevailed upon to couple with a donkey. Though regularly used as beasts of burden, mules still posed an enduring theoretical problem. As Aristotle had asked in the pages of *De Generatione Animalium* (On the Generation of Animals), why in the mule produced by a horse and ass "does there result something so dense that the offspring is sterile, whereas the offspring of a male and female horse, or a male and female ass, is not sterile?"[76] Porta had been less interested in why mules were infertile, but equally fascinated by the thought of kinds intermixing.

Breeders employing natural magic would, like Jacob, influence their animals through their vision and imaginations. Porta went beyond explaining the importance of a good environment to controlling the inputs that would shape animal imaginations, as "great is the effect of the mind or the power of the imagination."[77] He argued for a one-to-one relationship between what animals saw as they were copulating, what they imagined, and what that meant for their unborn offspring. Thus, when it came to sheep, anything except twigs should be removed from the ovine environs. A clean pasture made bright fleece. Peacocks and white chickens needed to mate in a birdhouse covered with white linen. The floors of their enclosure would be swept clean so no specks of dirt would undermine the color of the offspring. The ubiquitous horses of Naples appropriately provided more examples. Some breeders, wrote Porta, "urge the mares into intercourse and they fill up and decorate the stables where the mating takes place with tapestries and clothes in various colors."[78] If a breeder wanted to force the color black onto his animals, he might paint the mare with a black pigment before the hour of her romance with the stallion.[79] The stud would then fixate on the black covering as he mounted his mate, and this color in his eyes would impact the foal's color more than the mare's unpainted hue. Human women were also influenced by the environment surrounding them, and that environment influenced the shape of their children. Citing the pre-Socratic philosopher Empedocles, Porta suggested that pregnant women who often admired pictures and statues gave their offspring qualities similar to those works. Mothers in the making could game the system, giving their eyes only beautiful likenesses on which to meditate: Cupid, Adonis, or Ganymede might hang above the nuptial bed as paintings or appear as carved wood figures on the bedside table. The women also ought to be sure that they conceived in an orderly position, for it was in odd positions that "unnatural things" could be conceived.[80] There was a seamlessness to Porta's vision of how reproduction and sex worked among all different types of animals. Throughout Porta's animal kingdom, natural variation was partially the result of imagination. All creatures, from sheep to chickens to horses to humans, had a similar connection between their imagination and their vision. Sex needed to be staged, and the beauty of the surrounding world had a real impact on the appearance and behavior of future generations.

Porta promoted the artisanal knowledge of breeders as expertise in physiognomy, arguing that "for the hunters and the birders it is color that serves their craft, choosing those animals that are possessed with talents for hunting or the generation of offspring through certain particular signs."[81] This popular belief, which held that phenotype, or external appearance, provided signs that would allow one to predict an animal's qualities, was widespread in husbandry. Animal breeders and trainers often argued that color influenced an animal's character. Following the humoral logic common on the villa, duns (a dull brown-grayish color) and mouse-colored animals would be melancholic and cowardly. White horses would be phlegmatic and slow; bay horses (reddish brown or brown) hot blooded, happy, and agile. The horse trainer and author Federico Grisone argued that no color could be totally perfect, but some markings bore with them a greater likelihood of good behavior. White marks in the right places could make a fiery horse a bit calmer, "tempering the baldness which arises from dryness, that is to say from the true heat of his complexion."[82] A horse born with stockings on its right front and right hind legs was called a *travato*, and was deemed "dangerous and little-esteemed" by Grisone.[83] The theory was that such a horse developed with the two stockings joined closely together while in its mother's womb. After its birth, the animal wanted to keep those stockings together, and thus the horse's natural balance and gait were undermined with every step, making these feet easily entangled as the horse moved. Color was thus a predictor for the horse's inclinations and potential, shaping the animal's likely character through physiognomic features including the animal's head shape, form, and coloring.[84]

However, choosing a horse by color—or, indeed, breeding a horse for specific color markings—came with dangers. Natural magic worked with the grain of nature, unlike the demonic, which called on exterior influences to bend the natural world through illusions. But Grisone worried that those other influences might be afoot. Some breeders, perhaps, might make an unruly *travato* horse by harnessing less wholesome powers. Still, Grisone was certain: "There might be other reasons for either good or bad qualities in white markings," he wrote, "but since I perceive no truth in some of these occult opinions I prefer to omit them, and report solely what is seen clearly today, just as it was seen in the past."[85] Still, those fears Grisone left unwritten lingered behind breeders' work. Could they have power to manipulate or predict nature, as Bacon imagined the Atlanteans to do?

Predicting animals' behaviors in advance of their actions through physiognomic signs did not itself provoke inquisitorial attention because animals were not seen to have free will, as chapter 8, on physiognomy, will explore.[86] To be human meant to be able to choose between good and evil. For Augustine, writing in 388–390, the behavior Christ modeled in life displayed "no common rights between us and the beasts and the trees." In Matthew 8:28–34, Christ withered a tree that lacked fruit and dispatched devils into a herd of swine. Augustine wrote that "surely, the swine had not sinned, nor had the tree, for we are not so foolish as to believe that a tree is fruitful or barren of its own free choice." An inability to determine their own future fated plants and pigs to a status that meant killing them did not constitute murder. Whatever souls might reside in a tree or a pig, they did not become "wiser as does the soul of a man."[87] This view on animals was widespread by the Renaissance, and was generally maintained by the differing fragments of a rapidly splintering Christian world, although concerns about determinism and free will became increasing fraught throughout the Protestant Reformation.

Animal Education

Breeders created animals for a purpose. In the case of elite horses, that purpose was generally riding. To take a horse from paddock to arena required training. If Porta promoted the manipulation of nature, then another extensive tradition promoted nurture and animal education. Although their souls were not thought to become "wiser," horses, dogs, falcons, and other animals could all be trained, as myriad texts, especially the literature fo-

cused on training animals for the hunt, taught. To many, falconry epitomized this educational accomplishment: a bird of prey taken from the
wild, tamed, and taught to fly out into the wildness, kill for his new master, and then fly back to the man's arm and live a life in shackles.[88] Animals
born with a greater natural nobility—through beauty, physical aptitude,
or mental acuity—were generally thought to merit greater investment.

Federico Grisone built his career on the education of horsemen just
as much as the training of horses. Grisone had trained under the tutelage
of Cola Pagano, a riding master who had worked in the service of King
Henry VIII (1491–1547) before returning to Naples to educate the next
generation of equestrians via a riding academy of his own outside Naples
in the 1530s.[89] In 1550, Grisone published *Gli Ordini di cavalcare* (Orders
to Ride), which formalized the latest fashion in horsemanship, a rigorous set of training procedures aimed at shaping the ideal horse, in what
would become one of the most enduringly influential Renaissance equestrian books and the foundational text for modern dressage riding. It appeared in twenty-three separate printed editions between 1550 and 1620
and was translated into five different languages in the five years following
first publication, inviting a wider readership to participate in the elite networks of courtly masculinity held together through equine recreation.[90]
Grisone promoted Neapolitan riding and Neapolitan horses on the stage
of Europe. He influenced contemporary trainers Giovanni Battista Pignatelli, Cesare Fiaschi, and Giovannibattista Ferraro, who created an elite
equestrian network throughout European courts. Their work, in turn, provided central examples for the later contributions of Antoine de Pluvinel (1552–1620), who studied horsemanship in Naples under Giovanni
Pignatelli before writing the posthumously published (1623) *Le Maneige
Royal* (The Royal Stable) and instructing Henri III (1551–1589), Henri IV
(1553–1610), and the dauphin Louis XIII (1601–1643) in the art of horsemanship; the last was trained by Salomon de la Broue, who had also studied in Naples.[91] About two decades after its publication, Grisone's *Orders to Ride* had already traveled far and wide. In a manuscript inventory,
"Notes of books that are in the chest in 1569 in the Papal Palace here
in Viana"—which was dominated by learned, Latin works and humanist
tomes—we find a small group of vernacular texts pertaining to subjects of
daily concern in a noble's life.[92] First came fifty Latin texts like Apuleius's
The Golden Ass, the *Life of Cato*, and the poetry of Catullus, Tibullus, and
Propertius. However, nestled among Francesco Ferrosi's translation of
Vegetius's classic *On the Art of War*, M. Francesco Sforzino da Carcano's
The Three Books of Birds of Prey, and Giovanni Miranda's *Observations on
the Spanish Language*, one finds Grisone's *Gli Ordini di cavalcare*.[93] Books

on horses, birds, warfare, and the Spanish language encapsulated essential knowledge for an educated man of Renaissance Europe.[94] In 1575, Paolo Giordano Orsini brought Grisone's book and Francesco Sansonio's equine history, along with Pliny's *Natural History*, to Bracciano.[95]

The *manège* was a rigorous set of training procedures aimed at shaping the perfect horse. From pirouettes, where the animal rotated on the spot at a canter, to leaping caprioles, where the horse was required to kick out vigorously, to controlled *ciambette*, in which the animal supported all its weight on the hindquarters in a crouched rear, the *manège* movements were meant to highlight a horseman's total control. Like a leader in a partner dance, Grisone's riders were trained to control the tempo of their steed's movement. By counting in *mezzo tempo*, they slowed the horse's gait to an unnatural speed, counting time between movements with rests, just as in musical notation.[96] Through years of formal training involving many rapid circles, or *torni*, in mind-numbing succession, the goal was to create a perfectly submissive and highly athletic creature. Grisone's descriptions continue for many disorganized pages and are filled with details, such as where the rider ought to put his legs to signal different movements to the horse, how he ought to pull the reins, and how to best use his body to influence the horse's movement.[97]

The ability and willingness to learn were the horse's most important traits; by nature, Grisone's horse was flawed but perfectible. As Grisone put it, because the horse was "created by God to serve man and conform to his will," the horse needed "a capacity to learn that partially conforms to our intellect."[98] He acknowledged that the animal might be frightened when ridden through a city, past the butcher shop, or over a bridge, explaining that horses' fear resulted from youth, poor training, or underpreparation.[99] Rather than beating the animal on the head and neck, Grisone argued that one ought to encourage the horse to walk forward with a gentle pat on the neck. "You will notice," wrote Grisone, that little by little the horse "will become reassured of the thing that he fears. Then you will stop him near the thing that frightens him for a little bit, and at the time that he walks forward, remember to pat him on the neck."[100] Grisone saw horses as tools of transformation. The horse provided a venue for improvement, both of nature and in man:

> Horses have been and will forever be necessary for several levels and types of human professions—of letters, arms, and religion—a thing of worth beyond any might and a mark of honor above all other marks, a thing commanding admiration more than anything else, so that not only noblemen need them, but also men of lower ranks who through their efforts wish

to better themselves and become illustrious. Who cannot say that every Prince considers it glorious to call himself Knight.[101]

Grisone regularly used the language of "perfection," and explained that his goal was to perfect the natural horse through such sustained exercise, thereby making him the ideal companion for man; "What other animal is so sure and fearless, more equal to man than he?"[102]

The Fragility of Difference

To return to Russell's breeding strategies, I have shown how early modern breeders sought to impose positive breeding strategies by selecting animals for mating based on their appearance, sometimes with attention to their conformation and sometimes with attention to signs along the lines of physiognomy. In subsequent chapters, I will explore Renaissance projects to select by geography (breeding animals in environments like Southern Italy that were considered favorable) and to "grade up" their stock by importing foreign horses to create house *razze* of those animals. The maintenance of *razze* required human intervention. However, while most windows into the thinking of breeders emphasized enthusiasm about their projects, I will end this chapter with a breeder philosophizing about doubt and the degradation of Gonzaga family horses.

Most of the animal-related documents concerning the Roversella estate in Northern Italy are pragmatic affairs related to specific horses, or attentive to the general *otium* (pleasurable leisure) of the place.[103] At some point in the late sixteenth or early seventeenth century, Lodovico Nonio of Mantua encountered a dissonance between the health of the stable's horses and presumptions about their nobility. Whether Nonio was himself the breeding expert or consolidated the insights of other breeding experts is unclear from the fragmentary documentary record. However, what is clear is that his discoveries led him to be wracked by degeneration anxiety, frightened by the ways in which climate, food, and water could cause progeny to become completely different from their parents. In an incomplete memorandum, Nonio collected his "Particular Observations on the Illness that took place and is occurring in the races of Your Serene Highness in Roversella."[104] Nonio began by discussing the origins of the equine races that had found their way to Europe:

> It is necessary in the first part to reflect that, when from the past Sovereigns the following races were bought and like also with their fathers, these were conducted from those countries, from which they also have their names,

like the Turks, Barbs, Jennets, *Ubini*, and Courser horses, all of those come from countries of excellent climate and as a consequence of the exquisite foods, of perfect water so that it is necessary to surmise that they were having perfect health and in consequence a good nature which would be conserved for that space of time, but little by little they lost their robustness and also they lost their strength, and this weakened their complexion so that at present one can say, that nothing was saved but the names of their origins, and it conserved their nobility.

While the court breeders generally continued to call each type of horse by its place of origin, Nonio doubted the validity of this approach. He did not go on to describe the continued excellence, characteristics, and strengths of each type. Instead, Nonio noted that something seemed to have gone terribly wrong; some foundational assumptions behind the breeding project seemed to him to be incorrect. The breeders had a problem. The Northern Italian mists seemed not to sit well with animals imported from distant lands. Even as the stables swelled, complete with European variations of famous *razze*, Nonio noted a steady degeneration of their animals.

Even though he was clearly disappointed about the stock's proclivity to illness, Nonio did not suggest that the European emissaries who had imported them had been duped, left to purchase sickly or mediocre livestock rather than the finest of the bunch. Without any information about their health before the move, he supposed, as if to reference a perfect state of nature, that the equids had been healthy in their ancestral homes. "In consequence" of their original health, "a good nature" was conserved in the animals for a "space of time" after they moved to Europe, he wrote. However, "little by little they lost their robustness." Now, Nonio complained, nothing of these excellent origins remained in the horses except the names of their races, and those names "conserved their nobility." So far removed from the territories in which their original merits were earned, the horses became sickly steeds held up by their noble title. This observation is not so different from the critique of inherited nobility, leveled by Gaspare in *The Courtier*, that nobility could not, in fact, be inherited by blood, as discussed in chapter 3.

Nonio cast aspersions on the Lombard countryside's effects on animal bodies—an observation that matched an ancient Roman idea. Columella (4–70 CE) had also questioned the viability of moving working animals from one environment to another:

Indeed a native ox is far superior to one which comes from elsewhere; for it is not disturbed by change of water or food or climate and is not troubled by the local conditions. . . . When, therefore, we are obliged to bring oxen from a distance, care must be taken that they are transferred to country which resembles that in which they were born.[105]

Nonio took a similar position on the transfer of humans between regions, reasoning by analogy to explain the decline in the health of the imported European animals:

> Also in men there being many that if some accident brings them from one country to another, that which for their misfortune takes them from a good climate and goes to a worse one, one sees that, in the course of time, these healthy and robust ones that they were weakened and losing as it were the figure, the beauty, the complexion, and in large part their health.[106]

In Nonio's view, relocation not only posed a danger to individual animals, but also harmed subsequent generations. "Their offspring," he wrote, "preserve hardly any sign of their origin as, on the contrary, it is shown that they are of a bad country, and going to a good one they earn in brief good health and little by little they make a progeny completely different from the first." Having observed the Roversella stock, Nonio was convinced that the past sovereigns who imported these animals in the first place must have seen something fundamentally superior to the animals he saw. Over several generations, quality of current country—rather than race or heritage—was much more important in deciding the animal's general quality. When moved to a new region, horses' progeny came to reflect that region "little by little." Those "bad things that occurred in these races in the past and that occur in the present," wrote Nonio, came from the climate around the animal, the climate that created the food, water, and air it consumed.

Nonio's individual doubts fit into a larger seventeenth-century loss of optimism about transforming nature; by late in the century, it had become fashionable to complain, as William Cavendish did, that the Neapolitan breed and other Italian *razze* had decayed from their original greatness.[107] European patrons and breeders had managed to bring animals from around the Mediterranean world to the fields of Europe. Something, however, was lost over time. Perhaps such degeneration was the result of an illness, perhaps of the climate. Did Nonio's misgivings capture a moment when the seed was not tended with sufficient care and the resulting creature had grown wild? Neglect seems implausible in such intensively monitored stables. Wildness seems not to have been the problem either, but rather withering. For Nonio, *razza* was something fragile, and it could

not be simply transported from one place to the next. Early modern thinkers and practitioners wrestled with a complex reality in which interventions in breeding and training were only partially effective. From contra-
dictory theories and artisanal epistemologies of breeders, this leads us to
the question that motivates the next chapter: How did breeders seek to
make the work they put into *razze* more permanent?

CHAPTER 2

Razza-Making and Branding

Breeders helped popularize the language of race. Noble families patronized breeders and other experts to create specialized animals, along with nonliving commodities like wine and other craft products produced by specific houses and exchanged as gifts among them. Elite houses' internal records and correspondence concerning the exchange of gifts, which provided the fibers of social connections across European courts, amplified the use of the language of race, further promulgating a term and associated ideas about maintaining and distinguishing a population. Throughout the sixteenth century, Romance-language variations of *razza* referred to studs, breeds, and sometimes races, often emphasizing the artifice of the reproductive work that went into maintaining them. As we have seen, breeders of animal *razze* monitored generations of beings, raised them, altered them, guided their reproduction—often in tandem with astrological and environmental considerations—and then repeated the process anew. Cognizant of *razza*'s ephemerality, some feared that unrestricted breeding of animals would lead to a distressing erosion of unique variations. Perpetuating difference, in other words, required constant work. Animal branding provided a tool by which to render breeders' temporary work permanent across generations and distinguish bodies that appeared more alike than different.

Conceptual Blending of Razza

In Renaissance Europe, noble steeds often had a breed.[1] Throughout the sixteenth century in Romance languages, variations of the term *razza* were deployed alongside other terms to characterize lineage, stock, breed, and race. Like the overlapping conceptual packets of horse and nobility that defined the elite social position of knighthood, *razza* is the result of conceptual blending between terms related to domesticated animal breeding.[2]

As a vernacular term often describing animals, *race* has been traced to the late medieval period. By most accounts, *razza* and its variants have three possible origins. The first two are the Latin *ratio* or *generātiō*, although philologists have questioned this potential etymology.[3] More recent scholarship has emphasized origins in the Middle French term *haraz*, meaning "a stud of horses" and mares kept only for breeding.[4] *Haraz*, in turn, came from the fifth-century Latin *haracium*, a term that likewise applied to royal stables and studs of horses.[5] *Haras* was widely used; thirteenth-century English verses in the utopian poem "The Land of Cockaigne" describe *harace* or *stode* as an enclosure or establishment in which horses and mares were kept for breeding.[6]

In this study, I am interested in the use of race terms as they were applied to animal breeding in the late fifteenth, sixteenth, and early seventeenth centuries. This is a subset of how these terms featured in the full spectrum of meanings associated with variations of the term *race*. As this is a history of science focused on animal breeding and its analytical repercussions, I point readers to the many important works of scholars of medieval vernacular philology to pursue the precise origins of *race* across European languages. Given these parameters, I offer a few sentences to summarize a lively field of scholarship. Which vernacular Romance language came to the term first is hard to determine. What is clear, however, is that some variation of *race* spread through European languages like wildfire. Historians have recently sought to root the history of conceptual blends associated with race in metallurgy, fabric-making, and animal husbandry. The complexity of this semantic history seems to indicate that *razza* and related terms emerged from a confluence of overlapping thinking, as semantic fields converged without one singular origin or result. Related terms gained currency throughout the sixteenth century, by which time they were used to describe both humans and animals. New combinations of words generated new metaphors, which could facilitate new connections; those new connections created categorial boundaries that then distorted differences and likenesses.[7] Rather than disentangling these, I aim for this book to preserve their complexity, and I conclude that the term *race* itself has diverse origins.

Historians have offered many different readings of the history of race, beyond the term itself. In the historiography on race, one might see five strands of difference: ethno-religious-national identity, climate, natural slavery and embodied hierarchy, physiognomy, and selective breeding. Scholars focused on ethno-religious-national identity have framed inquiries around *ethnicity* and *otherness*.[8] A current thread of scholarship and activism contends that these differences might be better understood through attention to racism.[9] Scholarship focused on understanding climactic links to race has analyzed the early modern belief that the geohumoral nature of the environment helped determine inhabitants' complexion, health, and *calidad* (the sum of a person's embodied social station, including profession, dress, and behavior as well as familial heritage). The consumption of foods produced in each climate was also thought to determine bodily difference.[10] Works attentive to the history of Aristotle's concept of natural slavery consider classical ideas of servitude and engrained hierarchy, and how these notions bolstered the slave trade.[11] I see this book as examining the last two strands of difference—selective breeding and, to a lesser degree, physiognomy—as they pertain to the meaning of race in the early modern world.

Historians now have a strong sense that race, a term often used to designate selectively bred animals in the late medieval and early modern periods, became a descriptive term for visible human difference.[12] A tour of transmission hypotheses is helpful to think with. Vernacular speakers seized on this language and set of concepts differently in different regions. In some parts of the Mediterranean, such as Spain and North Africa, human ethnicity and animal breed were conflated around the Spanish use of the term *raza* by the fifteenth century.[13] Early mentions of race in Spanish—such as in Alfonso de Palencia's 1490 *Universal Vocabulario*—emphasize defects, stains, darkness, and damage in pottery, textiles, and gems; thus, Castilian *raça* was adopted early on as a metaphor for human imperfection and sin.[14] Writing about early Spanish, Ana M. Gómez-Bravo has therefore presented "*raza* as a culturally and linguistically situated metaphor built as a transfer from technical language into a coopted everyday vocabulary, facilitated by a common familiarity with the term through the pressures of religious and administrative language"; she proposes "that the early bipolarization of *raza* appears first in the conceptual transfer from the textile trade reinforced through semantic overlapping transfers from gemology and metallurgy and, to a lesser degree, veterinary lexicons."[15] David Nirenberg has argued that the Castilian word *raza*, with the significance of pedigree involving the passage of inherited characteristics, applied to horses in the first quarter of the fifteenth century. In the first half of that century, the term was applied to certain people to describe their origins, and sometimes an idea of taint associated with them.[16] Jews and Moors were among the first described with this language. Still, the primary use was equine. Caste, or *casta*—which was used in the Spanish context more than in the Italian documents with which this chapter is concerned, or in other European courtly contexts under Italian cultural influence—sometimes was a direct synonym of *raza* and sometimes carried a connotation of purity and nobility in reference to animals.[17] Other developing national traditions emphasized further meanings of "race" beyond defect. In France, by the early sixteenth century, "race," "blood," and "nation" were becoming deeply intertwined in discussions of humans and animals.[18] Charles de Miramon notes that the Middle French term *haraz* had transformed into "race" by the last quarter of the fifteenth century.[19] During the same period, German courts adopted the Italian term *razza* and arranged their animal collections to draw parallels between genealogical portraits of human nobles and the "noble horses in the stable," which, as Margócsy has suggested, "could also be considered as members of a genealogical chain."[20]

In Italian-speaking contexts, the word *razza* took on a mix of taxonomic and generative elements. John Florio's Italian-English *World of Wordes* considered *razza* "a race, a kind, a broode, a stocke, a descent, a lineage, a pedigree." In Italian, everything that was bred had a race, and everything had been bred. In any event, many early usages of the word explicitly combine the semantic fields of selective breeding and lineage, entwining a botanical language of roots with a language of stud farm animals, to characterize it metaphorically. This idea is substantiated by Florio's 1611 Italian-English dictionary, which mentions race under *principio*: "a beginning, a ground, or chief original, race and stocke of thing."[21] The verbiage of breeding has its own expansive varieties. Of these many terms, some were linked to *generáre*; Italians spoke of *allevare* (Florio 1611: "to foster, to breed or bring up") and Spaniards of *criar* (Covarrubias 1611: to invent something from nothing as God did; or, an alternative meaning, "often taken as engender, which is to say this land engenders valiant and robust men . . . or in Cordoba *se crian* good horses"). The fact that the word *razza* came to mean what we now think of as "race" does not mean that other words, such as the Latin *gens*, *stirps*, or *ethnos*, did not carry meanings associated with the modern notion of "race."[22] In Italian, as in Spanish, blood (*sangue*) carried similar connotations, although it was not prominently used in the breeding documents I have read: Florio defined *sangue* as "a kinde of blood. Also kindred, a stock, or parentage, a race or lineall descent. It is also taken for life, vigor, force or strength."[23]

Marking Noble Families

In the increasingly textual world of the sixteenth century, where property and status had to be affirmed through records, branding emerged as a specific and valuable tool for correlating text and object and conferring possession. Owners sought to write on the commodities that they possessed because that created a full circle between written text and physical life. It also provided shorthand, because one would not want to go through the

tedious process of recording details of a particular animal, person, or property, especially if one possessed many of them. Brands allowed owners to
multiply that information across diverse individuals or things, standardizing a wide array of bodies and property. Brands enabled distant supervision because they were readily discernible without localized knowledge.
They increased the commodity's exchangeability by conferring presumed
value through the brand itself, so long as the sign of the brand and the sig-
nifier of the commodity branded remained linked. Through brands, one
might confer property and status, and vice versa. To put it anachronistically, early moderns performed brand extension, putting their names on
associated things to extend the prestige of their houses to those objects
and to reciprocally benefit from those things' prestige.[24]

A growing branding economy standardized the localized achievements
of disparate European courts and made them exchangeable across wider
regions, complete with their own connotations of hierarchy and economic
and social value. A documentary tradition recorded specific details of
property with such care that the living world seemed that it might possibly be linked to the written word. Paper enabled this technology of record
keeping to become more widespread, more sophisticated, and cheaper,
such that individuals and families could generate their own records, creating a specific paper trail that became more important as printing accelerated it.[25] Likewise, the increased practice of double-entry bookkeeping
based on the rise of Arabic numerals was part of a broader increase of record keeping that fortified the brand economy, creating a legally binding
correspondence of things written on paper with who-owned-what and
who-was-what in the physical world, along with their rights.[26] A growing
bureaucracy, committed to specificity, transformed animals from named
individuals or general types to specific, branded kinds, making a named
horse a barb of a particular house race, as we will see in the case study
of the vast records of Neapolitan horse breeding. Indeed, these developments were part of the same movement that led to the growing interest in
identification cards, which provide specific bureaucratic details that identify individuals, but which are themselves deeply impersonal and unconcerned with individuality.

Even though most of the brands themselves, recorded on perishable
flanks and surfaces, are now lost, depictions of brands offer a window
into a specific type of economic system. Franz Fanselow's theory draws
a contrast between the bazaar economy and the brand economy. In a bazaar economy, featured goods were unbranded and ungraded. A prospec-
tive buyer would have little access to information to evaluate their quality before purchasing them. Rather, the buyer would be forced to trust

the individuals with whom he or she exchanged. If the commodities did not meet expectations for quality, the buyer would need to take aim at the seller directly or the larger bazaar of which the seller was a part. By contrast, in a brand economy, commodities are graded and marked with a seal. They are then exchangeable for similarly graded items. Goods are standardized and can be substituted for one another depending on their brands. In the wider community, consumers exchange information about various types of products. Because the individuality of the goods and traders is less important in this system, it is easier to quantify branded assets; types of things of greater or lesser value emerge.[27] The standardization of products through brands therefore prioritizes quality, exchangeability, authority, and ownership.[28]

In modern English, the word *brand* has many meanings. It is at once the act or result of burning, the mark made with a hot iron, a figurative sign of stigma or infamy, a trademark, a mark of ownership, or a particular class of goods. The scope of the English term hints at its legacy of conceptual blending through which burning, marking, ownership, and commodification came together to create such a multivalent word. In the Italian-language documents with which this chapter is concerned, branding was often described using variants of the term *mark*, which took on similar meanings of imprinting through fire and standardization. In Italian, a *marchio* referred to "a brand or marke upon horses, also a branding iron," according to Florio's 1598 dictionary. To *marchiare* was to apply the *marchio*, "to marke or brand horses," but it was also linked to marching in rank, as with soldiers. Likewise, in Spanish, a mark (*marca*) was "a sign that one makes in pieces of worked gold and silver and other things." The mark in such a piece would ensure its value of a certain amount, and the language slipped into referring to coins as *marcados*. Marked pieces had a guarantee of value. As Covarrubias's 1611 dictionary noted, *marca* also referred to "the mark or iron that they put on horses of *raza*."[29]

What were these marks, exactly? Often they were variations of the coats of arms adopted by elite families to emblazon their houses, furniture, and the standards they unfurled in war. For example, in a city filled with the Medici *palle* (the six-ball family crest), the Archivio di Stato di Firenze holds records of thousands of *stemme* (family coats of arms) related to various Tuscan families dating back to the twelfth and thirteenth centuries.[30] *Stemme* labeled families and noble houses, while a personal *impresa* was a visual token associated with an individual.[31] While sometimes in conversation, these symbols were quite different. The *stemma* was a stable symbol of the house, while the *impresa* was an individual symbol adopted by Renaissance elites. These latter symbols linked a picture associated with the body with a motto associated with the soul, and were used as witty tool of self-expression; as contemporaries explained, "An *imprese* (as the Italians call it) is a devise in picture with his motte or word, born by noble and learned personages to notify some particular conceit of their own"; this link will reemerge in chapter 8, on Southern Italian physiognomy.[32] If *imprese* were personal mantras of ownership and play, then family *stemme* built on a botanical allusion to designate inherited possessions. By 1611, John Florio defined the term *stemma* as "any stem or branching stalke. Also the stocke, the race or blood of a house or familie. Also a garland or chaplet of flowers."[33]

Across these many symbols, legibility was not a simple matter. The marks differed from the coats of arms but were related to them, sometimes as a simplified version of the family crest and sometimes as a separate symbol with linked iconography. The Renaissance world was filled with signs that needed to be read and decoded, replete with meaning that would have been clear to premodern viewers, but that we moderns struggle to see without instruction, as we lack the context and have been trained in a different intellectual style, as Michael Baxandall has argued in the case of art history.[34] As we will see in the case of physiognomy, animals, plants, and things took part in a symbolic language that often flowed seamlessly into the representational and allegorical languages of art and Christianity. In this visual morass, brand symbols carried specific meaning while also serving the general function of displaying ownership and standardizing commodities.

Branding as a Tool in *Razza-Making*

As it happens, breeders' efforts—to control the environment, influence animal imagination, and create offspring of ideal color and body shape— did not always prove effective. Signs of *razza* could be obscure to the untrained eye. Patrons, however, wanted to make sure their breeders' pricey work was legible. These Renaissance aristocrats lived in a world laden with signs, symbols, and correspondences. Some of these signs and symbols, especially those adopted by noble houses as family *stemme*, were supposed to represent actual features of nature. Others were elaborate puns that made artifice both close to nature and an improvement on it. *Stemme* and *imprese* were intensely important and meaningful devices in the Renaissance world, as expressions of family pride and visual play, which connected family to the system of signs and correspondences thought to organize the natural world. In a society committed to sumptuary laws, keen to impose order on human dress so that garments instantly revealed social

standing, the same needed to be true for animal companions. Iron branding created "distinguishing signs," to adopt Diane Owen Hughes's phrase,
akin to the earring that sometimes marked members of the Jewish community, or yellow veils worn as a sartorial mark, with an important excep-
tion: when impressed upon the skin, they were permanent.[35]

Branding proved especially important when breeding aspiration outran reality, as it did for horses. While the rhetoric of noble races increased
rapidly in the sixteenth century, the horses that breeders produced were
simply not that different from one another. Even in the idealized representations of horses from thirty European regions in Don Juan's *Equile*,
while the horses had distinct names and qualities ascribed to them, they
often did not look particularly different from one another.[36] Selective
breeding techniques known at the time were able to produce the multiplicity of equine *razze* that Renaissance aristocratic families claimed to
have. They could not, however, do so quickly. Aristocrats wanted immediate change, visible over the span of a few human generations. Such shifts
are possible with dogs, which can be reshaped with astonishing speed;
but the equine genome is simply less malleable. Thus, breeders turned to
branding: burning variations of their employers' family crests onto their
animals' haunches and cheeks using a hot iron.[37] By doing this, patrons
meant to render an animal's history, ownership, and characteristics clear
at a glance. As Duncan has written, "Permanent marks, burnt into the skin
of the cheek, flank, or shoulder of young horses, served as equine 'assay'
marks, giving proofs of the horse's value in the same way as an artist's signature might indicate the value of a sculpture or a painting."[38] Bred populations were therefore branded as the desire to render nature and commodities intelligible outstripped the ability to make them viably different.
Just as the novice gardener might leave the artificial descriptive tag on a
plant to remember its taxonomy and qualities rather than intuit them from
its growth, brands applied artificial marks to living bodies as an attempt
to render them more standardized and easily taxonomized. The process
of branding flesh was quite similar across mammals, though it always carried the danger of infection or hastened mortality brought on by the shock
of the wound. Animals to be branded would need to be restrained so that
the iron would make a clear impression on their skin. Before applying the
brand to an animal's flesh, the area's hair might be trimmed to prevent the
ignition of the oil. By the seventeenth century, cattle branding irons were
thicker than those applied to horses, indicating the elaboration and refine-
ment of a ubiquitous practice.[39]

The slippage between *razza* and brand, therefore, was entwined in several major historical arcs like the development of the bureaucratic state,

merchant capitalism, the rise of printing, and growing levels of literacy and numeracy. The sixteenth-century developments are no origin story, but rather an expansion and codification of a larger-scale and bureaucratic development. By the late Middle Ages, the rise of secular bureaucracies associated with courts (both royal and judicial) accelerated the rise of a written system replete with notaries, scribes, and the legal documents they prepared. In the late fifteenth and early sixteenth centuries, European culture experienced an acceleration of bureaucracy such that the difference in quantity became a difference in quality. The Renaissance became a world of record keeping and the writing down of everything owned, down to the last broken stool, to ratify who owned what. Elite families operated in a world where paper documents—whether to interact with the bureaucracy, move money, deal with courts of law, or defend rights of inheritance—were essential; families therefore made and retained records of both their own genealogies and their possessions to survive in an increasingly litigious society in which even nobles had to produce paper proof of generations having held that status. A section of the animate and inanimate European household was therefore formalized, rendered tradable, and marked for exchange through branding; this likewise applied to the active removal of human personhood as European social hierarchies extended into human slavery through the growing colonial system. Branding and trade saw social and juridical denial to humans of rights and their assignment of a status only marginally better (because manumission was possible, among other things) than that accorded to domesticated animals.

Katrina Keefer has argued that branding is an act of violence against an individual related to the process of commodifying a person or thing. Since antiquity, criminals had been branded as punishment so that their bodies would provide perpetual evidence of their deviance. Romans branded runaway slaves with the letters FUG on their foreheads to label them "fugitivus" to all who saw them. While punishment and commodification were by no means the same thing, many aspects of transatlantic slavery relied on the conceit of war captivity; thus, branding could mark criminality, resistance to slavery itself, and enslaved status. As companies took on an increasingly large role in European trade by the seventeenth century, slaves' brands tended to feature the company that had purchased them. However, if human cargo had different owners, then each individual owner's mark was impressed on their slaves; this is akin to what we see in the case of horses.[40]

Over the early modern period, the technology and look of brands changed little, although their application expanded. The marks in Renaissance livestock brand books resemble symbols from the Santiago alias Potosí, a Spanish schooner in the nineteenth century, and the human brands recorded in the register of liberated Africans in Sierra Leone.[41] Joanne Rappaport has examined the relationship between the branding of humans, physiognomic perceptions of a mixed-race physical aspect, and the category of *mulato* in seventeenth-century Bogotá: administrators were troubled not by the fact that a five-year-old slave had been branded on his cheeks with the letter S and a nail, but because the boy was a son of a Pijao Indian, a group that was supposed to be free.[42] The violent work of branding to render *razza* permanent and living bodies distinguishable as property directly and negatively affected Africans captured and people owned through the slave trade, as will be explored through the observations of Florentine merchant Francesco Carletti. These examples and others suggest that, while ideas about physical aspect and race diverged and crisscrossed during this period, branding remained a relatively stable, uncontroversial practice.

The Razza-*Makers of Naples*

To understand how breeding related to branding, I turn to Haniballo Musulino, a stable master, and his work to manage Southern Italian *razze*. An exploration of records of state-sponsored and private breeding projects in the Kingdom of Naples for the years 1541 to 1543 offers a window into records of livestock breeding. Under Spanish rule, the Neapolitan viceregal office of the Regia Camera della Sommaria collected records of breeding enterprises, including the birth, death, and sales of horses to the Spanish crown's patrimony; these records are now held in the Dipendenza della Sommaria in the Archivio di Stato di Napoli. These documents can be read as part of a larger narrative in which, because of the horse's military and social importance, early modern European states were invested in making sure they had an adequate supply. An English act in 1536 required deer-park-owning gentry to keep a specific number of mares that were taller than 13 hands to be covered by stallions of 14 hands; by 1540, King Henry required that mares 15 hands tall be kept for breeding on the English open commons.[43] For reference, today the threshold between a pony and a horse is 14.2 hands, or about 57 inches, at the withers (shoulder). In Castile, knights had been required to keep horses, and upwardly mobile non-noble knights had claimed their noble status through exemptions from service on horseback. Philip II revived the legal status of *caballero*, which emphasized the civic responsibility of horsemanship. His 1562 Iberian survey requested information about pasturage, local horse-raising practices, and the quality and quantity of horses of *casta*.[44]

Through the efforts of local animal experts, European courts and states managed the complex, geographically dispersed administrative enterprise of caring for an empire of animals sorted into *razze*. The troves of documents counting and organizing Southern Italy's animal capital grew from the same bureaucratic mindset that organized the galleys of Naples and the *tercios*, the Spanish infantry trained in Naples and marched up the peninsula to fight Spain's endless wars.[45] The range of documents at an official's fingertips was immense in both scale and variety, including information from monthly grain consumption to a particular horse's type, age, size, and sex. As master of the stables for Aniballo Ruffo, a low-level nobleman in Calabria, Haniballo Musulino's responsibilities extended from breeding horses to signing off on the books. His charges—including Dapple Grey Ferrandino, Prince the Bay, the Bay from Campania, Gusterra the Grey, the Sorrel and Grey Turks, Tirpaldo the *Ubero*, Muntire and Cosmano the Bays, and Sondaro the Sorrel—ate their way through bushels of barley.[46] A scribe recorded the horses' names and consumption, the stable master signed the records, and an accountant drew up the totals in Latin. Musulino personally verified the information: "I, Haniballo Musulino, Master of the Stables, faithfully promise that the following information is true and I attest to this through writing in my own hand."[47]

Large breeding enterprises had so many horses that, while the animals did carry individual names, these names alone would not suffice to identify them. For example, a 1542 record of the mares and foals of the royal race of Calabria divided the horses according to five different types: large, medium, small, jennet, and common.[48] In the entire Calabrian stable, there were only 33 noted stallions.[49] In Puglia, in 1541, 22 males had been gelded (castrated), losing their reproductive potential; just 17 stallions avoided such a fate.[50] Although these documents certainly shared general features, their organization and information depended on the priorities of the scribe or stable master who recorded them. Renaissance breeding, like so many other features of the period, oscillated between focused attention to detail that seems shockingly modern and deep inconsistency.

The *razza*-makers of the Kingdom of Naples were in demand. Italy's princes, dukes, marquises, ecclesiastics, and foreign elites all wanted their own designated *razze* and hired breeders to tailor animals to their needs. As we have seen through Grisone, Italian horsemanship was considered particularly superior to other European traditions in the sixteenth century, and so too were Neapolitan horses. Neapolitan and Mantuan horses were exchanged in 1520 at the Field of the Cloth of Gold.[51] The Neapolitan courser was particularly desired. When Thomas Blundeville listed English horse breeds, he included the Neapolitan courser and the Sardinian alongside Spanish jennets, Hungarians, Frieslands, Flanders horses, Turks, and Barbary horses.[52] Giovanni Bernardino Papa's manuscript "Treatise on the *razze* of horses" for the Tuscan grand duke described in detail "the most famous *razze* of the Kingdom of Naples," as well as those of Tuscany, in celebration of his patron's father. Papa argued point-blank that "above all the horses that I've seen and trained in Italy two are the most numerous and productive of in the large part good horses": the first were the horses of the grand duke of Tuscany, and the second were "those of the Catholic King of Spain," which were bred in the Kingdom of Naples; he clarified that "the principal *razze* that are in the Kingdom of Naples are those of the Catholic King of Spain."[53]

From Haniballo Musulino's and other breeders' records, it is clear that there was no one singular Southern Italian horse breed. Coursers for racing and jennets for riding, large, medium, and small, from Bari, Calabria, and Basilicata filled stables and pastures across the Regno. No one set of characteristics reflected all their varieties or abilities. Non-expert simplification erased regional animal cultures in favor of anthropomorphized animals fit to human stereotypes. For example, Stradanus's *Equile* portrayed the horse of Naples standing in front of the iconic view over the Bay of Naples with the ominous volcano Vesuvius in the background. The Latin beneath the image translates as follows: "Famous Naples and likewise happy Campania / Brings forth horses, suitable for wars, and powerful weapons, / And certainly for instruments to break up lumps of stagnant earth / And to endure cheerfully the heavy burden of the plow."[54] In truth, however, there was no singular horse Neapolitanus, and certainly not one lonely strain happily working in the fields.

Horses were branded according to an irregular set of customs associated with who owned them, who bred them, and what variety they were. In the Regno, horse brands were mostly equid coats of arms that designated the horses of various families.[55] These marks, characteristic of the noble houses that made the given *razze*, rendered the animals' bodies legible to an informed readership, helped by manuscript and print books detailing the origins of such marks. Illustrated manuscript books of brands created a collectible catalogue of the animals and the prized families from which they came. For example, a reader of "Los Hierros de los cavalos que ay hoy dia en Andalusia y Cordova, MDCXCVI" ("The Brands of Horses That There Are Today in Andalusia and Cordova, 1696"), now housed in the Austrian state archives, would find markings that identified horses of unknown breed (*casta no conozida*), the RE mark that labeled the king's breed from Aranjuez, and signs for breeds linked to private individuals such as Don Gaspar Serrano and D. Alaibar Pitus, whose horses bore a brand that looks like an armadillo.[56] Many similar books are in libraries and archives today, suggesting that they were a popular genre in private collections throughout Renaissance Europe. If it was worth creating a horse *razza* in the first place, the family would likely have been interested in a visually delightful mode of recalling their investment.

To trace these marks through books, consider the pocket-sized post-1530 *Merchi delle razze de Cavalli de Prencipi Duchi Marchesi Conti Baroni de tutte le Provincie del Regno di Napoli,* currently held by the Società Napoletana di Storia Patria. This manuscript acts as a map of Southern Italian horse ownership, including those animals marked by "Princes, Dukes, Marquises, Ecclesiastics, Private Houses, Foreign Powers with Absolute Monarchs, the Papal State, Cardinals, and the Ottoman Turks."[57] A reader of such a book could tell at a glance whether a horse before him was from the court of Mantua, Florence, or Naples by simply looking at its haunch and cheek, as shown in figure 2.1.[58] Because the Gonzaga bred multiple house *razze,* they used a different mark for each variety of horse. The Gonzaga crown appears on top of four brands for different *razze* bred by the duke of Mantua: *villani* (a type of riding horse) were represented with a C; jennets with CA in a diamond; turchi (Turkish horses) with a combined TC; and coursers (racehorses) with a sunburst.

Merchi delle razze offers a cross section of fractious local power relations. The manuscript organized the marks and their owners around hierarchy, like the society it was made to reflect. It began with the powerful princes and extended down to private houses. Princes had 69 recognized brands, dukes had 67, the houses of the marquises had 43, and ecclesiastics had 53. Foreign sovereigns had 17 marks, as well as separate categories for the papal state, cardinals (specifically Cardinal Di Lorenzo), and Turks, each of which had one. However, the marks from the highest-ranking houses were numerically surpassed by brands that characterized private houses, which numbered 143. The manuscript gave readers detailed social information to help them in navigating the Regno's stables and processions. In its final pages, the Neapolitan hierarchy that had been carefully dissected throughout the book was then integrated into Europe's system of social order.

The detailed marks and the notes written around them give us a sense of the importance of horse breeding and branding to local nobility. Horses were both representations and members of society, marked by their owners as both participants in and products of their domains. By looking at their hides, one could read a virtual who's who of elite families interested

FIGURE 2.1. Brands for the horses of Mantua. *Merchi delle razze de Cavalli de Prencipi Duchi Marchesi Conti Baroni de tutte le Provincie del Regno di Napoli*, MS segn. 22.D36 alle cc. 182r–v, 183r–v, 209r–v. Images published with the permission of Società Napoletana di Storia Patria. Any other reproduction or usage is forbidden.

in hoofed investments. Federico Grisone's house in Puglia raised an array of horses, most likely for the *manège* stables in Naples. Several other noble families with a prestigious place in the training world also had their own *razze* of horses: Antonio, Cesare, and Luigi Pignatelli all raised animals that they marked with distinct brands. The Cavalcante, an illustrious and extended house of Florentine origin, marked their own horses. The same was true with animals from the house of the Tufo, which bred horses worthy of export to Florence, selling to clients such as Ferdinando I de' Medici.[59] In Puglia, the Leonardi, Manti, Martori, Bank of Quarati, Antonio Rugiero, Count Bernabo of Basilicata, and Pardo Papacodo all bred horses marked with recognizable brands as part of a wider pattern visible throughout the Regno. Many of these *razze* continued into the seventeenth century. While Papa described *razze* of the local Spedale degli Innocenti in Tuscany, he also described *razze* of the house of Tufo in

Basilicata, the prince of Stigliano, the duke of Gravina, and the marchesa de Bienza, among others, and broadly celebrated the horses of Abruzzo and Bari.[60]

Brand books were relevant only if they contained current, useful information that kept the reader up to date with equine geography. When the prince of Ruoti was no longer satisfied with the wheel brand that punned on his town's name, he changed his mark to feature a fruit and the cross. Owners of brand books hastened to record such alterations, whether in manuscript or in print.[61] The link between brand and owner was often deceptively straightforward, especially far from established centers of breeding. But brands could be fraudulently imposed or bought out. Renton has shown that branding in the Americas was legally required but not systematically applied. By the 1560s, indigenous peoples in Cuenca were imitating the notches that marked horses.[62] More broadly, emendations to brand books in both manuscript and print can be read as the continued efforts of animal collectors to map their textual world onto the changing physical world around them, and we can imagine that these books were consumed by a larger group of readers interested in this world of animal capital from the outside.

Publicizing Italian Horse Brands

Among the many similar published brand books, we can capture horse breeding and branding in Southern Italy through two examples published in Venice: *Libro de marchi de cavalli con li nomi de tutti li principi et privati signori che hanno razza di cavalli* (Venetia: Nicolò Nelli, 1569) and Bernardo Giunti's later version of the same book, *Libro de marchi de cavalli* (Venetia: Appresso Bernardo Giunti, 1588) (fig. 2.2).[63] The citizens of Venice itself, a city of water and boats, had equestrian skills that were the butt of centuries of jokes.[64] It is not clear that a patron funded these volumes; noble houses did not need to pay to publicize their brands and *razze* to lay readers. These texts, written in the vernacular, were produced commercially, selling to a reasonably large audience by capitalizing on readers' demand to understand horse ownership. In short, both printers seized a market opportunity. While measuring the literate Italian demand for information about horsemanship remains challenging, numerous practical books on the subject suggest that in late sixteenth-century Italy, knowledge of equine brands was on the rise.

A comparison of the volumes issued by the Nelli and Giunti presses gives the overwhelming impression of similarity, yet the changes between 1569 and 1588 editions are revealing. Some marks appear quite regularly.

Pignatello family marks appear in the Neapolitan manuscript, the Nelli edition, and the Giunti edition, though the individual owners listed differed between the manuscript and the printed texts; the Nelli and Giunti editions give the brands of Cesaro, Fabritio, and Ferrante Pignatello.[65] Some houses had multiple marks. For instance, the Count of Potentia marked his race of coursers from Basilicata with either the letters "DO" or "CD" on a shield. Both books included comments about the utility or beauty of different types of horses. For instance, Zari Incencio's were reported to be "not very attractive but useful."[66] By contrast, Lord Antonio di Rugiero's animals were "small horses, but pretty and good."[67]

Each press relied on different plates, and Giunti removed some materials from Nelli's original edition and added others. The jennets of the race of the prince of Stilliano acquired an entirely new symbol: a diamond rather than a circle.[68] The race of the duke of Leone, which had been noted in Naples in 1569, was less important decades later; it does not appear in the 1588 edition.[69] Likewise, following his death in 1575, Count Ercule of Contrari, the Ferrarese gentleman, no longer had his mark recognized in 1588.[70] The race of the Count of Altavilla of the House of Caraffa had apparently also lost importance.[71] By contrast, in 1588, the princess of Salerno joined the prince of Salerno in having her own race of horse listed.[72] Lord Giovanvincenzo of Chiosan's horses passed their brand on to Zari Incenio, and Pierluigi of Parma's mark was extended to Pietro Alovisio Farnese, the duke of Parma.[73] In turn, several new Southern Italian brands found their way into Giunti's edition of the book, including those of Zuan Iacomo Detese, the Count of Ravasono, and the lords Gienari of the Kingdom of Naples, whose animals were identified as being raised in Basilicata, Puglia, and the Kingdom of Naples respectively.[74] Finally, for certain famous marks, the image could endure posthumously, even as its meaning changed. Although Bona Sforza, the queen of Poland and duchess of Bari, had died in Bari in 1557, her horses continued to be represented by her mark.[75] By 1588, however, Bona's name was no longer tied to the horse brand, yet the mark itself with the letters "BR" (probably Bona Regia) had not changed; it remained attached to a breed of animals roving the pastures of Bari decades after her death.

The placement of the brand signified different aspects of the animals' origins (see fig 2.2). Even when removed from their breeding farms, the horses' origins were emblazoned on their bodies: the emperor's jennets from Puglia had a brand on the right flank, while those marked on the left side were from Calabria. As archival records from the 1540s indicate, the horses were divided into different types, and a brand on the jaw designated whether they were of the *grandi*, *belli*, or *gentili* varieties. In the case of the

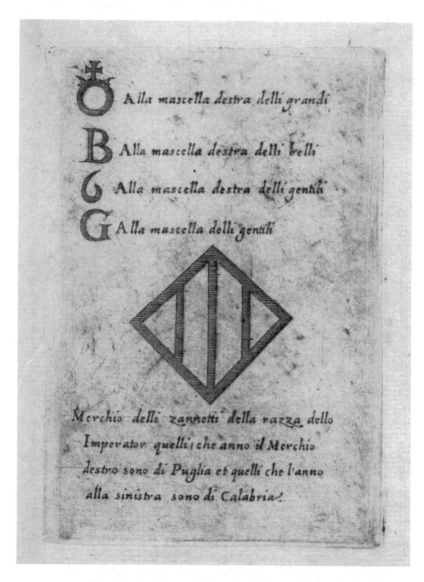

Ŏ Alla mascella destra delli grandi

B Alla mascella destra delli belli

Ł Alla mascella destra delli gentili

G Alla mascella delli gentili

Merchio delli zannetti della razza dello Imperator quelli che anno il Merchio destro sono di Puglia et quelli che l'anno alla sinistra sono di Calabria.

FIGURE 2.2. Branding the Emperor's horses. In *Libro de marchi de cavalli con li nomi de tutti li principi et privati signori che hanno razza di cavalli* (Venetia: Niccolò Nelli, 1569), 80r.

emperor's *belli zannetti* (beautiful jennet) horses, large jennets (*grandi zannetti*) would have a different mark. Although the Nelli and Giunti editions mention stock raised in the Este duchy of Ferrara, Don Franco da Este preferred to raise his race of horses in Basilicata in the Regno.[76] Even Don Ferrante Gonzaga kept a *razza* of horses in Puglia at the Serra Capriola, where these "handsome and good" horses were marked by an outsized "G" on their haunches.[77] By labeling animals on high-visibility body parts, the owners made their origins, quality, and ownership readily discernible.

Branding also mattered for the trade and gift exchanges that fortified diplomatic relationships. Close political contact between Habsburg Austria and the Spanish generally meant that the Viennese court had no problem acquiring Southern Italian steeds, whether from private houses, local princes, or the viceroy's stable. Horses poured into Vienna from many friendly Italian sources, including the Count of Condesa, Vicenzio de Ebuli, and Cesare Gonzaga.[78] In 1559, the marchioness of Gasto planned to send a few select horses to Emperor Maximillian II.[79] The examples of such gift exchanges are endless, all with attention to the quality of the animal, its potential use as a riding horse or something else, and the noble house that bred and gifted it.

In this system of barter and exchange, one finds caps, waiting lists, and carefully controlled population numbers. The fact that animals were often imported in pairs suggests what was presumed to be the likely outcome of exports. One required a combination of money and power to buy the animals. Buyers often wanted not only Neapolitan horses, but also equine expertise. In 1563, Cosimo I wrote to the viceroy of Naples that Paolo Giordano Orsini, duke of Bracciano and Cosimo's son-in-law, had sent Giovan Battista Pignatelli to Naples to provide a dozen adult horses and colts. Pignatelli was a successful equestrian specialist whose training methods had been noted by Grisone in *Gli Ordini di cavalcare*; his family was deeply embedded in the noble and equestrian worlds of the Kingdom of Naples. Pignatelli had recently come back to Naples from a trade mission in Constantinople, where he had been "impressed by the small but fast and agile" Arabian horses he had encountered in the eastern Mediterranean.[80] When Orsini requested a license to export six animals of the Regno's "Royal Race," he knew that Pignatelli would likely provide the best contact for realizing his ambitions.[81] Likewise, Francesco (later Grand Duke Francesco I) de' Medici frequently corresponded with Asciano Caracciolo, a relative of Pasquale Caracciolo, the Neapolitan nobleman who authored the colossal *La Gloria Cavalcare* (The Glory of Horsemanship) in 1566. On May 20, 1565, Asciano wrote that in three days he would send two colts to Prince Francesco by way of Fabrizion Summaio.

They were beautiful three-year-old colts of the King's Race, and he hoped that they would be a success.[82]

Under Spanish rule, exports of Neapolitan horses required a government license; securing one was no mean feat. Just as no one could simply travel throughout the Spanish empire without papers, and just as the legitimate and accepted forms of trade within its territories were limited, one needed permission to purchase Spanish horses. As the Kingdom of Naples fell under Iberian dominion following its conquest at the end of the fifteenth century, these restrictions also applied to Neapolitan horses.[83] Dressing horses with opulent tack, among other overt performances of luxury through horsemanship, was such a problem in Naples that in 1533, Charles V restricted it under sumptuary laws, and Viceroy Pedro de Toledo subsequently forbade gold and silver embellishments on horses beyond the headpiece, pectoral plates, and spurs.[84] In Neapolitan state budgets, the royal stables appear as part of the royal patrimony; they provided horses for the Spanish cavalry as well as the annual *chinea* given to the pope.[85] By the 1560s, Neapolitan stock had become difficult to obtain. In a letter from the fall of 1565, Alfonso Cambi Importuni described the considerable troubles he faced as he tried to export horses to Cosimo I de' Medici. Cambi did his best to keep the duke from paying the export fees.[86] Horse export licenses were controlled by the viceroy of Naples, and foreign princes made appeals all the way up to King Philip II. The consolidation of control over horse breeding was such that by 1560, even Maximilian, the future Emperor Maximilian II, had to write directly to the viceroy to obtain horses. He had requested a dozen colts from King Philip, which he expected to receive annually. Maximilian dispatched his servant Juan Antonio to collect the animals.[87] Maximilian was chiefly interested in "the race that your highness has in that Kingdom."[88] The following year, Maximilian again asked for the twelve horses Philip had allotted him and sent Juan Antonio to collect four from Naples. Maximilian asked the viceroy to send the rest up the peninsula in good time.[89] Later that year, he wrote to thank the viceroy, telling him that the animals that his servant brought back from Naples were some of the best, although among them there were more *ginetes* (jennet horses of Spanish or Arab origin) than he had expected.[90] Subsequently, Philip II's administration insisted on a basilica design for the Spanish Neapolitan stables, constructed in 1584, as his chief stable master, Diego de Córdoba, argued that the existing stables were inappropriate "for the foals and mares of 'his Majesty's Royal breed.'"[91]

Breeders, patrons, and horse traders mobilized a language of *razza* and perceptions of its social gradations. Emblazoned with house brands, their animal products advertised their resources for transforming the natural world into gifts and commodities. While the different propensities of the different types of animals mattered to potential buyers—note the disappointment in so many Spanish jennets above—the house that bred the animal and marketed its pedigree mattered quite as much. The selective breeding of animals, and the ideas of race it fostered, further connected social status, elite culture, and ideas of ownership.

Human Branding

The Florentine merchant Francesco Carletti (1573–1636) was as horrified about branding as Campanella was optimistic about the potential of selective breeding. Before Carletti traveled to Peru and questioned the classification of Andean camelids—an episode we will examine in chapter 6— he found himself entangled by a web of social and legal bonds. Nominally serving as the proxy of a Spanish woman in charge of a slaving expedition, Carletti was in fact subject to his father's authority in his family's illicit private venture.[92] He recounted that his conscience was wracked when he and his father traveled to Cape Verde to purchase slaves in spring 1594. Carletti wrote that

> the Portuguese, who kept [the slaves] like herds of cattle at their villas in the country, ordered them brought to the city so that we might see them . . . we bought seventy-five, two thirds of them males and the other third females, old and young, tall and short, all mixed together—as is the custom in that country—in a herd such as that from which, in our country, we buy a bunch of swine [armento di pecore], with all those precautions and circumstances of seeing that they are healthy and well set up without personal defects. Then each owner makes a mark [marca] on each slave—or, to say it more accurately, marks each of them with his mark, which is made of silver and is heated in the flame of a tallow candle. The tallow is used to anoint the burned place and the mark, which is placed on the chest or on an arm or on the back so that the slave can be recognized.[93]

As the multitude of the herd erased the identities of these individuals, human slaves were treated as swine, valued for their health with little care for their personhood. As Iberian empires expanded across the Atlantic, the branding of livestock had portended the branding of humans. Carletti later wrote that he had not participated in this branding ritual of his own free will. When he told the tale to the grand duke of Tuscany (probably between 1606 and 1608), he claimed that "this thing, which I remember having done under orders from a superior, causes me some sadness and confusion of conscience because truly it seems to be an inhuman traffic unworthy of a pious Christian." Did those humans, he asked, not likewise have "the same souls, formed by the same Maker who formed ours?" Nonetheless, Carletti participated in the banal evil of transoceanic empire, branding human bodies like livestock at the cost of not only their free will, but also his own.[94] The Florentine traveler concluded his rumination on branding with a turn to skin color as differentiation, and a desire to look past it, writing, "Although they are of different color and in the fortune of the world, nonetheless they have the same Soul formed in them by the same maker that forms ours."[95]

I conclude this section with a suggestive juxtaposition because animal branding was not happening in a vacuum. The branding of noble horses from ostensibly incommensurable *razze* unique to each house was used as proof of provenance. Carletti and other slave owners indicated possession using brands, too. Carletti, like Campanella, relied on an animal analogy, in Carletti's case, to emphasize the atrocity of the practice, likening the process of securing ownership over human slaves to the acquisition of swine. While he spoke of "jennet horses of the *razza* of Spain," Carletti did not employ the language of *razze* to characterize either the African slaves or the swine to which he likened them.[96] Packets of meanings were blended through the likening of ownership, provenance, and breeding in humans and animals, but situational nuances guided the extent of their overlap.

PART II

A Divergence in Breeding

CHAPTER 3

Razza-Making at a European Court

Wherever there is governance that spreads across territories, extracting resources from disparate peoples and consolidating them under the domain of a few families or institutions, or even when governance is confined to a single city that constitutes a political center, there are courts. These princely residences and their households—inhabited by a wide array of elites from regional sovereigns to local lords—accumulate wealth from surrounding areas, becoming a physical space of opulence for the successful consolidation of power and revenues. Throughout the early modern period, courts emerged in parallel across the globe, including in Europe and Mesoamerica, taking on different guises and aesthetics depending on local customs, but often sharing similar characteristics. At the center often sat a patriarch exalted with a title, whether duke or emperor, who reigned, surrounded by a wife or wives who, in turn, had the potential to wield considerable power; courtiers and aristocrats who liaised with the ruling family; and merchants and family networks connected to other courts of greater, lesser, or equal power. In some cases, female rulers took on a mediated version of this patriarchal role. A surrounding penumbra of artists, entertainers, and specialists—including animal breeders and trainers—worked to create a court aesthetic that physically manifested the power of the ruling family. An elaborate system of gift exchange provided a means of fortifying bonds of kinship and affinity through loyalty and obligation. In nearly all these court systems, power could be inherited through parentage and claims to ancestral clout rooted in the deep past. However, to access that family heritage, one needed to perform the status of an elite, behaving according to carefully prescribed cultural modes that could be learned.

Part II of this book approaches transatlantic intellectual history by juxtaposing ideas and practices developed in European and Mesoamerican courts. On either side of the ocean, imperial courts fostered concentrated breeding zones intent on creating versions of nature that revealed their cultural priorities and power. Agents collected natural curiosities and fostered their continued propagation. Both Mesoamerican and European courts emphasized the importance of selection for valued characteristics and a subtle hierarchy of both human and nonhuman animals. Remaining sources suggest that the court of Moteuczoma II in Tenochtitlan collected bits of the natural world; breeding practices in Mesoamerica sometimes involved ritual rhetoric associated with a process of selection that emphasized cyclicity and wholeness. Europeans, explored here through the Gonzaga court of Mantua, envisioned a hierarchical nature with elements that could be perfected. Through this chapter and the next, I argue that in Renaissance European courts, ideas of race diverged from global practices of selective breeding, setting the stage for a particular taxonomic hierarchy that developed in the social and political space created by European empire.

An Ideal European Court

"I tell you that one of the best *caballeros* of the world is dead," Holy Roman Emperor Charles V declared at the funeral procession of Baldassare Castiglione (1478–1529). Castiglione, a courtier, knight, diplomat, and nobleman extraordinaire, was buried in the cathedral of the Spanish court city of Toledo, where he met his end on February 8, 1529.[1] Called a universal man, an example of perfection, Castiglione had taught a generation how to conduct themselves in court, integrating European cultures with one common language of court etiquette transferable from one courtly center to the next. Castiglione would have known this etiquette well. Descended from the Gonzaga family on one side, he had been born in Mantua and educated in Milan. In 1491, Castiglione had met Isabella d'Este, the marchioness of Mantua, in preparation for Ludovico Sforza's wedding. Castiglione remained linked to the Sforza court until its abrupt downfall in 1499. He returned to Mantua to enter the service of Francesco Gonzaga, the marquis of Mantua, and regularly accompanied the Gonzaga entourage and represented their interests abroad. From there, Castiglione transferred to the court of Urbino in 1504, where he set the conversation featured in his later book. Diplomatic missions took him to England, to Bologna to meet the king of France, and to Rome, where he worked on behalf of the court of Urbino and the Gonzaga of Mantua. In 1518, in honor of Castiglione's marriage to Ippolita Torellini, Francesco Gonzaga had held a festival for him, complete with jousts, tournaments, and feasts.[2] From there, Castiglione's career continued abroad. Pope Clement VII dispatched him in 1524 as papal nuncio to Spain, where he joined the court of Emperor Charles V as it journeyed from Burgos to Valencia to Madrid.[3] Castiglione's political reports were a key source on Charles's policies, from the status of his marriage to his health to his emotional reactions to politics. His life consisted of traveling from one European court center to another. No wonder the emperor had thought him the best *caballero* (meaning at once horseman, knight, and courtier) around.

Although he was a remarkable leader who united and governed regions comprising all of Spain, the Netherlands, Bohemia, Transylvania, Milan, Naples, Sicily, and more, Charles V had never been much of a reader. Contemporaries wrote that despite his large library of untapped tomes, one of the three authors that he had read thoroughly—alongside Machiavelli on politics and Polybius on warfare—was Castiglione. A vivid conversation set at the court of Urbino, Castiglione's *The Courtier* (*Il Cortegiano*) taught "the foundations of civilized life" in a way that was legible and relevant even for the leader of Europe's largest empire in the sixteenth century.[4] When Castiglione had entered Charles V's entourage in Spain, the emperor had requested seventy copies of the book to circulate among friends and affiliates at court.[5] By the 1530s, the emperor had insisted on a Spanish-language translation, granting a privilege to print a copy in Barcelona in 1534, just as readership of *The Courtier* started to peak around the rest of Europe.[6] In the words of Geoffrey Parker, Charles V "did everything— walking, riding, fighting, dancing, speaking—with one eye on his audience," exemplifying what Castiglione had taught was the ideal natural performance of a member of court.[7]

Castiglione's popular book "professes to teach what cannot be learned, the art of behaving in a naturally graceful manner," as historian Peter Burke put it.[8] Amid his robust depiction of power in social habits, Castiglione recorded a vivid "how to" for becoming a savvy, relaxed member of a European center of power. Castiglione marketed *areté*, that Greek term for excellence that defined a fine horse's speed and a warrior's honor. Those with the most *areté* were aristocrats, and aristocracy was supposed to represent just that—the best of civility, polish, and good manners as themselves the paragon of excellence. Good manners could be contrasted with rural behaviors, literally *villania* (villainy) of the unrefined countryside. By publishing his book to a wider audience, Castiglione performed the belief that good breeding was the result of studied etiquette, perhaps as well as robust lineage; proper behavior was crucial to life at court. That wider audience consumed his message eagerly. Before it had even been printed, Castiglione was miffed to learn of a number of manuscript copies among the elite ladies of Naples. After its initial publication in 1526, the book was quickly translated into English, French, and Spanish. There were many references throughout Spain to *The Courtier*, and in the sixteenth and seventeenth centuries in Italy, sixty-two editions were published, along with many translations abroad. Sixteenth-century readers eagerly marked passages they wanted to remember, such as the instructions on riding that might set apart particular grace at court. *The Courtier* promoted the fashionable Italian culture in Spain, France, England, and elsewhere.[9]

If Castiglione captured a vision of the ideal courtier, then Mantua, where he was born and received the sustained patronage of the ruling Gonzaga family, was the ideal Renaissance European court. An innovative cultural center, immersed in fog and surrounded by rivers, artificial lakes, and canals constantly reshaped to fit changing needs, the city of Mantua in the Po Valley in Northern Italy became a capital of European culture in the sixteenth century. Like Innsbruck's Schloss Ambras to the north, it joined a network of court centers such as Ferrara and Urbino that attracted artists, doctors, scholars, and philosophers. First-rate artists came in droves, although Florentine Giorgio Vasari (1511–1574) was loath to admit that some of the greatest talents of the period bloomed on the banks of the Mincio rather than the Arno. Among them were Pisanello, Mantegna, Rubens, and architects Leon Battista Alberti (1404–1472) and Giulio Romano (1499–1546). Even after a catastrophic siege in the War of the Mantuan Succession and Napoleon's depredations, Mantua's fame persisted.

While this chapter focuses on Mantua, that city serves as an example and model for the patterns of behavior echoed elsewhere. In Italy alone, Ferrara, Urbino, Florence, Rome, Milan, and a myriad of other cities had their own variations on Mantua's culture and the approach to breeding it inculcated. The power of Castiglione's work, however, like that of the court culture he typified, was its increasing transferability as Europe's culture became more generalized or, indeed, Europeanized over the course of the sixteenth century.

Breeding Renaissance

When founding the field of inquiry we now call Renaissance studies, in which the study of Castiglione's text plays a prominent part, Jacob Burckhardt (1818–1897) traced key facets of modernity to the patchwork of Italian city-states. He claimed that "the house of Gonzaga at Mantua and that of Montefeltro of Urbino were among the best ordered and richest in men of ability during the second half of the fifteenth century."[10] Burckhardt saw Renaissance Italians' fascination with artifice as the defining feature

of their society, and of the modernity to follow. The state, war, and even
the self were all works of art, understood as an expression of creativity,
but also as an artificial creation. Though generations of scholars have complicated and contested Burkhardt's arguments, self-fashioning remains a
topic central to Renaissance studies.[11] While the Italian Renaissance is
celebrated for its belief that man could fashion himself to greatness, this
view was complicated by a lingering fascination with breeding. Just as one
strand of the Italian Renaissance celebrated a liberal view of human potential, another limited it by biology, reducing man to beast and prince
to stud.

When it came to Mantua, Burckhardt thought that the Gonzaga's commitment to artifice through breeding laid a cornerstone for the later biological sciences. He argued that "a significant proof of the widespread
interest in natural history is found in the zeal which showed itself at an
early period for the collection and comparative study of plants and animals." Menageries, gardens, and stables—complete with their lions, gi-
raffes, exotic plants, falcons, dogs, and horses—provided the circum-
stances under which "the foundations of a scientific zoology and botany
were laid." Patrons like the Gonzaga poured their resources into pro-
curing animal specimens from around the early modern world, breeding them, and using the resulting animals for entertainment, food, and
labor. If read through Nicholas Russell's schema of breeding strategies,
Mantuan breeders were particularly interested in the techniques of breed
replacement and grading up after chaotic forays into "multi-racial cross
breeding."[12] One can understand, then, how Burckhardt came to the idea
that "a practical fruit of these zoological studies was the establishment of
studs." In Mantua, "all possible experiments were tried, in order to produce the most perfect animals."[13] Burckhardt's contention—that these
breeding projects, themselves experimental, developed from zoological
observation—stands unchallenged and tacitly accepted, more than a hundred years later.[14]

What lingers in the Mantuan archives is extensive documentation of
systematic pairing and selective breeding as a part of the Gonzaga family's
campaign to promote their own nobility and court center. Writing in the
nineteenth century, Burckhardt saw the origins of his era's science of zoology in the Mantuan art of breeding. What he interpreted as "zoological
studies" resulting in "the establishment of studs" was the Mantuan razza-
making. Just as the Renaissance witnessed the transformation of artists
from anonymous artisans reproducing tradition to named figures famed
for creativity and individual excellence, some of the practitioners of the
ancient art of selective breeding influenced authors of books that theo-

rized reproduction, taxonomy, and natural history. Elites patronizing Italian breeders were deeply committed to performance—especially through hunting, racing, parading, fighting, and dancing—in a deeply fractured and highly competitive culture, and they invested in bureaucracy to record and support their successes.

The Emperor at the Mantuan Court

In the fifteenth century, Mantua had been intimately connected to other Italian and Alpine courts. The Italian Wars that swept across the peninsula, however, increased the challenge of self-preservation in the face of rising Spanish and French imperial interests. With Castiglione recently deceased in Toledo, Holy Roman Emperor Charles V's visit to Mantua in 1530 offered the Gonzaga the chance to pay homage to the lost *caballero* and enter into a mutually agreeable arrangement. Aware of the cultural impact of their breeding interventions, the Gonzaga aimed to win Charles V over by showing him their highly distinctive animal culture. This, they hoped, would at once impress Charles and help embed Mantuan animals (and by extension, the Gonzaga themselves) in the sphere of imperial influence without sacrificing Mantuan autonomy.

The Mantuans displayed to Charles V and his courtiers everything at which they excelled. The emperor was entertained with Mantua's superb animals, hunting opportunities, and palaces, almost from the moment he arrived that spring. Charles had hardly rested before he wanted to hunt; he called upon Federico II Gonzaga to ask about Mantua's famed hare hunting and falcon flying. Without further ado, the marquis set his smoothly running hunting machine in motion, summoning his hunting experts and their many excellent dogs, peregrines, and hierofalcons. The emperor left behind his guard and hunted with the marquis, using the latter's horses. The chase was successful; Charles killed a hare with a shot from his crossbow. The afternoon was left for birding, and they flew falcons of every sort. By the following day, Federico sent his hunting master to check whether there were any wild boar not too far away. On March 27, a mere three days after the emperor's arrival, they traveled to the lodges in Marmirolo for birding and boar hunting.[15] The emperor was offered the best of Mantua's noble animals in order to hunt the natural world.

Perhaps Charles was entertained in the Ducal Palace's banqueting hall, the first of Mantua's two Sale dei Cavalli (Horse Rooms), where dinner guests could attempt to identify trompe l'oeil horses prancing across the walls from glimpses of an eye, neck, fetlock, or hoof, winning the marquis's approval when they named his favorite steeds successfully.[16] Even the beds were dressed in golden brocades, perhaps won by Gonzaga horses in past palio races. Palazzo Te, Federico II's newly converted stable-turned-pleasure-palace just outside of the urban center, also featured prominently in the visit. En route there, the marquis praised the "beauty of the land and many other things."[17] As the Mantuan ambassador to Rome, it had been Castiglione who directed his friend, architect Giulio Romano, to the Mantovano to transform the Gonzaga duke's dreams of a pleasure palace into a reality.[18] In Vasari's account, when Romano finally arrived, "learning that Giulio had no horse, [Federico Gonzaga] had one of his own favorite horses, called Luggieri, brought out and gave it to Giulio." They mounted and together set forth, "outside the gate of San Sebastiano the distance of a shot from a crossbow to where His Excellency had a place and some stables called the Te, in the middle of a meadow, where he kept the stud [razze] of his horses and his mares." The stable for racing stallions stood near a specially constructed track for training palio horses. There, the marquis declared that he would like to build a "small place where he could go and take refuge on occasion."[19] On that pastureland where the horses resided, the marquis and Giulio Romano would build Palazzo Te. The palace was literally founded on a Mantuan horse breeding station.[20]

As the Emperor would see, the resulting suburban complex mixed architectural elements traditionally used in palaces and villas; it was like a Fabergé egg with decadent frescoes inside extensive gardens. Just as the mythical scenes in the frescoes were rendered with consummate realism, so too were the marquis's horses. The realism with which these animals were portrayed tied the mythical themes omnipresent in many other rooms in Palazzo Te back to the reality of life with the marquis, a life in which fine art was constantly accompanied by carefully refined people and animals. In the famed Sala dei Cavalli (yet another Horse Room in the city) loomed frescoes of horses drawn from life (fig. 3.1). As the original art historian, Giorgio Vasari, wrote:

> On the walls there are life-size portraits of all the most beautiful and most favored horses from the marquis's stock [razza], accompanied by dogs of the same coat or markings as the horses, along with their names; they were all drawn by Giulio and painted on the plaster in fresco by the painters Benedetto Pagni and Rinaldo Mantovano, who were Giulio's pupils, and to tell the truth they were done so well that they appear to be alive.[21]

The portrait of the "Morel Favorito" bears the horse's name; in it, he stands, bridled, ears pointing softly forward. On the other side of the ornate fireplace stands a light gray horse, similarly positioned, its exotic Turkish

FIGURE 3.1. A view of the Sala dei Cavalli at the Palazzo Te, by artists in Giulio Romano's workshop, in which horses from the Gonzaga stables towered over visitors. Wikimedia Commons.

breed on display through the dyed red color of its tail (fig. 3.2).[22] The red color of this horse's tail was produced using henna powder (*alcanna* or *archenna d'Oriente*) imported from Turkey; thus, the animal was meant to represent the broad range of the Gonzaga's diplomatic connections. Agents in the Ottoman Empire had sent this powder to Federico II, as recalled in a letter of September 4, 1521, in which Mantuan agent Giovanni Batista reported to Federico that he was sending a powder that would give his horses' tails a reddish tint. Jacopo Malatesta in Venice then received a request for three to four pounds of *archenna* so that the Gonzaga might color the tails of Turkish horses in their next race. The animal and its adornment embodied the Gonzaga's worldly diplomacy.[23]

Upon arrival at Palazzo Te, Emperor Charles V dismounted and entered a beautiful large room. Touring the palace, recalls the chronicle of his visit, Charles "stood for more than half an hour to contemplate, praising everything greatly." The party then went out to the stables, where the

FIGURE 3.2. Alongside a gray horse, a Turkish horse stands without a saddle, his tail dyed with red pigmentation that emphasizes his Eastern origins, in the Sala dei Cavalli at the Palazzo Te. Alamy.

"Illustrious Lord Marquis called out Master Vincenzo Guerrero, his master of the stables, and Master Matteo Ratto, his rider, on two great and superb coursers of the race of the Lord Marquis." All in all, the emperor was pleased with Federico II Gonzaga: "Sir, truly never had there been a city in Italy in which I was so well welcomed as in yours, nor so happily and honorably received as I was here."[24] By April 8, Federico secured what he had wanted all along. The emperor raised the Gonzaga marquisate to a duchy; Federico was now a duke.[25]

Isabella d'Este's Human Gifts

The Gonzaga populated their court with animal and human characters who would prove entertaining, be obedient, and capture the sense of wondrous, educated play that characterized Castiglione's world of poetry, jousts, festivals, and music. People, creatures, and other novel goods were exchanged across courts, too. Presented as gifts, they were intended to obligate the recipient and serve as a constant reminder of the giver's generosity. In discussions of their equine gifts, the Gonzaga deployed the language of *razze*, which they also described through metaphors of seeding, fruiting, grafting, and cultivation. The language of *razze* permeated their conversations from the garden to the stables, and even to the court's humans, including people now understood to have the medical condition called "dwarfism."[26] On September 11, 1532, Marchioness Isabella d'Este (1474–1539) sought the perfect recipient for a gift from her famed collections. The Renaissance patroness had in mind a carefully crafted human two-year-old, a gift that would demonstrate the wonders she was creating in the Gonzaga court of Mantua. Recalling past promises to Renée of France, who was married to the duke of Ferrara's heir, Isabella wrote to her cousin Diana d'Este to inquire whether the future duchess would still be interested in the gift. "Four years ago," Isabella wrote,

> I promised the Most Illustrious Madame Renée that I would give Her Excellency the first fruit to issue from my race of little dwarves [*el primo fruto che ucisse della raza delli mei nanini*], by which I mean a female. As Your Excellency knows, two years later a little girl was born. Though we cannot hope she will stay so small as my Delia, she will nonetheless without doubt remain a dwarf, and given her beauty, she deserves to be treasured. Since she is now at the point where she is beginning to speak and walk and is able to get around confidently all by herself . . . I thought I would send her to Madame.[27]

Fall had settled upon Northern Italy, the sweltering summer heat had abated, and Isabella was confident that the child could endure the hardship of travel. Cousin Diana's task was to investigate whether this was the right home for Delia's daughter, who, like generations of Mantua's fine horses, had a *raza* for which the Gonzaga claimed credit.

As a voracious collector who avidly pursued the finest artists of the High Renaissance, Isabella d'Este represents the archetype of a sponsor eager to influence the projects she patronized; her exhaustive instructions and persistence compensated for the relatively limited resources available to her as the marchioness of Mantua. Isabella managed to fill her *studiolo* with pieces by Mantegna, Bellini, Costa, Perugino, and Correggio and her *grotto* with antiquities. Through the measured allocation of power between them, she and her husband created a court that emblemized many of the political, military, and most of all, artistic priorities of the Renaissance. Beyond Isabella's artistic ambitions, her court's collecting likewise extended to exotic nature and artifacts from classical antiquity. Visitors to the court would find wonders—from gems and *pietre dure* to unicorn horns, coral, fish teeth, and whale bones—spread across several cabinets of curiosities.[28] Isabella inquired about placing Delia's daughter and getting her favorite clown back from Ferrara amid letters accompanying rose pillows, in thanks for copies of *Orlando furioso* and artichokes, and discussing plans for various tombs.[29]

In many Renaissance courts, people of short stature functioned as both servants and living curiosities, and were often depicted in artwork caring for exotic animals and noble children alike.[30] As Annemarie Jordan Gschwend's work on the Portuguese court of Catherine of Austria suggests, patrons regarded people with dwarfism as specialized servants. During Isabella's youth in Ferrara, the Este had employed dwarves as court entertainers and assistants to the ducal family. Atypical height at once excluded them from normal categories of advancement in Renaissance Europe, but also gave them access to the inner sanctuary of elite life through their visible difference.[31] The case of Petrus Gonzalvus (ca. 1537–1618) and his hair-covered descendants, well studied by Merry E. Wiesner-Hanks, might provide important clues to how Renaissance courtiers understood individuals with dwarfism. Gonzalvus was born in Tenerife and brought to Europe from the Canary Islands when he was about ten because of his hypertrichosis, which led to unusual hair growth over his body. He lived in Paris, the Netherlands, and Parma, married, and had seven children, three of whom were likewise covered in hair. Ulisse Aldrovandi referred to him as "the man of the woods," citing stories of wild men on the margins of society. Like servants with dwarfism, Gonzalvus and his daughters certainly mesmerized courtiers, doctors, and artists with their distinctive features. However, by most accounts, they were not mocked or displayed publicly, but rather incorporated into the court entourage. Isabella's note suggested that, when it came to human *razze*, selective breeding did not prohibit personhood, but complicated the line between exotic object and agentive individual. Little people likewise became part of the family in a subordinate but intimate role, and proximity to the sovereign potentially conferred on them power through the value of access. Renaissance courts were notoriously competitive, but those competitions were built around elaborately staged events requiring specialist roles for humans and animals alike. Feasting, hunting, dancing, and racing each required a specific entourage. Confronted with staffing these fixed roles, princes sought to perpetuate order by filling such positions with carefully prepared beings. More broadly, one vein of Renaissance Italian thought sought to optimize each element of society for its allotted role within the established hierarchy—each embodied manifestations of the links that made up the Great Chain of Being, from minerals to heavenly bodies, as will be discussed further in chapter 7.[32] The first "fruit" of Isabella's "race of dwarves" had been made to play an allotted part.

Delia had clearly known other people of short stature, who could be found in courts and noble households throughout central Europe, where they maintained the collections of curiosities of which they were a living part. In Milan, a person of short stature named Biagio had served at the court of Francesco Sforza (1401–1466); later, Janachi similarly worked for Ludovico il Moro (1452–1508).[33] Between 1465 and 1474, another Mantuan individual with dwarfism found herself prominently immortalized among Marquis Ludovico Gonzaga's (1412–1478) family in the frescoes that line the Ducal Palace (fig. 3.3). Finely dressed, she stands beside Barbara of Brandenburg (1422–1481), another marchioness of Mantua. A dog, perhaps the beloved Rubino, glares out protectively from beneath the seat of Ludovico Gonzaga, Isabella d'Este's grandfather-in-law. One might go so far as to suggest, as Dániel Margócsy has, that European courts extended their interest in curiosities to *razze* of horses—which might be extended to other human and nonhuman animals that were collected and displayed as wonders.[34] To put it bluntly, cabinets of curiosities took living form in the shape of courts, which in turn housed *wunderkammer*. Coming full circle, by the late seventeenth century, a custodian of short stature cared for the museum of Marchese Ferdinando Cospi (1606–1686),

FIGURE 3.3. Court scene in the Camera degli Sposi, Palazzo Ducale di Mantova, painted by Andrea Mantegna between 1465 and 1474. The scene depicts Ludovico Gonzaga and his family, as well as a loyal dog beneath his chair, a dwarf woman, and a counselor who has been interpreted as Leon Battista Alberti. Wikimedia Commons.

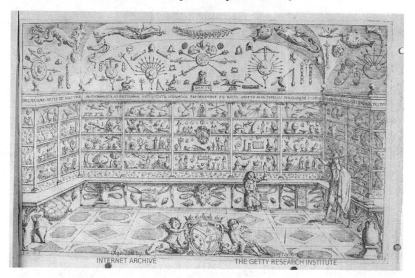

FIGURE 3.4. Sebastiano Biavati, custodian of the Museo Cospiano, points to curiosities of nature surrounding him. In Lorenzo Legati, *Museo Cospiano* (Bologna: per Giacomo Monti, 1677). Getty Research Institute, Los Angeles (3023-784).

which had been annexed to the *studiolo* of acclaimed naturalist Ulisse Aldrovandi (1522–1605) in Bologna (fig. 3.4). Cospi hired Sebastiano Biavati, along with his sister Angelica Biavati, who had not yet reached the height of 30 Roman inches at fifty-five years of age.[35] People of short stature created a common thread between court and cabinet as living exotica prized in a culture transfixed by the wondrous.

Isabella's discussion of her influence in Delia's life was just a more explicit version of what happened throughout the court. Indeed, because noble lineage defined the inheritance of property and status, lords and their networks policed the choices required to make both the beasts and the humans of the next generation. The restrictions of court life were as abundant as its privileges; individual volition was largely ignored. Renaissance elites had low expectations of autonomy, assuming that everyone, from the court entertainers to the duke himself, would have their partners chosen for them with an eye to the resulting offspring.[36] Isabella's actions, though high-handed and coercive, had the potential to guarantee Delia's daughter a lifelong position in a rarefied world of relative comfort. As it happens, Renée and her court at Ferrara did not become a home for Delia's daughter. Despite this complication, Delia's daughter did not have the chance to stay in Mantua with her mother for long. Isabella must have quickly extended the offer to another family member. About a year later, on October 25, 1533, Isabella heard from her daughter-in-law Isabella di Capua (1510–1559), princess of Molfetta, that the girl with dwarfism had arrived. "I was glad to hear it, since I longed to know that she had made it safely," Isabella wrote, "and my pleasure was doubled when I learned from Your Ladyship's letter how happy you are to have her with you and what amusement she is providing you."[37] This joy, however, was fleeting. The exchange quickly shifted to the surplus of milk yielded by Isabella di Capua's left breast. Hints we get from letters like Isabella's reveal a female world of breeding, one that implicates women alongside men in the making of razze. In contrast to her husband's animals—which were prized for size, speed, and exotic origins—Isabella's description of her raza delli nanini emphasized domesticity and infantility. Making a new line of beings to coddle and adore, whether they were civet cats or humans of short stature, extended the realm of her domus.

The Gonzaga court likewise sought out black Africans, who were trained to work and often placed in positions of trust. In April 1515, Isabella wrote to Rigo Carpesano that "we understand that our bitch, which you have, is pregnant as we desire to breed a pair of dogs"; Isabella then wrote, "Our Black, the master of the Wardrobe, is commissioned the task of making them breed."[38] Despite his position, she identified the master of the wardrobe only by his skin color. However, he was certainly not the first black servant or slave to work in the Gonzaga court. In 1491, Isabella, then the newly married sixteen-year-old marchioness of Mantua, had continually pestered her agent in Venice to locate a very young black girl ("moreta") and encouraged her agents to search for a child between one-and-a-half and four years of age, about the age of Delia's daughter. Isabella specifically requested that the little girl be "as black as possible."[39] Her attempts to purchase the four-year-old daughter of an African gondolier were thwarted by Isabella's mother's move to invite the entire family into her service; Isabella then acquired a two-year-old black girl from a Venetian orphanage.[40] To modern readers, Isabella's project of collecting people with skin deemed black might seem to be among the most racialized of her dubious ventures. However, to my knowledge, the term razza does not prominently appear in her discussion of the black Africans.

The Intimate Lives of Companion Animals

Isabella d'Este's sense of entitlement to human bodies came from the animal culture that distinguished court life. When Mantegna painted the Gonzaga family between 1467 and 1474 and rendered a dwarf servant standing among the courtiers and family members, the dog Rubino sat at Ludovico III Gonzaga's feet, glowering from beneath his chair, at once a half-threatening secret weapon and a beloved companion (see fig. 3.3). In this case, art followed life. Stories of Rubino's real adventures lingered on in Mantuan lore. Once, when the dog wandered away, the marquis instructed his wife, Barbara of Brandenburg, to organize a search party. When swollen glands threatened to take the dog's life, he once again ordered Barbara to secure the animal's comfort, this time in death. Ludovico requested that Rubino's body be buried in a casket and his interment marked with a tombstone, as Rodolfo Signorini has traced.[41] The grave bore a commemorative poem:

> Rubino the Dog.
> Having been esteemed for long and faithful love for my melancholy
>   master,
> killed, I lie here watched over, my lineage [stirpe] protected,
> the master has thought me worthy of the honor of this grave.[42]

Not just a dog, a commodity, or a hunting companion, Rubino had been an adored part of Ludovico's family; his epitaph recognized both his character and his intimate relationship with the marquis. However, even in death, the gravestone asserted that Rubino's "lineage [stirpe] was protected." A successful burial meant not only laying the body of this beloved dog to rest, but also securing a comfortable future for his descendants by preserving his line. In a sense, Rubino had been the head of a household, too.

Along with other long-lost dogs whose lives had been commemorated in verse and stone, Rubino appeared in Ulisse Aldrovandi's natural history of dogs in De quadrupedibus digitatis, an encyclopedic natural history first published in 1637. The acclaimed Bolognese naturalist built on a humanist tradition of collecting stories and descriptions of nature. In addition to exotic creatures like dragons, his natural history featured more familiar domesticated animals such as dogs, chickens, and horses.[43] For these descriptions, Aldrovandi drew much of his research from animals whose histories were deeply interwoven with humans', whether they participated in long-forgotten battles or belonged to famous contemporary households. Spanish dogs, Gallic dogs, and hunting dogs came with their own citable textual remains, not unlike Rubino's epitaph. Among many others, his illustrations included a lean hunting dog bred by Sebesten Mamilla akin to a modern Italian greyhound, a French dog, and two types of Spanish dogs with distinct ears, tooth placement, color, and body shapes (fig. 3.5). In his chapters on quadrupeds, Aldrovandi captured a dizzying array of animal

FIGURE 3.5. Dog breeds in Renaissance Europe, including Spanish dogs, a French dog, and a hunting dog bred by Sebesten Mamilla. Ulisse Aldrovandi, *De Quadrupedibus digitatis viviparis*, libri tres, et *De Quadrupedibus digitatis oviparis*, libri duo (Bonon: Apud Nicolaum Tebaldinum, 1645), 545, 560, 561, 563. Wellcome Collection. Public Domain Mark.

types, which in turn highlighted the success of Renaissance ambitions to create recognizable animal breeds. Different dogs were their own sort of wonder to be placed alongside monsters and exotic animals in different works of natural history; Aldrovandi collected them as a juxtaposition of nature's artifice and the marvels of the contemporary breeding projects that had made such animals.[44] The publication of Aldrovandi's writings, as with those of other natural historians explored in later chapters, fixed in print the temporary differentiations of the domestic animals that surrounded him. After all, a line of spaniels might be differentiated from a line of mastiffs only so long as their populations were maintained by keeping such groups of differentiated animals apart.

To return to Russell's breeding strategies, I have suggested that Renaissance Italian noble houses sought to make distinct animals through "positive" breeding strategies: selection based on phenotype; selection and relocation based on geography and a belief in a strong environmental influence on stock; sire or group rotation; and hybridization.[45] The Gonzaga were particularly committed to supplementing their local stock with foreign animals. Closely clustered small courts throughout Italy fostered competition but were constrained by a relatively narrow local base of flora and fauna. To win in this competitive hothouse required raising the stakes, as Giancarlo Malacarne's foundational scholarship on the Gonzaga's hunting, courtly animal, and palio-racing culture explains.[46] Mantuan lords therefore turned outward to other regions of Europe and the Mediterranean to gain access to other varieties of domesticated animals.

We can see this interest in international trade in court dogs through the Gonzaga's efforts to secure Dalmatians and Turkish hunting dogs. To do this, the Gonzaga relied on a network of agents and local experts to source and care for the best stock. In the case of Dalmatian dogs, for example, Gonzaga agents located prized animals across the northern Adriatic to enhance stock bred in Mantua. In January 1530, Iacopo Malatesta, Marquis Federico II's (1500–1540) lead diplomat in Venice, wrote of his experience in Dalmatia (present-day Croatia):

> I have been trying for the female Slavic dog that your Excellency wanted, and I have been working a great deal in order to have one from Schiavonia. Master Petro says that they have a lovely race [*razza*], but they are small. Master Zohanni says that in the mountains of Vesentina one can find a very good sort, and that he might . . . send one to your Excellency. From here, I will undertake every practice possible to make a large one, and I will see that one comes from Schiavonia.[47]

Even amid the high drama of the Italian Wars, with an imperial army besieging Florence and Holy Roman Emperor Charles V preparing for his coronation in Bologna, the Gonzaga mobilized their diplomatic network to make big Dalmatian dogs and bring them back to Mantua. The Gonzaga were also interested in Turkish hunting dogs; by November 5, 1498, "two large dogs" had arrived from Constantinople.[48] Gonzaga agents claimed that they imported only the "most beautiful and highest quality" from various houses throughout Europe.[49] Once Dalmatians and Turkish hounds arrived in Mantua, they would have been received by the Gonzaga before being consigned to the care of a specialist. Sarah Cockram has shown how these specialist handlers, especially those who traveled with exotic animals, "could act as brokers and mediators between the worlds they traversed: between cultures, bringing foreign creatures and techniques to far-away places."[50]

Like the manager of the stables with whom the Gonzaga lords maintained constant contact, the *canatéro* (master or keepers of dogs) often enjoyed a favored position at court for their connection to hunting.[51] In addition to the *canatéro*, the Gonzaga lords also regularly corresponded with the *strozzieri* ("a killer of wilde beasts") and *falconieri* ("a Faulkner, a hawker").[52] The master of dogs tended to the hunting dogs' health and provisions, and sounded the horn during a hunt. With the dogs' relatively short lifespans and rapid reproduction cycles, knowledge of canine gestation played a particularly important role in the *canattiere*'s daily work. Hunting with a pregnant female dog posed considerable risks for both the mother and her pups; thus, the *canattiere* remained attentive to his charges' needs so that years of the dogs' training, care, and breeding would not be lost on one ill-timed hunt.[53] When a dog contracted rabies or behaved oddly, he was the first to report it to the lord. When a recent canine mother suddenly ate her offspring, it was Zo Francesco's responsibility to describe the whole scene in lurid detail.[54]

Why would the head of state and his consort pay such close attention to the intimate lives of animals? Simply put, the Gonzaga were committed to good breeding, broadly defined. Isabella d'Este often had precise plans for how to match her dogs and for where such pairings ought to occur. In February 1493, she wrote to Leonelle da Baesio that he should "send to me by this ship our dog called Moretta so that she may be covered by Puchetto, of the Excellency of madonna our mother."[55] Isabella instructed Leonelle to be careful with the timing and guard against the chance that Moretta would "go mad with another dog" while in heat before she could be bred to Puchetto, the dog of Isabella's mother, Eleanor of Naples (1450–1493), the duchess of Ferrara. As with dogs, local family networks alone could not provide sufficient feline variety for Isabella's ambitions. In the late fifteenth century, Isabella acquired a Persian cat from Damascus. Decades later, she sought to import a civet cat to mate with her domesticated cats.[56] Since the Gonzaga understood that breeding made the most exceptional individuals, it was appropriate in such a hierarchy for the ruling couple to make the essential decisions, encoding the choices of each generation on the future. As this chapter has shown, noble members of court sometimes had their animals' portraits painted and their names recorded. From eulogies for dogs to portraits of horses, the Gonzaga's favored animals had their individuality praised and treasured.

Through the ancient tradition of gifting exotic animals, the Gonzaga's international reputation grew. In turn, diplomatic gifts frequently recognized and contributed to the Mantuan project to collect exotic animals and create domesticated animal varieties. Social capital became animal capital that became social capital anew. Throughout the 1520s, the Gonzaga received puppies from England and the Ottoman Empire, sometimes as spontaneous gifts, other times as requested ones.[57] Once foreign animals arrived, they were ranked subjectively by the lord's assessment of their quality, often using physiognomic characteristics like their visage, general shape, or other bodily characteristics to inform conclusions about their character. Isabella's correspondence betrays a focus on animals determined to be "the most beautiful and best in the house."[58]

Family Resemblances and the Gonzaga's Horses

As we have seen, Renaissance Italian courts worked to collect, improve, and display a wide array of human and nonhuman animals, including but not limited to dogs, cats, falcons (though these were trained in captivity after being removed from the wild), humans of short stature, and exotic servants. While each of these beings had their own life cycle and use, they were increasingly described using the shared language of breeding, which emphasized the selected and cultivated nature of their bodies. Horses — the distinguishing animals of European nobility—dominated this discourse and noble interest, as they were among the most expensive, valuable, and high-status domesticated animals at court.

Horse breeding programs flourished under Isabella's husband, Francesco II (1466–1519), and their son Federico II (1500–1540), who cultivated close relationships with other rulers to import steeds from the Ottoman Empire, Africa, Arabia, Ireland, and Spain as part of an effort to create Mantuan equine *razze*. This effort was sustained through a network of emissaries abroad who funneled animals to stud farms in Mantua. When setting up the stables and breeding facilities, Ludovico III had dispatched his agents to Southern Italy, to the Aragonese court in Naples, for its acclaimed horses. When it came to selecting appropriate animals, breeders and traders used physiognomic observation to determine animals' character, the potential character of their future progeny, and their ability to reproduce effectively. In 1470, Antonio Donato di Meo, Mantuan ambassador to the Court of Naples, was in search of possible mothers for the herd. In language that resembled Alberti's description of an ideal mare's conformation, he requested,

> first, that they have a good body and that they are not stretched, the high back and not sagging, and above all that they have large wombs [*la natura grande*], that those that have the biggest parts make the most beautiful horses. It would be very dear to be able to have such mares because . . . our race [*la raza nostra*] . . . would be able to reuse them much better.[59]

A mare that had foaled successfully before would be a better investment. Four days later, Donato di Meo searched for a founding stallion, "a good stallion that had a good mouth and good back . . . because our *raza* must begin well to succeed well, and having good stallions we are certain every day to succeed better."[60] Seven years after Donato di Meo's mission, the Mantuan marquis received a letter from Antonio de Calcho, who was tracking down Sicilian horses on his behalf and reported on the selection of horses—barbs, Turcomans, and Neapolitans—and how to settle accounts.[61]

Breeding Gonzaga horses required, first, a place to keep the studs, meaning the flocks or broods used for breeding. A wide array of correspondence notes that the Gonzaga's many brood mares were driven from one pasture to another depending on the time of year. At high summer, they were herded from Lake Garda to a town near Tyrol in the shadow of Mount Baldo, where Alpine grass grew. By the time Francesco II was marquis, in 1484, two hundred mares grazed on the pastures of the Pietole breeding establishment in Virgilio; most of these animals were barbs and coursers. Grooms, trainers, and stable masters ran stables and stud facilities in nearby Gonzaga, Governolo, Marmirolo, Pietole, San Sebastiano, Sermide, and Soave; stable masters from these territories regularly wrote directly to the marquis. Near the canal dividing the old Roman city from the Isle of Te was San Sebastiano, a well-known stud facility that held the best horses not used for racing. In Governolo, the Mergonara stable housed the Gonzaga's racehorses "of the breed of the house" (*razza della casa*) and accommodated many of the most acclaimed stallions and palio racers.[62]

Breeders' practical experience raising horses led them to observe family resemblances firsthand, although, as suggested in chapter 1, they did not often dwell on the theoretical reasons for the likenesses and deviations. Consider the records from 1490 regarding Zirifalcho, a very fine horse that lived in Mantua. Zirifalcho's virility was by no means in doubt. That year, three legitimate offspring were born to him by different females: one was born healthy, another's birth was so difficult that her mother died in the process, and the third was monstrously malformed. As mares were foaling in late March, the *vicario marchionale* of the Mergonara region, Martino Maria de Anguissola, wrote to the marquis, "Last Friday a mare named la Cornachina, sister of Renegato and covered by Zirifalcho, had a handsome colt, dark bay with a bit of a star in front and white sock, like his father."[63] Despite how frequently he sired Gonzaga horses, their health record was not perfect. "Impregnated by Zirifalcho," the stable manager wrote of another horse,

> yesterday she had a filly who looks just like Zirifalcho, but having the feet and legs rotated in form such that she could not get to her feet, and having

a short and twisted snout above and the part of the said snout below much longer than needed, and in that way she could never take her mother's teat; yesterday she was dead. The mother is doing well.[64]

Even with the finest horses and stables, breeding remained an uncertain affair; family resemblances held in most, but not all, breeding interactions. Although regular reports from the stables detailed their charges' early lives, they contained no sustained discussion about why the birth record was so variable and what might have gone awry in each case. To know animal breeding meant to be willing to experiment anew, not to theorize the origins of monstrosity.

Seeding Razze

While neither the Gonzaga patrons nor their hired breeders were particularly introspective about their projects, they regularly adopted botanical metaphors to explain their process of creating *razze*. In a 1514 letter, Lorenzo de Medici's son Giuliano told Francesco II Gonzaga's brother, Cardinal Sigismondo Gonzaga, that he was planning to challenge the Gonzaga's success in palio racing by "sending agents to the Barbary coast of North Africa to find the first class palio horses." Francesco replied that he might look for horses in Spain, Naples, Sicily, Turkey, and Phrygia, as his family had.[65] By the sixteenth century, the Gonzaga had honed their reputation as high-end horse traders uniquely connected to the world's best equine markets. In the 1521 *Chronica di Mantua*, Mantua's court humanist Mario Equicola explained the Gonzaga lord's approach to horses:

> Nature, above every other brute animal, was inclined to love horses.... Not content to have the seeds and races [*seminarii & Raze*] of the Kingdom of Naples, of Sicily, and of Spain he sent to that part of Asia and Africa that today is called Turkey and Barbary. Therefore, he seeded [*seminò*] very fast horses in his country, where they became such, that in little time there were born in Mantua the Asians, which have left a large piece behind.[66]

For the Gonzaga, Western horses from Iberia and Southern Italy were neither exotic nor fast enough alone; they sought to cross them with animals from Turkey and the Barbary Coast. Imported animals were seedlings from which the Mantuan equine race was supposed to flower, to perpetuate Equicola's metaphor.

Over the sixteenth century, the Gonzaga stabled and bred a variety of different types of horses created from their international exchanges and local stock, as Andrea Tonni has traced.[67] For example, a numerical inventory charted the horses covered by a given stallion in a year when "the horses of the big race come as beautiful as the world" (fig. 3.6). Mares were branded, herded as a group, and kept away from stallions, except for intentional breeding. Using the same categories employed throughout many European courts, names for human ethnic groups or warriors were folded into popular animal names. The barbs (pl. *barbari*, sing. *bárbaro*; "barbarous, incivall, also a Barbary horse"), imported from North Africa (Barbary Coast), were reputed to be fastest.[68] Jennets (pl. *ginette*, sing. *ginétta*; "a iennet or Spanish horse"), used for riding, had long been carefully bred in Andalusia; their light bone structure was supposed to come from their Moorish ancestors, which included North African barbs. The Turcomans (pl. *turchi*, sing. *turcó*; "a Turke, also a Turkish horse") came from the Ottoman Empire and were known for stamina. For hauling people and goods, some Mantuans kept heavy, strong *virgiliana* horses, named for the poet Virgil's ancient home in Mantua; for farm work, they maintained the *villana* breed, which was supposed to be strong and hardy. The coursers (sing. *corsiere*; "a steed or courser of Naples"), often Neapolitan in origin,

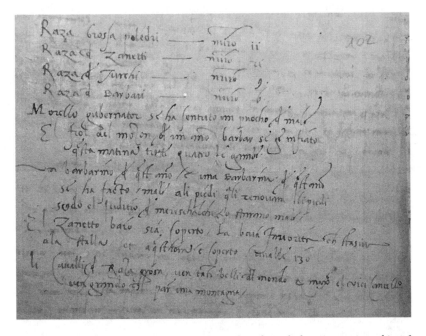

FIGURE 3.6. Inventory of the Gonzaga house breeds, including *zannetti*, *turchi*, and *barbari*. Archivio di Stato di Mantova, Archivio Gonzaga 2475, fol. 102, May 8, 1509. Image courtesy of the Archivio di Stato di Mantova.

were acclaimed for their elegant proportions and power, and were used in parades, tournaments, and *manège* riding. Over time, the word *barbari* came to refer to fast horses regardless of the animals' breed or origin. However, originally the Gonzaga and their stable masters generally applied it primarily to the barb horses that they had imported or bred in the Mantovano region. *Barbari naturali* referred to the horses imported to found the breeding program; they were born in places outside of Mantua. In some records, their descendants, *barbari* (or *cavalli*) *de la raza de la casa*, seemed to lose their "natural" status, even if they were produced from two imported horses. As soon as the horses were selected by the Gonzaga breeding system, they became representatives of the "race of the house." Some of the Gonzaga's *barbari de la raza de la casa* were crosses between *turchi* and *barbari*.[69] Local breeding (in an expansive sense meant to include both animal generation and training) made it possible for the Gonzaga to impress their brand on stock with foreign origins, literally and figuratively. Performance and *razze*, including the labor of attending to horses' mating habits, environment, individual health, and training, might have made some Gonzaga horses "noble," but that was not an adjective to which Gonzaga horses were entitled by birth alone. While the language of *razze*—that is, a language of breeding—pervades correspondence and administrative documents, an emphasis on horses' purity of blood does not. When pedigrees were collected, as we will see in the case of the book of palio victories by Francesco Gonzaga's horses, these records focused on particular characteristics or victories ascribed to the unique, champion horses.[70] Pedigrees celebrated notable animals' successes, but the remaining archival evidence does not suggest that these charts were extensively used by breeders to compare the family trees of different successful studs or mares with an eye toward the success of their progeny generations down the line. This approach appears to have been quite different in kind from, or at least less systematic than, the breeding strategy of pedigree analysis that Russell analyzes in seventeenth-, eighteenth-, and nineteenth-century England, in which the detailed parentage of an individual animal is recorded and "the breeding value of an individual is determined by its parentage over several generations."[71]

As we have seen from contemporary Southern Italian studbooks, many *razze* of Gonzaga horses were held in high esteem; however, Mantua was most famous for its racehorses. The palio horse racing circuit, a series of events organized across Italy, offered the Gonzaga a reason to sink considerable resources and time into their horses. A highly competitive and fractured city-state culture fueled intense competition, the winners of which were decided by measurable outcomes. Like human height, a horse's speed

was a measurable outcome of breeding projects, and one that was easier
to categorize than a complex set of characteristics that might approach an
ideal of equine perfection. With their simpler genetic correlations, fancy
characteristics responded more quickly to breeding interventions; early
modern breeders debated to what extent speed fit into this category.[72]
Winning races conferred such popular support that courts were incentiv-
ized to reward even incremental improvement derived from training reg-
imens or breeding changes. The Gonzaga imported horses internationally
in part because they valued winning races locally.

Palio races coursed through the heart of the city before the roar of a
crowd. Through their research on palio racing, Galeazzo Nosari, Franco
Canova, and Elizabeth Tobey have painted a vision of race days: the piazza
filled with attendants standing on the palio cart (*carro trionfale*), grasping
ropes to secure the banner, which was both the finish line and the prize in
one. Only a substantial gust of wind would have made the heavy fabric—
velvet, brocade, or damask—ripple, since it was likely lined with hundreds
of ermine skins. The winner was the horse that reached the palio banner
first, and occasionally jockeys had to touch the banner as they passed. The
race itself was often a direct shot, run *alla lunga* on a straight track through
the city center, starting off at a city gate and ending in a piazza. Florence's
famous San Giovanni palio was 1¼ miles long, while the palio of Bologna
was a longer race at 2 miles. Local authorities often poured sand and earth
from the stables to cushion the blows of horses' hooves upon the stones.
Nobles stood on upper terraces to watch the race, while commoners lined
the street and peered around stone corners to catch a glimpse of the pack
as they charged by.[73]

Victory mattered to the Gonzaga lords and their team, who acquired,
maintained, trained, groomed, and rode the horses into competition.
Their victories were lovingly recorded in the 1512–18 *Libro dei palii vinti
dai cavalli di Francesco Gonzaga*, an illustrated manuscript that depicted
the Gonzaga's champion horses above a list of their victories.[74] The book
was organized around individual horses, whose portraits graced the top
of the page, followed by details about their victorious races. "A Mare of
the House Race" (fig. 3.7) managed a slew of victories in Roman carnival
races, and on Bologna's Saint Peter's, Saint Raphael's, and Saint Martin's
Days from 1511 to 1512. Despite this formidable list of ten wins, she apparently remained named "our mare" from beginning to end. An equally
banally titled male, "Dappled Horse of the *razza*," was depicted on the
pages after her. He did not achieve such an illustrious career, with only
one victory for an Alexandrian velvet in Rome in 1506. However, he managed to win that race even though it was "the first time he had ever run a

FIGURE 3.7. The illustrious "Mare of the House Race." Silvestro Da Lucca, *Il libro dei palii vinti dai cavalli di Francesco Gonzaga, 1512–1518, mineato da Lauro Padovano e forse altri due miniatori*, s. 16. Falk Family, Palazzo Giustian Recanati alle Zattere, Venice, Italy, private collection.

Palio race."[75] Veteran palio horses were more likely to win, and imported barbs were the animals to beat. Winning at such a prestigious competition was an important boon for the Gonzaga diplomatically, financially, and culturally. It fueled their reputation as masters of nobility and made their animal gifts more prestigious when offered to sovereigns, both European and Ottoman. As the Italian Wars raged throughout the peninsula and the fates of many small states hung in the balance, the palio circuit remained an important expression of social power. It offered a real, if informal, way in which the Gonzaga could challenge their Northern and Central Italian neighbors, the likes of which included Giuliano de' Medici and the Este of Ferrara.

Animal Imports and Making a Natural History

As eastern Mediterranean politics shifted over the sixteenth century, it became easier for the Gonzaga to access Ottoman markets directly. By the time of Federico II, Naples under Spanish rule had diplomatically turned away from its historical connections to Greece and the Ottoman Empire; instead, it was jointly ruled by the Council of Italy in Simancas, Spain, and a Spanish viceroy. In his time, Francesco II Gonzaga had relied on agents in the Ottoman Empire and North Africa to provide horses to sate his thirst for palio victories. The Ottoman Empire was so wide that its sultan had access to stock from the Balkans to Syria and beyond. A Venetian ambassador to the Ottoman court wrote that the sultan had racing stallions from Bursa and Adrianople, as well as animals given as tributary presents from Cairo, Damascus, and Baghdad. Suleiman the Magnificent (r. 1520–1566) controlled horse breeding in Syria and could access an important horse fair in Aleppo.[76] Thus, what constituted a "natural" Turkish horse was itself variable. Despite their diverse and ambiguous origins, when Ottoman-bought horses made their way to Mantua, the nuance of their history was subsumed under the breed name *turco*. Upon Gonzaga victories in palio races, the crowd apparently cried, "Turcho, Turcho." Perhaps this was because the Gonzaga had received horses from the Ottomans, or because Ludovico Gonzaga had been nicknamed Turco, or perhaps because his family had earlier developed a diplomatic relationship with Sultan Bayezid II (r. 1481–1512). Regardless, it was clear to the crowd that Eastern powers influenced Mantuan success in horse racing. Mustafà Begh, governor of Herzegh, also had a diplomatic relationship with Francesco and, in 1492, sent him horses as a token of friendship.[77] The Ottoman connection with the Gonzaga was not unique, however. The Hafsid rulers also exported horses to the Gonzaga, and in 1492, the king of Tunis had allowed a Gonzaga agent to export mares.[78] Despite, or perhaps because of, the persistent conflict and competition between Europeans and the Ottoman Empire, the Gonzaga were eager to incorporate Ottoman stock into their horse breeding initiatives.[79] In sum, while Iago meant to insult Othello by equating him to a Barbary horse, the insult had meaning only because so many Renaissance Italian nobles did want coursers to give to their cousins, jennets to impress the Germans, and Barbary horses to breed new speed into their local mares' offspring.[80]

Myths emerged around the Gonzaga breeding projects, which in turn gave their brands, discussed in chapter 2, greater value. Pasquale Caracciolo, author of *La Gloria del Cavallo* (1566), argued that the Gonzaga produced the most horses on the Italian peninsula.[81] However doubtful that claim may have been, given the scale of Southern Italy and its commitment to horse breeding, Mantuans were interested in cultivating that reputation. For centuries, historians have debated the role that Gonzaga horses played in the foundation of the Thoroughbred racehorse breed.[82] Like their dogs, the Gonzaga horses were held up as examples by no less an authority than the Bolognese naturalist Aldrovandi in *De quadrupedibus solidipedibus volumen integrum*: "In our age in the home of the Mantuans a noble breed of horses is reared with great passion which live in the regions full with marshes and swamps that go back to the ancient customs and character of the Venetians."[83] Aldrovandi continued, "The most highborn and noble progeny of horses [*generosissima equorum sobole nobilissima*] are seen in this city surrounded by a lake." He noted that

> no other king in Europe has developed military horses of all kinds, more valuable and more noble, than those created by the Mantuan leader. He developed them at great expense directed to the propagation of the off-spring by selected mares kept for breeding from the entire overseas province, with amazing and fertile success, from which it is certain that after a while other leaders also created their own.[84]

Thus the fame of Gonzaga breeding was preserved in the burgeoning field of natural history.

Good Breeding and the Renaissance Courtier

Conversation and debate were competitive sports at the Renaissance Italian court, although they were in some ways less public facing than the palio circuit. In the treatises that represent these closed-door conversations among elites, descriptions of animal excellence were used to buttress, question, and explain the origins of human nobility. Was nobility the result of breeding? Must the ideal courtier be born of noble blood? James Hankins has argued that "in humanist literature legitimacy of exercise contrasted above all with legitimacy in the most basic sense of the word: legitimacy of birth."[85] Thus, humanists argued that a moral virtue could be crafted outside of lineage, therefore stressing the importance of education in galvanizing human potential. This idea countered the more prevalent notion that "breeding was key"; as Hankins explains, "Renaissance Italy, despite its relative urbanization, was still overwhelmingly an agricultural society, and even city folk knew that the stock of plants and animals could be improved by breeding."[86] Still, as new families of elites first acquired, then defended, their noble social standing, the qualifications of nobility shifted; the role of blood inheritance was up for debate.

Castiglione's *The Courtier*, which had contributed to his status as Charles V's best *caballero*, provides a particularly revealing window into the qualities of this debate and its penchant for animal analogies. In the optimism of the early sixteenth century, nobles were free to discuss the bases of nobility in the protected courtly centers of Italy; such debates would become increasingly politically charged over the course of the century, censored by Church and inquisitorial forces. *The Courtier's* conversation was set in the Palace of Urbino in March 1507. The cast of characters featured real nobles and courtiers, including Elisabetta Gonzaga (duchess of Urbino), Margarita Gonzaga (the duchess's niece), Lord Ottaviano Fregoso (doge of the Republic of Genoa, lived 1470–1524), and Cesare Gonzaga (another kinsman). Count Lodovico Canossa (1476–1532) was one of Castiglione's relations and friends in life, as well as an interlocutor in *The Courtier*. A Veronese noble, Canossa had grown up in Mantua and Urbino, then served as a diplomat for the papacy and King Francis I. Through these characters' dialogue, readers could share the privilege of sitting comfortably alongside these elites and listening to their elegant conversation. Still, even as Castiglione was writing his treatise in the 1510s, the dynamic court of Urbino that he had glorified was dissipating, as its ruling dynasty was dispersed and its population subjected to war and occupation. While the wittiness of the conversation and the principles of civilized living that his book espoused endured, its specific context was elegiac, treasuring a memory that harked back to an earlier time in his life before he had met Charles V.[87]

In the fictional dialogue of *The Courtier*, a life surrounded by well-bred nobility led Lodovico Canossa to argue that the apple rarely fell too far from the tree. In those of high birth,

> nature has implanted in everything a hidden seed [*occulto seme*] that offers a certain strength and property of its principle to all that derives from it, making it similar to itself. As we see this not only in breeds [*razze*] of horses and other animals but also in trees, whose offshoots nearly always resemble the trunk; and if sometimes they degenerate, that proceeds from the bad farmer. And so it is with interventions of men, who, if they are born of good quality [*se di bona crianza*] and they are brought up [*sono cultivati*], nearly always they are similar to those who preceded them, and often they improve upon them; but if one fails to look after them well they will become wild and never ripen.[88]

Canossa held up Isabella d'Este's brother, Don Ippolito d'Este, Cardinal of Ferrara, as an example of the combined powers of excellent seed and careful cultivation in his person, aspect, words, and movements. Whether that excellence was born or created, however, might have been straightforward for Canossa, but was up for debate with others.

Not everyone agreed with the heritability of noble characteristics. Castiglione portrayed young Gaspare Pallavicino (1486–1511) as contending that "in the courtier, to me it does not seem necessary to find this nobility." He went on to cite that there were many people with "most noble blood" [*nobilissimo sangue*] who were wicked, while many of those of ignoble birth earned the admiration of posterity through their own virtues. He argued against the hidden seed hypothesis, noting that "if it is true what you have said, that is, that the hidden strength of the first seed [*seme*] is in everything, we would all be in the same condition for having had the same start [*principio*], nor anyone more noble than the other."[89] Variations in nobility, Gaspare argued, primarily came from fortune. While those who were equipped with greater talents and beauty succeeded better and more easily, Gaspare thought that had nothing to do with nobility, as it is equally present among people of humble origins. So, Gaspare concluded, nobility of birth is nothing but a name, "acquired neither through ingenuity nor through force or art, and being a matter for which to praise our ancestors rather than ourselves."[90] By Gaspare's argument, a rose by any other name would still smell as sweet; a dog of any breed could be a noble dog if it behaved excellently. Still, Gaspare's contributions were neither marked by their consistency or by their continuing liberal approach to nobility. When the conversation turned toward what would please women, Gaspare bolstered his experience with ancient philosophy: "What would please most women for having well-disposed and beautiful children will, in my opinion, be the community that Plato wanted in his *Republic* and in that way."[91]

When Gaspare insisted that human virtues were "granted by God to men and cannot be learned, but are natural," Lord Ottaviano Fregoso countered that Gaspare saw men as overly unhappy and perverse. Had they not "industriously found an art to tame the meek wits of the beasts, bears, wolves, and lions, and by it are able to teach a pretty bird to fly at the will of man and return from the woods from his natural freedom voluntarily to snares and servants?" If they had that skill, why could they not "by the same industry find arts to benefit themselves, and, with diligence and study, make the human soul better?"[92] Nature, Ottaviano contended, was trainable, and the humans who trained it were even more perfectible than their hawks.

Training and shaping potential were essential for both humans and animals, but training was generally afforded only to good stock; one could not transform a sparrow into a falcon. While mutable, neither the natural world nor the prince's mind was a tabula rasa (a blank slate) to cultivate. No matter how good his courtiers and companions, if a prince were not inclined to learn, then all labor would be in vain. Lord Ottaviano took an example from husbandry to prove his point:

> Just as any good farmer [*agricultore*] also would labor in vain if he were to begin cultivating sterile sea sands and sowing it with excellent seed, because that barrenness is natural in that place; but when the diligence of

human culture is added to good seed in fertile soil, and to mildness of climate and rains suited to the season, very abundant fruits are always seen
to be widely born; nor is it on that account true that the farmer alone is
the cause of them, although without him all the other things would avail
little or not at all.[93]

Everything hinged on what constituted good soil. Recall that, in Lodovico's vision, nobility was inherited. On the other hand, Gaspare had contended that there was no noble seed; capricious fortune determined one's
opportunity to excel. In Ottaviano's argument, however, seeds represented
training, soil stood in for talent or potential, and crops were the fruitful products of that combination in the form of personal excellence. The
analogies to agriculture and husbandry took on different analytical work
in each example; in all cases, breeding practices, farming, and courtly animal culture more generally informed intellectual debates about the nature of inheritance.

Castiglione's set-piece conversation highlights an ongoing debate about
breeding, training, and the nature of nobility, along with the well-studied
themes of *sprezzatura* (that art of living excellently effortlessly), music,
ethics, classical education, and discursive power. On one hand, Christian
origin stories, the theology of free will, and the experience of badly behaved nobles and talented commoners militated against encoding nobility
in blood. On the other hand, an unquestioned assumption about hierarchical society, a recognition of the reality that offspring share characteristics with their progenitors, and the experience of selectively bred animals suggested a more determinist possibility. The same debate continued
throughout the seventeenth century, as Silvia De Renzi has explored in the
1620s arguments between Ludovico Zuccolo and Alessandro Tassoni.[94]

From Courtly Debates to Imperial Hierarchy

Castiglione's wildly popular book itself became controversial outside of
Italy over the sixteenth century, though not for its remarks on nobility
and inheritance. Following Charles V's death and the growing militant
Catholicism and isolationism of Iberia under his son Philip II, *The Courtier* raised hackles in the Spanish empire. No editions were published in
Spain from 1573 to 1873, although many copies in Italian continued to circulate.[95] The Spanish Inquisition allowed only expurgated editions of the
book following 1612, and questions about its contents were raised as early
as 1576. In Peru, Fray Juan de Almazar, an Augustinian from Lima as well

as a professor at the Universidad de San Marcos, used censorship of Castiglione to help ignite the Peruvian Inquisition. Almazar sought to censor books that were suspicious, heretical, or forbidden to prevent the spread of religious dissent in the Spanish Americas. In 1582, he appeared before the Inquisition to condemn *The Courtier*; the book taught free thought (*enseña mucha libertad*) and raised "scandalous propositions" against cardinals.[96] Almazar's condemnation of *The Courtier* in Lima more than a half century after its initial publication in Italy highlights the importance and circulation of Castiglione's ideas and the potential threat they posed.

Italian court culture was increasingly seen as anticlerical, liberal, and overly enthusiastic about free speech, especially when it came to jests at the expense of cardinals. Although Spanish courts flirted with Italian culture, they never committed to its norms. A new generation of informal advice texts, such as the manuscript "Instructions for Negotiations in the Court of Spain for Lord Ludovico Orsini Sent to His Catholic Majesty by the Duke of Bracciano," taught courtiers coming from Italy how to change their behavior to better fit with Spanish customs. The Italian visitor ought to maintain "gravity in his person in every action with the Nation of Spain, the great aspect of the *Cavaliere*." In Spain, one should "demonstrate religion in every action" and maintain that region's dress code, which, to the Italians, most closely resembled that of the cardinals. One should refrain from laughing, be sure that one's family dressed modestly and was all "well bred" (*ben creata*), and check that one's company all maintained a good reputation. In Spain, one ought not walk too much, but ride in a carriage or on horseback instead. Most of all, one ought to "speak little and with great consideration, because the Spanish are great enemies of words."[97] To these Renaissance Italians, the Spaniards seemed a bit like the Romans did to the Greeks: austere, militant, and pious.[98]

Meanwhile, across the Atlantic, the Spanish were promoting their own viceregal court system that consolidated colonial power in centers like Mexico City and Lima. Spanish colonizers built up governments on indigenous foundations, intending them to be rivals to European centers of power.[99] In the next chapter, with attention to the tensions between imperial hegemony and local autonomy, I use the metaphor of agricultural grafting to explore this practice of overwriting indigenous power channels. With a radical shift in geography, I turn to the Nahuas' ideas of fecundity and inheritance, explored through corn, seed, and blood during the same years that the Gonzaga poured resources into their breeding projects. Parallel—but importantly different—approaches to agriculture, animal husbandry, and theoretically linking one generation to the next appear in a Mesoamerican court system folded into Spanish imperial control over the sixteenth century. Through this geographic juxtaposition, I seek to destabilize a presumption of novelty around European approaches to breeding and cultivation and explore what, if anything, was unique about the European interest in *razze*.

CHAPTER 4

Corn, Seed, Blood
in Mesoamerica

Tlaloc and Cultivating Futurity

A stone figure, identified by the Johnson Museum of Art as "Seated figure with Tlaloc mask and maize," from Aztec (Mexico), 1440–1520 CE, grasps an ear of corn (fig. 4.1). The figure's right hand rests upon his knee as if about to strip the kernels—presently separate and whole, like teeth in a smiling mouth—from the cob. The phallic maize ear is disproportionately large compared with the male figure, its cob running the length of his calf and its pruned husk reaching to his jaw. Nahuas associated Tlaloc, a rain or storm god, with sustenance and fertility. Spanish-Nahuatl texts produced in the sixteenth century noted that Tlaloc had been honored, among many other places, at shrines of the main temples of Epcoatl in Mexico-Tenochtitlan's ceremonial precinct.[1] The name Epcoatl was one of the alternative divine titles for Tlaloc, and it was used to refer to his ritual representation. During the feasting and blood sacrifices of Atl Cahualo (a ceremony whose name means "[when the] water leaves" at the first month of the year), children were addressed as Epcoatl, dressed as representations of Tlaloc, and sacrificed at the Pantitlan. Their deaths were a debt payment for rain and fertility. Their tears portended future rain. If they fell asleep at the vigil, no rain would come.[2] Children, the fruits of human fecundity, were offered to appease the forces that made agriculture fruitful.

This chapter explores ideas of inheritance and fecundity in Nahuatl-speaking Mesoamerica, and how those ideas shifted in the decades following Spain's conquest of Mexico in 1519–21 and the installation of a new colonial regime. In 1492, just as Renaissance Italian patrons collected living marvels and sought to improve their *razze* of animals, Nahua agriculturalists celebrated their reliance on cultivation. Domestication and selective breeding shaped Mesoamerican nature, from maize plots to floating chinampa gardens to pet-keeping practices. In the highlands of Mexico, maize (*Zea mays*) had been transformed from the ancestral crop *teosinte*.[3] Beans, chocolate, chili, rubber, squash, and vanilla had all been

FIGURE 4.1. Seated figure with Tlaloc mask and maize. Aztec (Mexico), 1440–1520
CE, volcanic stone. Collection of the Herbert F. Johnson Museum of Art, Cornell University. Acquired through the Membership Purchase Fund; 73.013.002. Image courtesy
of the Johnson Museum.

honed through selection to serve human interests, as had cochineal insects, turkeys, dogs, and honeybees.[4] While archaeologists have debated
to what extent people were conscious of the changes they were effecting,
many scholars now emphasize deliberate selective breeding practices over
the past ten thousand years rather than reading these changes as results
of inadvertent selection.[5] Sixteenth-century Nahuatl-Spanish sources are
replete with a vision of nature intermingled with human intervention,
including discussions of intentional selection based on the language of
"seeding." To see selection in the "Seated figure with Tlaloc mask and
maize," we might note that kernels came in many colors and sizes, and
that human agriculturalists chose which to plant and which to consume.

The concept of *xinachtli* (a term that meant seed, plant seed, or, when possessed, semen) slipped between plant, animal, and human bodies.

After the Mexica (popularly known as the Aztecs) followed the vision of Huitzilopochtli and founded Mexico-Tenochtitlan in the center of Lake Texcoco (ca. 1325–45), Mesoamerican agricultural traditions were ritually reconstituted in the new imperial center. As the polities of Tenochtitlan, Tetzcoco, and Tlacopan joined together in the Triple Alliance following the Tepanec War of 1428, their new empire standardized the calendar, religion, law, and tribute extraction; tribute included quotidian agricultural products (like maize), higher-value items (turquoise and quetzal feathers), and foreign slaves to be sacrificed in the central precinct.[6] I follow Matthew Restall's suggestion that we read Moteuczoma Xocoyotzin as a collector; his royal precinct featured a "zoo-collection complex" of flora, fauna, and other objects that were intentionally acquired, maintained, and observed.[7] By this reading, Moteuczoma welcomed the Europeans into his court in part in an attempt to collect them. In turn, his power was comprehensible to Europeans like Hernán Cortés because American courts, like European ones, enculturated a fascination with the wondrous and a desire to collect it. After the Spanish conquest, European ideas about selective breeding mapped imperfectly onto existing Nahua notions. Here, I evoke a botanical metaphor in my representation of that history, which suggests that, in the first decades of contact and conquest, the Spanish grafted their own traditions of husbandry and agriculture and its related discourse onto indigenous notions of heritability.

Nahuatl Sources and Nature Studies

A central conceit of this chapter is its use of Nahuatl-Spanish documents produced in the sixteenth and early seventeenth centuries both to access the changes taking place in New Spain during the period when they were produced and to glimpse cultural practices from the century before the European invasion.[8] Not everything changed with the Spanish conquest, and Mesoamerican approaches to cultivation and breeding did not slip away without a trace. To navigate the intricacies of this shift, some scholars of Nahuatl have elaborated James Lockhart's periodization of language changes to parse how life might have been before the arrival of the Spanish in 1519, despite the relative paucity of written documents compared with what we have for the sixteenth century.[9] Lockhart suggested that there had been few changes to Nahuatl from 1519 to 1545–50, a period defined by early Spanish colonialism and by the *encomienda* system in which whole indigenous states were assigned to one Spaniard. The language shifted with extensive noun borrowing between 1545–50 and 1640–50 under the new colonial order distinguished by the *repartimiento* system and Spanish-style towns.[10] Attention to these shifts allows us to analyze the emergence of neologisms and the borrowing of Spanish grammar, husbandry practices, farming, and associated vocabulary.

For changes in language and the increasingly hispanicized vocabulary of cultivation, breeding, and lineage, I rely on Nahuatl-Spanish dictionaries such as *Vocabulario trilingüe* (ca. 1540) and Alonso de Molina's 1555 and 1571 versions of *Vocabulario en lengua castellana y Mexicana*.[11] These dictionaries were based on word lists compiled by the Spanish humanist professor Antonio de Nebrija (1444–1522) in his *Vocabulario* or *Dictionarium*. Nebrija's *Vocabulario español-latino* (1495) included twenty thousand entries and translated between Castilian and Latin. By comparison, Florio's Italian-English dictionary and Covarrubias's Spanish dictionaries from the late sixteenth and early seventeenth centuries defined fewer terms, but offered more expansive definitions.[12] In Nebrija's *Vocabulario, generacíon* was linked to Latin *generatio*; in reference to lineage, it was also linked to Latin *genus, stirps*, and *propago*. In turn, *Vocabulario* linked Castilian *genero por linage noble* to Latin *genus*; *casta*, in respect to "good lineage," was likewise linked to Latin *genus*.[13] To my knowledge, *raza* (or *raça*) in relation to animals was not defined.

The many translations of Nebrija's word lists changed as word usages changed across the early modern world. Byron Hamann has shown that "these changes in content allow us, now, to reconstruct a *genealogy of vocabularies*—a tangled family tree in which earlier editions spawned descendant generations both legitimate (legally licensed) and bastard (pirated printings)."[14] The anonymous manuscript *Vocabulario trilingüe*, now held at the Newberry Library, was likely made by Nahuatl speakers who copied out a Nebrijan Castilian-Latin dictionary and added Nahuatl translations in red ink. Other missionary dictionaries were extensively based on lists from Nebrija's dictionary. Their compilers, then, requested, or insisted, that indigenous-language speakers provide them with definitions for these words.[15] Molina's dictionary was built around the 1536 Granada expansion of Nebrija's *Dicionarium*, which Alonso de Molina and his collaborators themselves expanded through the enlargement of existing categories and the listing of new Spanish words to capture ideas from Nahuatl.[16] In different instances, Nebrija's categories were uncritically transferred to Mesoamerica or expanded to meet missionaries' linguistic needs, or (more rarely) Castilian categories were modified to capture the subtleties of indigenous words.[17] To a large degree, however, European humanistic vernacular vocabulary infused the dictionary definitions of terms translated from the sixteenth-century Mesoamerican world. I have not found the term *raza* in refence to animal breeding or human difference in the *Vocabulario trilingüe* or Molina. However, as we have seen elsewhere, by the early seventeenth century, Covarrubias's dictionary included the term. In lieu of searching for animal *raza* in the Americas—for which endeavor future researchers might productively turn to sources on ranching—I turn to language used in discussions of lineage and cultivation in an early estate map, dictionaries, the Florentine Codex, and the writings of a Nahua chronicler.

The collaborative and encyclopedic nature of the Florentine Codex (1580) makes it a particularly rich source for reconciling older Mesoamerican practices and changing approaches to nature, agriculture, and husbandry.[18] Unlike Nebrija's word lists, it did not command a wide early modern readership. This compendium was produced by around two dozen Nahua artists and collaborators under the supervision of Salamanca-trained Fray Bernardino de Sahagún (1500–1590) in the College of Santa Cruz de Tlatelolco outside of Mexico City.[19] Written in Spanish, Nahuatl alphabetic, and pictographic texts, the Florentine Codex combined the European and indigenous American ways of knowing plants, animals, medicines, traditions, and people that made up a Nahua world.[20] I particularly rely on sections of book 11, *On Earthly Things*, that describe maize, turkeys, and in the next chapter, dogs. Scholars have noted that the structure of this book contains influences drawn from Aristotle's *Historia Animalium*, Pliny the Elder's *Natural History*, Isidore of Seville's *Etymologies*, Bartholomeus Anglicus's *On the Properties of Things* (ca. 1240), and Jacob Meydenbach's *Hortus Sanitatis* (1491).[21] After its completion, the codex was gifted to the Medici, and it is now held by the Biblioteca Medicea Laurenziana in Florence, which makes it a particularly appropriate source on the connections and siloes of knowledge between Italy and the Spanish empire. It was not widely read in Italy, but an Italian translation of books 1–5 was produced in the 1580s.[22]

Sorting Out Seeds

Maize (*cintli*), seed (*xinachtli*), and dried maize seed (*cinxinachtli*) were key terms for Spaniards seeking to understanding sixteenth-century Nahua customs because they were crucial to both agriculture and ritual. Participation in farming did more than mediate cultural admission: it also provided a dominant metaphor for Nahua practitioners and intellectuals to theorize what was passed from one generation to the next, and to link human life to a wider nature. This link between metaphorical system and agricultural practice has been widely studied by historians, archaeologists, and anthropologists of modern indigenous Mexican communities.[23] When writing of the modern Nahuas of Amatlán, Alan Sandstrom observed that "the Nahuas, in their abiding wisdom, say of themselves, 'Corn is our blood.'" These words link them to their horticultural way of life, the powerful spirits in their pantheon, the animating principle of the universe, and their vision of the beautiful green earth that has given them life."[24] In his description of a modern-day fertility ritual in which shamans cut out the images of seed spirits in paper dressed in girls' and boys' clothing, Sandstrom characterized villagers' views of Tonantsij, the fertility mother, as she ruled over her children, the seed spirits (*xinachtli*)—of which corn is paramount—that "represent the life force or potential for fertility of each crop."[25]

The process of selecting corn seeds appears in *On Earthly Things*, in a chapter that focuses on sustenance and begins with maize. However, the cultural centrality of corn was intentionally undermined, or at least deemphasized, by the Spanish author writing this section in parallel. While the Nahuatl text in the right-hand column of the section, starting on fol. 246r, emphasizes different colors of corn glossed with short Castilian subheadings, the Spanish text in the left-hand column, ostensibly a translation, notes the "diversity of maintenance," meaning the traditions of agriculture and food culture, "that have nearly nothing in common with our own." It notes that indigenous Nahuas ("these people") are unaccustomed to "tame animals that we use, those that come from Spain and other parts of Europe." In the trope of difference-making that Acosta would repeat, the Spanish text emphasizes that there is no local Mexican equivalent to the Old World's wheat, horses, oxen, donkeys, sheep, or goats. The Spanish-language text then wanders into discussions of evangelization, Sahagún's personal experience teaching the Christian faith, and the immediate threat of pestilence; all of these topics have even less to do with the Nahuatl attention to corn variants in the adjacent column.[26] As the Nahuatl text turns to the specifics of maize planting, emphasizing what local agriculturalists are actually creating rather than opining about the European bioproducts they lack, the Spanish text recounts a prophesy. In this vision, recounted with slips into the first person, the few Spaniards who had settled the land would cease to be and the land would become infested with local beasts and wild trees. Once the Spanish were gone, they would multiply and populate (*multiplicando y poblando*) the land. Sahagún concluded that, because of this inevitable local fecundity, "there would always be a large number of Indians on these lands."[27] In this telling, the Mexican landscape was set to grow wild as soon as Spaniards were no longer there to prune it.

This disparity of emphases fits into a wider tendency of European and subsequent Eurocentric historiography to emphasize only the technologies that Europeans developed or wielded versus those developed by indigenous American peoples. Marcy Norton has asked what might be gained with a "a more suitably generous definition of technology, a definition that would allow for the inclusion of cultivated or even foraged plants as well as prepared foods, one that also allows for communication devices and literacies, auditory and kinesthetic arts like music, dance and prey stalking, and building and furniture technologies such as thatching and hammocks."[28] In sum, even in this encyclopedic text, the Spanish authorship undermined the perception of the Nahuas' agriculture as technology, depicting their selected flora and fauna as missing the essentials that made up those of Europe.

Both the Nahuatl alphabetic text and the corresponding images emphasized the labor and intention behind maize. When it came to creating maize varieties, a planter was charged with the ceremonially important task of "choosing a seed," as art historian Helen Ellis has explained.[29] The illustrators of the codex depict this figure bearing a bundle of cobs, kernels, and seeds to sort (fig. 4.2). Kneeling down, he sorts through the kernels extracted from cobs like that held by the seated figure with Tlaloc mask. Maize kernels were diverse in appearance and included white, yellow, reddish, tan, and other varieties.[30] The codex's description of these practices records a set of repetitive phrasing that would be useful to evangelists hoping to recognize Nahuatl variations on a given topic: "I cause

FIGURE 4.2. Sorting maize kernels for ritual planting. In Fray Bernardino de Sahagún, *General History of the Things of New Spain: The Florentine Codex*, bk. 11, *On Earthly Things*, chap. 13, "Which tells of the kinds of sustenance." Florence, Biblioteca Medicea Laurenziana, MS, Med. Palat. 220, fol. 249r. Biblioteca Medicea Laurenziana. By permission of Ministero per i Beni e le Attività Culturali e il Turismo. Further reproduction is prohibited.

it to have a slender cob. I cause it to form first. I make it like a gourd seed. I cause it to have a slender cob. I make the cob slender."[31] Actions such as harvesting, clustering, and husking were recorded in the first person, recalling the repeated human intervention involved in agriculture. The next Nahuatl section elaborated how "maize in the ear [is] selected for planting. The best seed is selected. The perfect, the glossy maize is carefully chosen. The spoiled, the rotten, the shrunken falls away; the very best is chosen. It is shelled, placed in water. Two days, three days it swells in the water. It is planted in worked soil or in similar places."[32] The kernel body is described with language like that describing a pregnant woman: swollen and fecund. *Cinxinachtli* was "precious, our flesh, our bones."[33]

Xinachtli linked reproductive potential in the present with hopes for the future. In the Florentine Codex's discussions of body parts, *xinachtli* featured in the language of human sex, as Alfredo López Austin has shown. While most sections of the codex have Spanish equivalents written in the left column, book 10 lists bodily organs in Nahuatl without Castilian equivalents.[34] The paragraph on semen follows terms related to testicles. It begins with *omicetl*, which López Austin understood as "the bone that coagulates," revealing that the Nahuas believed that bones were filled with vital forces. It moves quickly to *xinachtli* (seed) and *tlacaxinachtli* (human seed).[35] Nahuas believed that both parents produced liquids with generative powers. Just as one had to place the right number of seeds in the hole to cultivate a plant, too much *xinachtli* could lead to problems in childbirth: the baby could get stuck in the woman's abdomen because of this gluey adhesive. With such generative power, too much seed risked becoming a bad thing.[36]

Human seed grew into humans. As scholars of human deep time have shown, medieval Eurasian manuscripts include variations of the Tree of Jesse, representing the genealogy of the Messiah, and texts from the Abrahamic tradition "are replete with lists of men begetting and begotten."[37] Likewise, Nahuas relied on genealogies to link living individuals and past ancestors, which were often pictorially represented as lines stretching across time; these links were used to explain kinship ties and defend hereditary titles.[38] However, over the course of the sixteenth century, it seems that the notions and language of European kinship were stamped onto indigenous relationships. In her study of Nahuatl kinship terminology, Julia Madajczak revised the interpretation of *tlacamecayotl*. By 1571, the term was defined by Molina as "abolorio de linage o de generacion"; subsequent scholars have interpreted this term to mean "a descent group, a lineage" and have linked it to the terminology of rope.[39] Madajczak

suggests that the original meaning was not descent, but "a family understood as people connected by means of consanguineal relations."[40] This
research suggests that lineage meant something particular in sixteenth-
century Southern Europe as compared with kinship networks in Southern
Mexico. European marriages increasingly derived a sense of purity from a
marriage with one wife with Christian ancestry (we might think back to
the peacocks in the introduction). Meanwhile, however, some members
of the Nahua nobility practiced polygyny. Beyond providing advantages in
finances and labor, multiple wives could bear multiple heirs, which could,
in turn, increase the likelihood that a lineage would prosper, as Lisa Sousa
has described.[41] Tensions between European and Nahua ideas of inheritance had material stakes.

Turkeys-Eggs-Turkeys

Mesoamericans raised ducks, turkeys, and dogs for consumption.[42] As in
many other cultures, Nahua breeders learned about living animal bodies
through selective breeding around the home. Intimate knowledge of turkey reproduction was possible because turkeys, unlike their wild relatives
(the *cuauhtotolin* or wild turkey) resided in the home.[43] While the Spanish
text describes them as "domesticated and thus known," the Nahuatl highlights personhood and human labor in their production: "It is a dweller in
someone's home, which can be raised in someone's home, which lives near
and by someone."[44] Women—illustrated in the Florentine Codex with a
distinctive feminine hairstyle—played a leading role in the breeding and
raising of turkeys.[45] Like dogs, whose malleability defined them, domesticated turkeys were bred in different colors, from a smoky gray (*pochectic*)
to black (*tliltic*) to white (*itztac*) and more. Unlike most of the animals in
the Florentine Codex, which are described on their own terms, human action pervades the description of turkeys, as it did the description of maize.
Again, the Nahuatl text switches to the first person in a set of repetitive
phrases on how to prepare turkeys for consumption: "I pluck it," "I remove its wings," and "I pull out its bill."[46] Keepers even used seeds from
the *teopochotl* tree to cause the birds' bodies to swell.[47] The codex's illustrations detailed the birds' reproductive cycle (fig. 4.3) from copulation
(in the upper right) to egg laying (center right) to incubation under the
mother's warm body (center left). A human presence is visible in the little
house on the bottom left of the illustration. Through the turkey, Nahua
breeders and observers saw a life cycle that created new generations inside
a shared human-animal home. Life transferred from one generation to

FIGURE 4.3. *Totoli,* or domesticated turkeys. In Fray Bernardino de Sahagún, *General History of the Things of New Spain: The Florentine Codex,* bk. 11, *On Earthly Things,* chap. 2, "Of Birds." Florence, Biblioteca Medicea Laurenziana, MS Med. Palat. 220, fol. 57r. Biblioteca Medicea Laurenziana. By permission of Ministero per i Beni e le Attività Culturali e il Turismo. Further reproduction is prohibited.

the next through living rocks (by which I mean turkey eggs or *tōtoltemeh* from *tōtolin* ["turkey"] and *tetl* ["stone, egg"]) that, like seeds, carried the next generation of birds inside them. As with the visual text of the *totolin-totoltetl-totolin*, cycles defined Nahua visions of growth and generation.[48]

The language of seeding and selection is not uniform across early colonial sources written in Nahuatl, so it would be as much an overstatement to say that all breeding ran through discursive variations of *xinachtli* as to say that European selection functioned solely around *razze*. Rather, these discourses slipped into and out of use, varying with communities and their relations to nature and with the bureaucracies that left records about those relationships. From turkey pens to chinampa gardens, Nahua agriculture required the active manipulation of a sexed nature to create desired outcomes.[49] Thinking back to chapter 3's depiction of Renaissance Italian breeding, we might note some parallels. Despite the domain difference (horses and humans of short stature versus maize), a metaphorical system interwove agriculture and cultural priorities. Both connected human, nonhuman animal, and botanical bodies, particularly through an emphasis on seeding. Both provided space for human nobility. It is possible that the emphasis on male seed in written Nahuatl-language sources came from the early influence of European terminology and a Galenic medical system that was already prominent in Mexico by the 1540s. Nonetheless, the ritual importance of seeding combined with the dispersion of descriptions of male and female sex require interpretations to maintain space for pre-Columbian American theories of reproduction that differed from the European norms. Some instances suggest a shared source: European beliefs impressed onto the inhabitants of New Spain. Other cases point to the parallel coming-into-mind or coming-into-text of a set of long-standing embodied practices of breeding.

Collecting in Moteuczoma's Court

Hernán Cortés (1485–1547) arrived on the coast of the Gulf of Mexico in 1519, having sought out a preexisting empire that could be conquered, its elite replaced, and its commoners kept creating wealth for their new masters.[50] The conquistadors off-loaded their horses, which indigenous onlookers thought to be *mamazah*, or deer, having never seen such animals before. Cortés's letter to the Holy Roman Emperor Charles V stressed the impressive scope and nobility of Mexico-Tenochtitlan, highlighting the infrastructure, collecting, and architecture that were legible and impressive within a European courtly context (fig. 4.4). For his own part, Moteuczoma Xocoyotzin (ca. 1466–1520) was a collector who had created a

FIGURE 4.4. Human and animal entertainment at the Nahua court. In Fray Bernardino de Sahagún, *General History of the Things of New Spain: The Florentine Codex*, bk. 8, chap. 10, "On the Pastimes and Recreations of the Lords." Florence, Biblioteca Medicea Laurenziana, MS Med. Palat. 220, fol. 19v. By permission of Ministero per i Beni e le Attività Culturali e il Turismo. Further reproduction is prohibited.

zoo complex that "contained hundreds of species of flora and fauna and thousands of people (some as part of the collection complex, many more as workers)," as Restall has stated.[51]

Cortés described the palace of Moteuczoma in an architectural vein, walking the reader through its precincts and gardens, past pools filled with domesticated birds; he would later portray them as representing each *linaje* of birds in the region.[52] He emphasized the expense of the space, writing that birds living in the pools ate ten *arrobas* of fish each day, and that Moteuczoma employed both regular animal caretakers and experts in animal health. The palatial corridors and galleries, filled with collected objects, animals, and humans, had been designed for Moteuczoma's pleasure. In one passage, Cortés darted from Moteuczoma's collection of albino people—"men, women, and children, whose faces, bodies, eyebrows, and eyelashes are white from birth"—to cages of eagles and other birds of prey. Cortés asserted that three hundred men cared for these birds, along with lions, tigers, wolves, foxes, and other cats, who were likewise fed fowl. Near this palace menagerie was yet another palace that contained "a number of men and women of monstrous kinds, and also dwarfs, and crooked and ill-formed persons, each of these had their separate apartments and their separate keepers."[53] Years later Bernal Díaz del Castillo recalled similar memories of Moteuczoma's court and the jesters and buffoons who entertained the lord. Like Cortés, in his next breath Díaz discussed other wonders: palatial rooms filled with books, weapons, and idols. There were even poisonous snakes kept in earthen vessels, fostered from eggs and fed human and canine flesh. Díaz recalled the fierce noise of the jackals, foxes, lions, tigers, and serpents. Gardens were filled with trees, flowers, and medicinal herbs. "There was as much to see in these gardens as everywhere else," wrote Díaz, "and we could not tire of seeing his great power."[54]

A court scene in the Florentine Codex likewise depicted jesters, musicians, hunchbacks, little people, and "monstrous men," as well as animals, entertaining the lords.[55] The collections were both human and animal, with "dwarfs, cripples, hunchbacks, and servants" alongside "eagles, ocelots, wolves, mountain cats, and birds."[56] Older women taught younger women of the household to make cloth, take charge of food, and prepare chocolate; their handmaidens included "hunchbacks and dwarves [*tzapameh*] who sang and played the drum to amuse them."[57] *Tzapameh* fell into a wider category of individuals that William Gassaway has termed "exceptional bodies."[58] These individuals—often described loosely as *in tzapameh in tepotzohmeh*, or "dwarves and hunchbacks"—appear in stories of elite households.[59] This practice, too, seems to have been akin to parallel court collections in Mantua, although I do not know of evidence comparable to Isabella d'Este's letter about Delia's daughter to suggest the manner in which these people were collected.

Restall argues that, to the Europeans, Moteuczoma's zoo-collection complex meant wealth, suggested Moteuczoma's grandeur, and confirmed Aztec barbarity (especially insofar as the inhabitants of the menagerie hissed, howled, and yelped), just as it provided the fodder for European collections that would appropriate some of the same objects.[60] He goes so far as to suggest that the collections were Moteuczoma's attempt to attain universal knowledge akin to that of the creator deities: "It was therefore imperative that Montezuma collect the Spanish newcomers in order to know them."[61] Whether or not Moteuczoma admitted the Europeans to collect them, the European invaders ultimately seized his power. About two months after they landed, Cortés and his men took Moteuczoma prisoner in his own palace on the pretext of a battle near Vera Cruz between Totonac soldiers and a Mexica force. Exactly what happened is contested; each side tells the story differently. Ultimately, Moteuczoma was killed, whether by a Spaniard or by an onlooker's stone. Over the next year, Tenochtitlan fell to the Spaniards and their allies. By 1521, the Mexica had fallen.

Grafting a New Spain onto the Valley of Mexico

Early modern horticulturalists, European and Mesoamerican alike, understood variation, fecundity, and domestication. However, the practice of grafting, which had been common in Afro-Eurasia since antiquity, seems not to have been shared. The olive tree, which historian Fernand Braudel heralded as the calling card of the Mediterranean, was a favorite model for the widespread use of grafting.[62] Liz Herbert McAvoy, Patricia Skinner, and Theresa Tyers have shown that the practice long provided a language for conversion and mixed marriage, as recalled in Saint Paul's Letter to the Romans. Grafting reproduced plants already known to be strong, while seeds introduced variability.[63] Agricultural practice slipped into metaphor, which in turn slipped into social interactions and practical science. For example, sixteenth-century Ottoman scholars described how the Ottoman intellectual class of the Rumis grafted themselves onto the purest Islamic stock, acquiring a lineage of nobility that was particularly pure, "as if two different species of fruit-bearing tree mingled and mated," and thereby obtained the right to rule.[64]

In this section, I turn to the metaphor of grafting and its real use in New Spain to explore changing agricultural practices and the new colonial system. The social-climbing young men who made the Spanish empire violently harnessed the power of demography.[65] Often these men came from lower social positions in Andalusia, Extremadura, and the Kingdoms of Naples and Sicily. They dreamed that in the New World they would become land-owing nobility. To that end, they sought to create and acquire estates and aimed to engage in the same kinds of breeding programs that were the mark of nobility back home in Europe.[66] In what follows, I read a garden of grafted plants as something of a stand-in for Spain's vision of the ideal structure of its empire. Like the Mediterranean farmers they were, the Spanish cut down branches of indigenous trees. Then, they attached European branches and cultivated a composite plant that was supposed to grow European fruits, with mixed results. This is not to say that the Spaniards thought the result of their reproduction with native people would be European fruits, but more broadly that they would come to perpetuate a Christian system and a Christian nature inextricably linked to Europe while also maintaining a local rootstock.

Soon after Mexico-Tenochtitlan fell to Spanish power in 1521, Nahuatl speakers began to write about unfamiliar Spanish fauna; not surprisingly, they used words for things familiar to them. As James Lockhart noted and Chris Valesey has extensively researched, Spanish cows, bulls, and oxen were termed *cuacuahuehqueh* (horned ones), calves were *quaquauheconetl* (young of horned ones), goats were (*quaquauhe*) *tentzone* (bearded [horned] ones), a pig was *caoyametl* (peccary), sheep were *ichcatl* (cotton), a rooster was *Caxtillian huexolotl* (Castile turkey cock), a hen was *Caxtillan totolin* (Castile turkey hen), and a cat was *mizton* (little cougar).[67] During this moment when old categories struggled to hold new fauna, documents like the *Tira de Don Martín* pictographically told a history of the conquest showing a conquistador battling an indigenous warrior (fig. 4.5).[68] The European knight is astride a horse that looks rather like a deer with its small rostrum, thin legs, and short neck, suggesting this early moment of transitions in language and animal culture in which Nahuas called horses "deer" (*mamazah*). This depiction captures Lockhart's characterization of Nahuatl during the first generation of the colonial encounter, which was distinguished by little linguistic change and the hesitant adoption of Spanish loanwords such as *caballo* (horse), *vaca* (cow), *pollo* (chicken), and *carnero* (livestock or mutton).[69]

Agriculture and animal-human relations shifted alongside language. During this period, Texcoco was part of the terrain contested between conquistadors and their heirs (who were eager to set themselves up as a new aristocracy) and Charles V's court (which did not want to create unruly aristocrats in the New World and had genuine pangs of conscience

FIGURE 4.5. A conquistador battles an indigenous warrior with shield and obsidian-bladed sword at a place later glossed as Tenanpolco (Place of the Large Walls) in 1519. *Tira of Don Martín* (Codex Tetlapalco/Codex Saville), National Museum of the American Indian, Smithsonian Institution, 13/6913.

about the ethics of the treatment of indigenous people).[70] As Howard Cline has shown, the conflict surrounding the estate of Don Carlos Ometochtli Chichimecatecatl provides a window into this transition; the events were recorded in a map and Inquisition documents shortly after Don Carlos was executed by the Inquisition sometime between 1539 and 1544. Don Carlos was a local lord of indigenous background who had been raised as a Christian in the household of Fernando Cortés since 1524. In exchange for his family's initial loyalty to the Spanish, Don Carlos was granted control of the Oztoticpac estate, where he settled around 1536. During that time, Cline suggests, he shifted his agricultural practices, raising cattle and employing new European methods of grafting.[71]

On December 31, 1540, Pedro de Veraga petitioned the Inquisition to return fruit trees that, he claimed, had belonged to him under a contract that had been concluded with Don Carlos years before. Since Don Carlos was now dead at the hands of the Spanish authorities, he required his trees, or at least his half of their fruit yields. These trees were more precious because they had not been planted, but rather were surgical hybrids, stitched together to support new European foliage. The grafter inserted part of one woody plant (the scion) into the rootstock of another (the stock) to create a genetically composite plant. Unlike seeds, which often carry unique genetic material, grafted branches reproduce the characteristics of the plant they were taken from. As shown in the Oztoticpac map produced after Don Carlos's trial (fig. 4.6), pears, quinces, apples, pomegranates, and peaches grew on the shoulders of indigenous apple and cherry trees. The trunks of *cochitzapotl* (*casimirora* or Mexican apple) and mountain black cherry trees provided the stock, to which branches were attached using a whip and tongue graft. An arboreal Frankenstein had cut down the Mesoamerican flora to provide the rootstock onto which an exotic scion bearing foreign fruits could be grafted. A single new circulatory system united the two plants, and a cambium layer of woody exterior created a new skin around the transplant. The mapmaker recorded the trees in Spanish with a Nahua number to specify quantity. To my knowledge, this is the first documented instance of grafting in New Spain.

Grafting bore little resemblance to Nahua visions of growth and generation, which, as we have seen, emphasized seeds. By contrast, an oft-reprinted text by Columella advanced the typical European assumptions when it came to summoning fruits from the earth: avoid seeds and their variability whenever possible. Columella advocated grafting (*institionis*) as

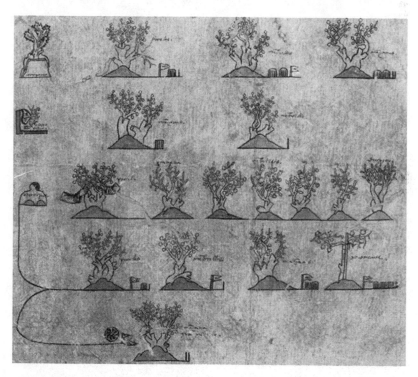

FIGURE 4.6. The Oztoticpac Lands Map, ca. 1540. 76 × 84 cm. G4414.T54:209 1540. 09, Library of Congress, Geography and Map Division, Washington, DC.

the way to regularize fecundity: "A tree which is ingrafted is more fruitful than one which is not, that is, than one which is planted in the form of a branch or of a small plant."[72] In Columella's view, botanical organs readily matched, it seemed: "Any kind of scion can be grafted on any tree, if it is not dissimilar in respect of bark to the tree in which it is grafted; indeed, if it also bears similar fruit and at the same season, it can perfectly well be grafted without any scruple."[73] Long after Don Carlos, in 1582, Texcoco remained a land of grafted plants as administrators reported their growth of cherries and Castilian apples to the Spanish government. Across the sixteenth century, a long list of Spanish fruits were imported to Texcoco and grown there, including peaches, apricots, and pears. Pomegranates, figs, olives, and grapevines fared badly in comparison to oranges, limes, and other citrus varieties.[74]

Meanwhile, European agricultural practice spread. While pigs ruled the islands, livestock husbandry featuring cattle, sheep, and horses flourished in the Valley of Mexico. By 1580, the province of Atlitlalaquia was thought by residents to be "a very healthy province without notable illnesses to speak of. There are some ranches of cows and mares between the rivers of Acambaro and Apatzeo; there must be more than 100,000 that graze between the aforementioned rivers."[75] The successful transfer of Castile's rural economy to Mexico allowed the Spanish to re-create an idealized version of their homeland. Those with enough resources could cloak themselves in imported European lifestyles, based on European agriculture, animals, and foods, while in the Americas. "The Spanish have their farms, by which they live and are sustained via local goods and those from Castile," wrote an observer from Coatepec in 1579 in one of the *relaciones geográficas*. "These goods included a variety of wheat, corn, barley, and flour, which they traded, and they also have ranches of colts, young bulls, sheep, young goats and pigs that they gather and fatten."[76] Even Spanish dogs did well, as I will explore further in the next chapter; in Tepexpan, villagers "raised both dogs of Castile and other wild ones and they grew in number."[77] Pedro Gutiérrez de Cuevas and Juan de Écija of Cuiseo de la Laguna emphasized the similarities between Spanish agriculture and nature and that found in their province of Michoacán. They wrote that locals

raise their sheep with such good order and harmony like in Spain. In addition, they raise a good number of Spanish hens and the wild ones. . . . There are also quail like those of Spain, crows, hares, rabbits, deer, starlings in great number. There are doves, falcons, wolves, foxes that are not so different from those of Spain. There are pigs of Castile and many of the natives have begun to raise them.[78]

From the corn, seed, and blood of Nahua Mesoamerica, a New Spain was cultivated through the introduction of European food products and animals, Spanish agricultural practices like the *mesta* system and grafting, and associated ideas of heredity and nobility.

A Changing Language of Heredity and Agriculture

We might read Spanish ideas of heredity, agriculture, and population as grafted onto Nahua ones, in part through how terms for animal breeding, nobility, and lineage were translated in dictionaries. The vocabulary lists that Europeans used to organize Nahuatl responded to a predetermined set of words and concepts deriving from the Latinate European tradition. After 1540, Nahuatl-Spanish sources underlined connections between the root *xinach* as both seed and semen, and nobility and race through iterations of *xinachtli* and *xinachtia*.

The Nahuatl terms for future generations of nobles and horses under Spanish colonialism incorporated the language of seeding. As we have seen, the horse, which the Nahua had first called *mazatl* (pl. *mamazah*), took on the Spanish loanword *cahuallo* (pl. *cahuallomeh/cahuallos/cahuallosmeh*). Around 1540, the authors of the *Vocabulario trilingüe* had translated Nebrija's "*Caballo bien proporcionado. Equus.us.*" into the Nahuatl for deer (*maçátl*). They added red-ink Nahuatl translations to "small horse" and "dwarf horse," but offered no equivalents for other horse terms like "Moorish horse" and "light horse."[79] In the *Vocabulario trilingüe*, a studhorse became *cahuallo xinachtli*, which translates literally to horse seed.[80] In these early translations, a botanical root linked to seeds slipped into human fecundity. *Tlacaxinachtli* connoted the seed of a man or woman.[81] The translators linked Nebrija's category of *generacion* to *tlacamecaiotl* and *tlacahioaliztli*, and similarly linked the Latin term *genus*, which, as mentioned, Nebrija translated as *generacion como linage*, to *tlacamecaiotl*. Regardless of the preexisting Nahuatl categories for kinship, this translation exercise sought likenesses for European ideas of lineage and linked the broader Latinate category of *generatio, genesis* to the same term that was increasingly employed to translate Spanish ideas of lineage.[82]

Flipping between Castilian-Nahuatl and Nahuatl-Castilian definitions, Molina's 1571 *Vocabulario* likewise defined *tlacaxinachtli* as the "seed of a man or woman" and as "seed of man."[83] *Xinachtli* meant "seed," "vegetable seed," or "herb or grain seed," or alternatively, seed of linen or onion.[84] An olive tree—that quintessential Mediterranean fruit—was not generally planted, but rather was cut and grafted onto a rootstock. According to

Alonso Molina's 1555 dictionary, in colonial Nahuatl the process of grafting an olive branch became *quauhxinachtli*, or wood seeding; this neologism literally meant "wood-like seed" or "tree-like seed"; likewise, the seeds of plants like chia and amaranth were defined using the term.[85]

Through the process of translation and seeking analogies, early colonial dictionaries employed agricultural metaphors that linked seeds, human origins, and a few nonhuman animal origins.[86] Correspondingly, Molina defined a noble man as *tlaxinachchotl*—or man of seed line; a "genera-tion of noble *caballeros*" became *tlatoca tlacaxinachotl*.[87] The translation of the Christian phrase "Be fruitful and multiply" (*in nexinacholoz in net-lapihuiloz*) included the same metaphor of seeding.[88] Molina's variations of Nebrija's word lists included Nahuatl translations of other terms and phrases such as "generous, of good lineage," "genealogy of noble lineage," "having natural nobility or dignity," "degenerate, or not doing the duty according to one's nobility or lineage," and "emblazon the nobility of his lineage falsely."[89] Molina's dictionary noted that *tlacaxinachtin* meant "the beginning of the human kind [*generación*]" as "Adam and Eve, or others somewhere in the world, they begin to multiply and raise children."[90] In short, these phrases accommodated existing vocabulary designed for a specific purpose: to speak about European-style hierarchies and Chris-tian origins. In terms of animal breeding, just as in the 1495 Nebrija *Vo-cabulario*, *raza* does not feature in either Molina's 1555 or 1571 dictionary.[91] Molina connected "casta" to purity; to live purely, one lived with *casta*.[92]

Describing "Respected Lineage"

This chapter ends by tracing the use of these terms associated with lineage in the work of Nahua chronicler Don Domingo Francisco de San Antón Muñón Chimalpahin Quauhtlehuanitzin (1579–1660), who scholars call Chimalpahin. In analyzing this episode, I build on the work of Camilla Townsend, Daniel Nemser, and María Elena Martínez.[93] In his discussion of a 1612 riot in Mexico City, Chimalpahin employed a Nahuatl translation of *limpieza de sangre* in a passage discussing public concerns that men of African descent would prey on Spanish women to intentionally color their lineage. A Nahua annalist from Chalco, Chimalpahin recorded history and ancient words (*huehuetlahtolli*), conveying Nahua events as part of a con-nected, cosmopolitan world. The Chalcan kingdoms had been settled by Nahuatl speakers from 1261 to 1303, and Chimalpahin recalled that both his mother and father descended from noble lineages. By 1464, the Chalca had become tributaries of Tenochtitlan; a century and a half later, Chi-malpahin would narrate a multigenerational tale as the Mexica lost their own independence to the Spanish.[94] In the previous century, the Spanish had transformed the former Mexica capital into Mexico City, a viceregal center from which Spain ruled Mesoamerica and the Philippines.

On May 2, 1612, twenty-eight black (*tliltiqueh*) men and seven black women were hanged in front of the palace for inciting a rebellion with the intention of killing their Spanish overlords.[95] Chimalpahin's account of the event, written in Nahuatl, was corroborated by an anonymous *Audiencia* report written in Spanish.[96] First, the account: New Spanish investigators became convinced that the black rebels had targeted Maundy Thursday and intended to use the celebratory procession of Holy Week to kill their masters and, reportedly, install a black king and Morisca queen named Isabela. The plan was to transform power throughout New Spain, as the elite Spanish would be replaced by blacks, just as the Spanish had imposed themselves in place of indigenous leaders from Moteuczoma downward. Details about the rumors of the plot, doubtless inflated by fear, and the gory execution and dismemberment of the rebels dominate remaining chronicles. Chimalpahin expressed some doubts about whether the plot was legitimate or simply a Spanish conspiracy theory. Regardless, three years after expelling the Moriscos from Spain, administrators believed that, as Lockhart, Schroeder and Namala have translated it, "if [the Black rebels] had been able to kill their masters, a Black King would take over with a mulatta woman, a Morisca, named Isabella. They would become rulers of Mexico."[97] Chimalpahin recorded that the rebels allegedly plotted to kill the male Spaniards and their mestizo male children. The rebels also allegedly planned to make the Spanish women sex slaves.[98]

To make sense of this alleged plot, Chimalpahin invoked the changing language of social categories, breeding, and inheritance. He described all of this in Nahuatl by combining Spanish loanwords for Moriscos and Africans with Nahuatl *xinachtli*. For example, the rebels were specifically concerned with the Spanish women of *mahuiztic xinachtli*, or respected lineage.[99] According to the translation by Lockhart, Schroeder, and Namala, Chimalpahin wrote that the rebels aimed to make a new population: "reportedly if [the rebels] should engender females by the Spanish women called mulattas and Moriscas, reportedly they would not kill them; they would live and the rebels would bring them up, because when they grew up the blacks would take them as wives so that their procreation, their lineage, their generation [*innepilhuatiliz yn intlacamecayo yn intlacaxinacho*] would turn black."[100] In this language, lineage traits could either be white and clean, or black.[101] Physiognomic language featured in the reported plot, too; Chimalpahin wrote that they would kill any women "if they weren't really good looking and with comely faces" and thus suitable to be taken as wives.[102] Chimalpahin wrote that his fellow indigenous people were particularly worried that the new black rulers would brand *indios* on the outsides of their mouths like animals, marking them as bestial inferiors.[103] He narrated the terror of local people with the phrase "we commoners," including himself in the anxiety that they would have become "vassals to these new black overlords, to whom they would have delivered tribute and services."[104] To be branded was normally a mark of ownership and servitude reserved for black slaves and livestock. In short, Chimalpahin wrote, as he completed his record of this plot to turn the colonial world upside down, "all the different things that the Spaniards were responsible for arranging they reportedly were going to take from them so that the blacks would have access to them, would thereby make the rules of behavior."[105]

PART III

A Brave New Natural World

CHAPTER 5

Canine Mestizaje

Living Animals, Thinking with Animals

This chapter and the next trace the ways in which New World nature prompted a reevaluation of how breeding could transform populations, human and nonhuman alike. Whereas part II of this book considers parallel cultures of selective breeding and European ideas of managing *razze*, part III moves to the grand experiment of the transatlantic encounter to trace ideas of mixing across populations and the imposition of Christian nature. The demographic, environmental, and cultural shock of Spanish imperial conquest and settlement in the Americas made ideas of difference urgent. I trace two opposing inclinations—to recognize the porosity of natural types and to isolate them through taxonomy—through dogs and camelids.

The moment of contact in 1492 was the beginning of the mixing of the continents, ushering in a new epoch of global history, one defined by contact and conquest, but also by creation. During that time, people in the Americas were both labeled as "mestizo" and self-assigned the term.[1] In her analysis of race in the colonial era, Joanne Rappaport has argued that "the concept 'mestizo' functioned as much as a metaphor as a social category, standing in for a range of types of mixings, not only between people of different sociracial categories, but also between those of different social statuses." She continues that the term *mestizo* referred to the mixing of "pure" and "impure" blood in crossbred animals like mules; "to say that mestizos constituted a fluid sociracial group in the colonial period is insufficient."[2] A term for hybridity derived from Latin *mixtus* and originally used in zoological discourse, *mestizaje* became a term of identity that characterized the human mixing in the Americas.[3] In this chapter, I suggest that *mestizaje* is likewise useful for thinking about the collision of European and American nature-cultures.

In Spanish America, several animal terms came to be central to describing proliferating human diversity.[4] Sometimes these terms retained their animal valence. Sometimes that bestialization was lost in reuse and the words became social categories. While the word *razza* emerged as a neologism from horse breeding, other terms came from special attention to dogs. *Criollo* originally meant "little bred thing," but came to refer to people born in the Americas, particularly people of African and Spanish origins.[5] By 1550, it had become a bureaucratic category so widely used that it appeared in church records. Terms like *mestizo* and *mulato* also began with animal usage and became descriptions of human difference. In his 1611 dictionary, Covarrubias defined *mestizo* as "that which is engendered from different species of animals, from the verb to mix." *Mulato* first applied to mules, infertile hybrids born of donkeys and horses. But *mulato* also developed a human meaning.[6] As Covarrubias wrote, it referred to "the son of a black woman and a white man or the reverse, and for being an out of the ordinary mixture, they compared it to the nature of the mule."[7] The examples continue: words for wild dogs, like *lobo* (wolf) and *coyote*, were derogatory slang for thieves and tricksters in the early modern period. Over the course of the colonial period, *lobo* came to refer to a child of Indians and blacks, and *coyote* was used to describe children of Indians and mestizas.[8] As cultural and socioracial mixing continued, the language of animal hybridization proliferated.[9] Although this chapter focuses on Spanish and Nahuatl speakers in Mexico, it bears mentioning that few Italian speakers interacted with these terms: Florio (1611) does not feature definitions of *mulatto, mezzo sangue,* or *creolo*. Arguments based on absence are hard to sustain, but this seems to suggest that the Spanish world's deployment of these terms was, to some degree, self-contained.

Many pre-contact American and European mammals differed so greatly that they could not effectively interbreed. However, members of dog populations and human populations could, and did, find mates in their species across the Atlantic. Europeans and Mesoamericans alike had long been domesticating dogs and keeping them as pets. But at the moment of contact, these dogs served very different functions. Both canine malleability and culturally specific breeding had spawned European dogs that frightened Mesoamericans and Mesoamerican dogs that seemed relatively harmless to Europeans. Europeans used dogs as enforcers of violence; Spanish courts had lapdogs like spaniels (literally *espagnol*) and Maltese; large Spanish sheepdogs called *mastín* dotted the countryside and guarded flocks of sheep.[10] By the fifteenth century, both Mesoamericans and Europeans had developed a mosaic of distinct domesticated dogs, including the hairless Mexican *xoloitzcuintli* and the European mastiff. Old World dogs brought over by the Spanish interbred with New World dogs. As I will discuss below, by the late sixteenth century, the *xoloitzcuintli* had become a mestizo dog that recent genetic studies reveal to be infused with genetic material from shepherds of European origin, just as their human companions of the New World developed increasingly mixed heritage. This mosaic of mixing posed a challenge to natural historians and colonial administrators alike, who sought a clear order to nature.

This chapter traces the changing definitions of dogs in the Florentine Codex, the writings of Dr. Francisco Hernández, and the subsequent manifestations of Hernández's notes in the Mexican Treasury, the compendium on New Spanish nature completed by the Accademia dei Lincei by way of the Neapolitan Nardo Antonio Recchi and published in Rome in the seventeenth century. I introduced the Florentine Codex as a source in the previous chapter, so here I will focus on the latter two sources. Dr. Francisco Hernández was a Spanish royal physician, or *protomédico*, dispatched by the crown to gather information on the useful plants, animals, and medicines of the Americas; he lived in Mexico from 1571 to 1578.[11] Hernández was an immediate contemporary of Sahagún, and his travels through Mexico in search of botanical information have been recorded by local *relaciones geográficas*.[12] Given this mission, Dr. Hernández remained less interested in the growing predominance of European nature in Mesoamerica; rather, he set out to "differentiate the indigenous plants and animals from those we Europeans brought from our mountains by sea on the long voyages to these lands."[13] Hernández derived knowledge from his own experience, ancient texts, and conversation with local experts.

The print history of Hernández's work is notoriously complex. Some of his work circulated in Mexico. Here, however, I will follow that strand that returned to Europe and came to be known as the "Mexican Treasury." Hernández himself returned to Spain with extensive notes and collections, and these were eventually studied by the Neapolitan apothecary Nardo Antonio Recchi. In turn, Recchi's notes were acquired by the Accademia dei Lincei, to which we will turn in this book's final chapter. German physician Johan Faber (1574–1629) annotated the text, and the Roman prince Federico Cesi added botanical tables. The project was brought to publication through the labors of Roman collector Cassiano dal Pozzo as the *Rerum medicarum Novae Hispaniae thesaurus seu plantarum animalium mineralium Mexicanorum*. As readers continue from here, please note that I seek to represent how each source described dogs in their own terms, rather than to standardize them according to either classical Nahuatl orthography or modern scientific ideas of breeds. Like all creatures and all scientific knowledge, neither the dogs of Mexico, nor naturalists' authoritative written descriptions of them, remained stable.

Canine Columbian Exchange

The Columbian Exchange is a historian's category that Alfred Crosby coined, in his eponymous 1972 monograph, to describe the flow of plants, animals, and microbes between the Old World and the New after Christopher Columbus's voyage in 1492.[14] In the last half century, historians, geographers, and biologists have turned to both written and physical remnants of this interchange to trace the collision of Eurasian, American, and African natures and cultures over the course of the early modern period. The tomato, originally Nahuatl *tomatl*, slowly made its way into Italian cuisine, horses galloped across the Americas and grew wild as mustangs, and chocolate became a fashionable beverage in Europe.[15] Wheat, cattle, and smallpox traveled to the Americas, while maize, tobacco, and potatoes returned to Afro-Eurasian networks of commerce and consumption. Viewed alongside other invasive species, scholars have noted that colonizing Europeans had a great deal in common with other nonhuman animals as they consumed their way through the New World, transforming it through their appetites.[16] Following Crosby, historians have unearthed myriad archival examples that have further illustrated that the exchange was bilateral, but by no means equal. Dogs offer us a new vantage point for understanding the implications of the Columbian Exchange, the most drastic incident of mixing in the early modern period.[17] Many accounts of the exchange have focused on how the absence of interbreeding and the limited genetic pool of stock transferred over long distances led to lower genetic diversity in the distant populations.[18] Humans and dogs, however, constitute exceptions to this rule. Both interbred with isolated populations of their own species.

Mesoamerican agriculture, as we know, emphasized the cultivation of maize. Scholars have argued that the Nahua and other peoples did have domesticated animals—primarily, though not exclusively, in the form of dogs, turkeys, and bees—but that none of these animals was raised on a large scale.[19] However, I argue that these differences from European agricultural traditions have biased historians toward focusing on the divergences between European and Mesoamerican norms rather than their similarities. From the perspective of the history of science, if not of economic history, the volume of outputs generated by breeding in a society is less relevant than the implications that a system of controlled reproduction had for that society's metaphorization of agricultural processes. Regardless of scale, Nahuas and other Mesoamericans, like Europeans, employed systematic breeding that had real implications for both their animals and their theories of nature.

Just as Spanish Catholicism and indigenous religious practices eventually syncretized to create a distinctively Mexican Catholicism, so too did a syncretic canine *mestizaje* emerge over the course of the colonial period. By analyzing evidence from natural historical accounts written throughout the early modern period, we can see how the Mesoamericans' effort to sustain differences in dogs was reallocated to other projects. Following the imposition of Spanish colonial power, its importance waned as disease threatened the health of human agents and animal yields. Generations of Mesoamerican merchants had once made a business of selling dogs, principally for meat. These practices were considered problematic by Europeans, who did not eat dogs and sought labor for their stock farms, where they could raise Spanish livestock with their own nuanced, European gradations of breed. The colonial period facilitated a gradual destruction and replacement of a whole cultural system that had sustained Nahua dog breeding, and as a result those breeds began to disappear. At the same time, natural historians created a vision of fixed nature that allowed Mesoamerican breeding projects to slip through the documentary record. Influenced by early narratives that stressed the inferiority of American societies that lacked large Eurasian livestock such as horses, camels, pigs, cattle, and sheep, these accounts often de-emphasized the artificial variation in Mesoamerican domesticated animals. The resulting shift was used in turn as proof of the wild, uncivilized qualities of Mesoamerican societies.

Before Europeans came to the Americas, there were at least five different types of dogs already there. Archaeological evidence confirms that there were small Chihuahuas and Chinese crested dogs (a breed indigenous to the Americas, their name notwithstanding) in North and Central America; a hairless variation of the *xoloitzcuintli* in central Mexico; a hairless dog variety in Peru; and the mute dog of the island of Santo Domingo.[20] These dogs performed an array of functions. Some pulled carts, some helped hunt, some were raised for human consumption, and some for companionship. The archaeological evidence of their spread throughout the continents highlights the reality of regular migration, conflict, and interchange among the numerous societies in the Americas prior to Columbus's arrival.[21] The original dogs of the New World were once thought to have been almost entirely replaced following European contact.[22] Recent genetic evidence shows decisively that this was not so. As a 2016 genetic analysis demonstrates, while Mexican and Peruvian dogs certainly mixed with others, their respective clades may each retain the "aboriginal New World dog genomic signatures intermixed with the European breed haplotypes, similar to the admixture among European, African, and

Native American genomes that can be found in modern South American human populations."[23] Just as Amerindian genomic signatures persist in populations where Amerindian communities were thought to have been destroyed, American dog genomic signatures survive.[24]

Nahua Dog Breeding

Long before the Mexica, dogs had been a central part of Mesoamerican culture. Tlatilco ceramics from 1200–900 BCE have been identified as depicting a dog nestled into a female figure's lap.[25] Art historians have noted depictions of dogs from the Classic Maya corpus (600–799 CE).[26] The line between wild and domesticated was vague for canines, for all could be tamed to some degree. Coyotes had long been interbred with female dogs to create hybrids noted for their loyalty and reliability as guard dogs. Although Bacon would later hold up intermixing across types as a signature of new breeding practices, it did not require an expert to make such hybrids: one simply tied up a female dog by a mountain for a few days, kept her well fed, and waited for her to be impregnated by roving coyotes.[27] In Teotihuacan between 700 and 1100 CE, over 455 dog sacrifices have been identified. Of these, twenty were possible wolves, likely dog-wolf hybrids fed on vegetable diets that show they were domesticated.[28] For the late postclassic Nahua, dogs were living manifestations of Xolotl, the canine twin aspect of Quetzalcoatl. Some of the most iconic illustrations of Xolotl, a pair of monumental sculptures that formed the western gate of the sacred precinct of Tenochtitlan, show the loose facial skin of *xoloitzcuintli*.[29] Each night, Xolotl (as the evening star) descended into Cihuatlampa (the west) just as sacrificial dogs guided the dead to the underworld.[30]

What European languages defined as "animals," Nahuatl viewed as "living entities." Molly Bassett has shown that "like modern-day Nahuatl speakers, the Aztecs understood life and being as existing along a continuum that did not (and still does not) draw hard-and-fast distinctions between the (in)animate, the (super)natural, or the (super)human."[31] Bassett elaborates this idea by charting how Nahuatl speakers characterize the existence of a spectrum of animate entities through two "to be" verbs they acquire—*itztoc* and *eltoc*. She contends that "modern speakers recognize two categories of human beings, *macehualmeh* (Nahuatl speakers) and *coyomeh* (nonspeakers), both of which belong to a group of animate entities. *Tecuanimeh* (wild animals) and *tlapiyalmeh* (domestic animals) compose the next category. Finally, some modern speakers class *xihuitl* (grass, plant life) as animate because it exhibits movement in the form of growth. All of these entities take the animate 'to be' verb *itztoc*."[32] Bassett argues that most modern speakers characterize plant life, rocks, and stone through *eltoc*. Other scholars question whether this spectrum extends yet further beyond modern taxonomies of animals to include a great variety of other animate ontologies in plants, and even minerals—all of whose animacy is expressed in images by emanations.[33] For the purposes of this study it is enough to note that the language of being is shared between animals and divinities without separations; one might profitably compare the spectrum of animacy that Bassett defines with the Christian notion of the Great Chain of Being, which I will address further in chapter 7.[34]

Notably, Fray Alonso de Molina's 1571 dictionary included a definition for "wild animal or living thing" under the Nahuatl *yolqui*, which also meant "resuscitated from the dead, hatched egg, or thing risen from death to live."[35] Christopher Valesey and other scholars of Mesoamerican human-animal relations have focused on how the Nahua's spectrum of animacy influenced interactions with animals. Valesey has shown that early colonial Nahuatl blurred the division between animal and human by using the same language to describe human and nonhuman anatomy.[36] There was, however, a marked difference between wild and domesticated animals. In both modern and older Nahuatl, *tecuani* meant wild animals, or "those who eat people."[37] Canines walked the line between the wild and domestic, with coyotes and wolves on one side and dogs on the other.[38]

Among the Mexica, breeding dogs was a symbol of wealth, success, and fecundity. A passage in book 4 of the Florentine Codex describes the good fortune of the dog seller born on the day Nahui Itzcuintli (4 Dog):

> If he bred dogs, he whose day sign was the dog, all would mate. His dogs would grow; none would die of sickness. As he trafficked in them, so they became as numerous as the sands and so they barked. It was said: "How can it be otherwise? The dogs share a day sign with him. Thus the breeding of dogs resulted well for him."[39]

The text notes that the Nahui Itzcuintli–type person smothers his dogs, or kills only stolen dogs. If they happen to become a dog seller in the marketplace, they will never have to work a day in their lives because all their dogs will breed and people will buy all of them, regardless of what the dogs are like.[40] In both the Spanish and the Nahuatl columns, the text describes the market as filled with dogs that were muzzled so they would not bite and the breeder's role in making dogs with short hair and long hair; the Spanish text noted the "pleasure of seeking that which was best."[41] While the calendrical name Dog, which one received if one's parents selected that day from around the time of one's birth, did not fate one to a career as a dog breeder, it did support that option.[42] In the Cuernavaca region, it would have been a sign of good fortune for a woman to be named Little Dog (*Chichiton*) between 1535 and 1540.[43] Passages in the Florentine Codex differentiated between those who killed their own dogs and others who killed only stolen dogs. After digging a hole in the ground, the executioner smothered the dog in the mud with their hands, suffocating the animal.[44] The owner would put a cotton cord around the neck of the dead animal, stroke it, and say, "Stand guard for me there! Thou shalt pass me over the place of the nine rivers."[45]

Textual evidence from the Florentine Codex suggests that breeding dogs led to elite social status, which influenced a vision of dogs as a symbol of capital. Aware of dogs' malleable forms, it appears that Nahua breeders directed nature's dynamism. The compilers of *On Earthly Things* featured several dog varieties, three of which might be interpreted as akin to distinct breeds (figs. 5.1, 5.2, and 5.3). The *tehuih* was described as "shiny, hairless, flat, and smooth" in Nahuatl; around the large figures, the Spanish text noted that "bred in this land were dogs without any hair" and "the hair they have was very little." Similarly, the *xoloitzcuintli* was "a dog with no hair at all; it goes about completely naked." To keep a *xolo*, one required a cape on which the animal could sleep. The authors of the codex imagined how and why the dogs were hairless and had so much dark black skin. Those who were less successful simply needed to smear the dog's body with *oxitl* resin when the animal was a puppy and its hair would fall off entirely. The Spanish text noted a difference in opinions about the origins of hairlessness: many thought that "dogs are not born like this" while "others said that they are born without hair in the towns of Teutlixco and Toztlan." Finally, *tlalchichi* were simply *chichi* (dog[s]) that were short and squat—that is, close to the ground (*tlalli*); these "ground dogs" were small, squat, and round, not unlike the otters that swam in the nearby waterways, to which they were compared.[46] Even the Spanish text noted that this type was "very good to eat." After these types, the Florentine Codex's Nahuatl column lists pet names for dogs: they were called *chichi* or *itzcuintli*, or lovingly addressed as *zochcohyotl* (flower-coyote), *tetlamin*, and *tehuitzotl*.[47]

The authors of the codex repeated the term *cequi* (some, a separate part) to emphasize the variety of animals through repetition. Some dogs "are black, white, ashen, smoky, dark yellow, dark, spotted, and some spot-

ted like ocelots; some are high, tall, and towering; some medium; some
are smooth, some shaggy, others woolly . . . their ears pointed, concave,
hairy, or shaggy."[48] They captured all types in this description, emphasizing the multiplicity of forms and names.[49] This syntax, central to ritual speech, subjugated many characteristics to describe one creature: the
dog.[50] It was echoed in the Spanish translation, which simplified things,
writing that "the dogs of this land have four names: *chichi, itzcuintli,* also
xochxocoiotl, also *tetlamin,* also *teuitzotl.*" In the end, the authors and illustrators delineated a Nahua dog by two key features. First, the dog was an
eager companion and was allowed into the house. It was a "tail-wagger." It
loved to consume revolting foods.[51] Second, dogs came in many shapes,
colors, and sizes. However, across the spectrum, they all shared an essential sociability. The Spanish text translated this as "they are meek, they are
domestic, they accompany or follow their owner, they are collected, they
wag their tails as a sign of peace, they bark, they lower their ears to their
necks as a sign of love, they eat bread and green maize, and raw and cov-
eted meat, they eat dead bodies, they eat meat."[52] Like other sixteenth-
century natural histories, the Florentine Codex uses the dog as a comparison to make other animals more comprehensible: the otter (*aitzcuintli,*
which immediately follows the dog in the codex) is likened to a *tlalchichi,*
and the description of the mythical ahuitzotl uses *tehuih.*[53]

Other portions of the Florentine Codex mention the cooking of dogs.
In describing the banquets of rulers and merchants, the Nahuatl text reads,
"They killed, they singed, they prepared dogs."[54] The authors mention dog
meat (*itzcuinnacatl*) in their description of the Tepeilhuitl festival, noting
that it offered "fruit tamales and sauce perhaps with dog meat, perhaps
with turkey."[55] Celebrants at the merchants' feast, Tealtiliztli ("The Bath-
ing"), ate dog meat at the bottom of a sauce dish; the celebration would
have been so large as to require between twenty and forty dogs and eighty
to a hundred turkeys.[56] In book 10, "Of People," the authors list *itzcuinnacatl* among the low-quality goods that the bad meat seller offers, noting
"he claims dog meat to be edible."[57] It is possible that this insult is tinged
with the aspersions that the Europeans cast on eating dog meat.

European dogs bred for their ferocity in war were a weapon unlike anything being used in Mesoamerica at the time.[58] The Spanish mastiffs—
alanos, large game hunters and bull baiters—of the conquistadors were
fabled tools of war, bred into their form across generations of spectacu-
lar battles and hunting. Both European sources, like Theodore de Bry's
image of Vasco Nuñez de Balboa (1475–1519) unleashing his large dogs
on the Panamanians, and rock art in Colombia suggest that dogs were
used to massacre indigenous people and allowed to eat their corpses.[59]

FIGURES 5.1, 5.2, AND 5.3. Variants of dogs in the New World. In Fray Bernardino de Sahagún, *General History of the Things of New Spain: The Florentine Codex*, bk. 11, *On Earthly Things*, chap. 1, "On Four-Footed Animals," Florence, Biblioteca Medicea Laurenziana, MS Med. Palat. 220, fols. 16v–17v. By permission of Ministero per i Beni e le Attività Culturali e il Turismo. Further reproduction is prohibited.

Beyond letting slip the dogs of war, Europeans used various breeds to guard property, hunt, herd, and provide companionship, as John Beusterien has shown.[60] Selective breeding for specific functions, then, shaped dog breeds in both Europe and the Americas. Their differences came down to both their variable genetic inheritance and their purpose. For instance, Europeans did not generally eat dogs, and had no incentive to do so in a culture of large mammal and fowl husbandry. Likewise, Mesoamericans had no need for herding dogs, since there were no sheep or cattle to herd.

Animal Mestizaje

By the time the Florentine Codex was composed, after a generation of Spanish rule, indigenous peoples were familiar with European dogs. Many would have encountered them in person and would also have heard accounts of European dogs used as weapons of war throughout the Americas. Decades after the Siege of Tenochtitlan, the authors of book 12 of the Codex emphasized how the introduction of these new, terrible animals marked a difference between the invaders and themselves. They had to negotiate how to even describe dogs that were at once so familiar and yet so unlike Mexican dogs. They described them alongside firearms: "very large, their ears folded, with great dragging jowls. They had fiery, blazing eyes, flanks with ribs showing and gaunt stomachs. They were very tall and nervous as they went about panting with their tongues hanging. Their hair was spotted and varicolored, like ocelots."[61] By contrast, the Spanish text devotes much less attention to the horror of this description, writing simply that the messengers "also gave an account of what the Spaniards ate, and of the dogs they brought, and the way they were, and the ferocity they showed, and the color they had."[62] What made them so intimidating was left out of the language easily read by European eyes. Similarly, the images of dogs were legible to European sensibilities: in the visual language of book 12, the authors depicted in color a tall European dog with a curving tail as one of the intermediaries between European knights astride horses and indigenous peoples.[63] Perspectives on normal nature was radically contingent on one's home culture. Just as the Spanish saw manatees as "sea cows," Nahuas originally called horses "deer." They saw the vicious dogs of war through the pelts of their own large, ferocious predatory cats, like the jaguar; these occasionally carried off humans, just as canine invaders mauled their human enemies.

While the War of Mexico marked an immediate rupture, deeper change in Mesoamerican culture occurred by a slower process of erosion, driven by illness, forced labor, and a changing set of values. Edible dogs appear to have been an important part of the Nahua diet, and that centrality endured throughout the sixteenth century. Along with insects, fish, and birds from the lagoon and Lake Texcoco, dogs featured prominently among the wares in the marketplace. In the fifteenth century, war had meant that the Triple Alliance forces would loot cornfields and kill domesticated turkeys and dogs; people in the towns regularly hid their foodstuffs, including both corn and dogs, when they feared an invasion.[64] From the first encounter, the Mexica's eagerness to consume dogs fascinated the Spanish.

In his second letter to Charles V, describing events from the fall of 1520 to the spring of 1522, Hernán Cortés reported that Mexica markets offered "small gelded dogs which they breed for eating."[65]

The tradition of dog consumption persisted throughout the sixteenth century, despite Spanish opposition. In Chilchotla, Michoacán, Pedro de Villela and Francisco Gorjón Toscano noted that "they had dogs that they would fatten up like pigs."[66] Likewise, inhabitants of Texcoco "did not have any [other] kind of animal for service or eating if not a genus of dog, retriever size, which grew fat for the common people to eat."[67] In the towns, local breeding and consumption of small dogs (*perrillos*) shocked observers who embraced European dietary norms. Priests like Dominican Diego Durán (1537–1588) insisted that Nahuas ought to be prohibited from eating "filthy things" like "dogs and skunks, badgers and mice."[68]

Local reports suggest that sixteenth-century Mexican dog breeders sometimes developed populations of Spanish and Mesoamerican dogs in parallel, although the numbers of the latter started to fall. Some areas with few Castilian animals were replete with "the small dogs they had already before," as Juan Gutiérrez de Liébana wrote from Cuatro Villas.[69] Other regions with greater Spanish influence already struggled to maintain local dog breeds. Reports from Tepexpan and nearby Tequizistlan recorded that "they bred a great quantity of dogs of those sent from Spain and some of those from the land, although those are few."[70] As the Nahuatl language started to erode to fit Spanish vocabularies, so too did dogs' role in food culture suffer under the colonial transition.[71] In the wake of initial contact, concentrated Mesoamerican canine culture slowly waned.

In a February 18, 1579, report from Mérida, Martín de Palomar, citizen and alderman of the city, suggested that intermixing was well underway. The region had been peaceful until an uprising around 1546, in which locals killed more than thirty Spaniards, pronouncing their loathing of all things Spanish, "and even the dogs and cats and the trees of Castile were uprooted, along with anything else that belonged to the Spaniards."[72] When the Spaniards allied with the people of Tutul Xiu, Hocaba, and Ahcanul, who had not joined in the rebellion, they, in the language of the *relación*, "pacified" the region. After that conflict, intermixing continued, as Palomar explained, through animal *mestizaje*:

> There are dogs native to this land that do not have any hair and do not bark, that have thin and sharp teeth, with small, stiff, and raised ears. These are fattened up by the *indios* to eat and they have them as a great gift. They come together [*se juntan*] with the dogs of Spain and they breed [*engen-*

dran] and the mixed pups [*mestizos*] that come from them bark and have hair. The Indians also have other kinds of dogs that have hair, but they do not bark either and are the same size as the others.[73]

Labeling mixed pups born of the union of dog kinds from Europe and the Americas mestizos showed the generality of this term at this moment as it simultaneously became a socioracial category for humans. Thinking back to the conceptual blending hypothesis advanced earlier in the book, and Rappaport's discussion of the multiple uses of *mestizo* and its continued metaphorical valences, perhaps the Spanish observed this instability in populations, were alarmed by it, generated metaphors from it, and the metaphors seeped back into ideas about human reproduction.

The canine genome is famously malleable, meaning that new breeds of dog can be produced in relatively few generations. Thus, dogs in the New World in 1490 probably would have looked markedly different from their descendants in 1520, even without concerted human intervention. However, as the cultural ideals of Mesoamerican societies changed over the course of the sixteenth century, so too did their animals morph along with these new specific needs. A "plague" of European sheep swept across the Mexican landscape, to use Elinor Melville's term; it should not be surprising, then, that an array of sheepdogs with herding skills came with them.[74] The newly introduced animals were, unsurprisingly, better accustomed to and bred for this work than native dogs. Even so, European and indigenous breeds mixed. Suddenly, short-haired or hairless dogs with surprisingly strong herding capacities appeared, even though they had not previously been developed to herd.

Modern geneticists have noted that German shepherd dogs hybridized with both Peruvian hairless dogs and the *xoloitzcuintli*. Canine genomic expert Heidi Parker and her collaborators have described this overlap, noting both the impact on the indigenous population of dogs and how "geographically distinct subsets of the same breeds show that some degree of admixture" occurred in the imported European dog breeds when they were introduced to the Americas. Parker et al. continue by arguing that "these data suggest two outcomes of breed immigration that mirror human immigration into a new region: the immigrant population is less diverse than the founding population, and there is often admixture with the native population in early generations."[75] Mexico's national dog, the modern *xoloitzcuintli*, has thus been a mestizo dog for nearly half a millennium, which seems fitting for a country that defines itself as a mestizo nation.[76]

The Puzzle of Difference

This section transitions to considering the writings of European naturalists as they sought to enumerate and characterize Mesoamerican dogs. When faced with the transformations of the physical world that would come to be known as the Columbian Exchange, early modern natural historians often preserved nature in their writings as it had been when they encountered it.[77] The dynamism of the canine body became a set of fixed forms that obscured its potential for change.

During the exploration that characterized the first decades of contact, dogs emerged as a model for comparison. Dogs provided a means of measuring relative difference. The Maltese was such a standard feature of Spanish households that its size became a benchmark for describing New World otters, cats, and dogs.[78] In another vein, dogs' famous malleability made them ideal examples for describing variations among other animals, as Jeremy Paden has stated.[79] Strange American dogs posed a problem for European expectations of the boundaries of species. This is best exemplified in the debate concerning the mute dogs of the Caribbean and Central America, as Enrique Alvarez López has shown.[80] The "dogs that never barked" fascinated Columbus and other European visitors to Taíno lands.[81] In 1495, Columbus's companion Michele da Cuneo had described "quadrupeds which live on land" on the islands of the Caribbean, including "dogs which do not bark."[82] Likewise, Nicolò Scillacio reported that "there are so many dogs, however, though none that bark and no madness" (which seems to imply the absence of rabies).[83] The silent American dog became a trope, not unlike the trope of the "silent Indian" viewed as dumb and mute rather than self-contained and thoughtful in preparing speech.[84]

The next generation of critics doubled down on the importance of barking to the definition of *dog*. Could a dog be a dog if it could not bark? Spanish chronicler Gonzalo Fernández de Oviedo (1478–1557) characterized barking as a "natural thing for mongrels and dogs of all sorts." The wide circulation of Oviedo's work inspired many others to take an interest in the silent dogs that were "really remarkable, compared with those of Europe and most parts of the world."[85] Spanish experimenters tried to understand the dog's silence using three methods: exposing the dog to pain, translocating the dog into a new environment, and vivisecting the dog. The first solution seemed most obvious and was probably attempted most frequently. The more pain European dogs experienced, the more noise they would make. Not so with these Hispaniola dogs. In his *General History of the Indies*, Oviedo noted that "even when they were beaten with sticks or cut with knives, they never whined, just emitted a certain low throttled growl, barely audible."[86] No matter how poorly they were treated, "all these dogs here and on the other islands were dumb."[87]

Next, the experimentalists tried to connect dogs' behavior to a particular location by reasoning analogically. Would the inner nature of a dog change if it was relocated to a new place? Oviedo noted that Pliny the Elder, an ancient Roman natural historian, had mentioned the mute frogs of Cyrene and Seriphos. Pliny's text recorded that when those latter frogs were moved to another place, they could croak.[88] They also knew that some animals quickly died when brought from their homeland to another territory, and thus these frogs were purported to have lost their special characteristics when relocated. Might the same be true for dogs? Oviedo wrote that he "wanted to test whether these dumb dogs, if removed from their country, would bark in another." He took a silent dog from Nicaragua and transferred it to Panama, three hundred leagues away. Unfortunately, his plan was not successful: "Here too it remained dumb." Oviedo did not give up on his classically founded theory, however. Perhaps the problem was Panama's seaside geography, "because [Panama] is all one coast and mainland and, as I have said, in all of these parts and these islands the dogs that are native thereto are thus dumb." He tried to continue his experiment by bringing his canine companion to Europe, but "when I left for Spain he was stolen from me and I had brought him up and he was very tame."[89] While Oviedo's dog might not have made it to Europe, other examples of these silent dogs did.[90]

Francisco Hernández's Dogs

Given that their behavior did not change with pain or relocation, naturalists considered how vivisection might solve the question of American canines' silence. After traveling throughout Mexico as part of a royal mission, Spanish physician Francisco Hernández considered how the dogs he encountered in the Americas had thrown a wrench into Pliny's extensive descriptions of canines. In his comments on Pliny's *On Terrestrial Animals*, Hernández noted that "in New Galicia there is a breed of hairless dogs, with a smooth hide, mottled, like a setter although somewhat bigger, and they have a bark dissimilar to the rest."[91]

Hernández sought to prove that a dog's inability to bark was the result of its anatomy. Rather than experimenting on the mute dogs of Santo Domingo, he instead tested European dogs to see if they could be made silent. If changing a silent dog's environment did not make it any less mute, then the cause of European dogs' barking and Hispaniola dogs' silence must be based in their bodies. Hernández described an experiment to remove a dog's bark, writing that "in past years, as an experiment, Nicolas de Vergara—architect, painter and excellent sculptor from Toledo—and I cut the back nerves, and thus we deprived [the dog] completely of bark and voice."[92]

Most of Hernández's writings about nature were not centered on vivisection. Instead, they depicted a rich, urban Mexican culture that cultivated animals of economic importance. He emphasized how the indigenous people of New Spain chose to cultivate their land- and waterscapes with the most useful varieties of flora and fauna. For Hernández, Mexican nature was controlled and civilized, not chaotic or accidental. Therefore, when he set about describing the dogs of New Spain, Hernández focused only on those he thought were completely indigenous, leaving aside "the dogs known in our Old World, and brought almost all by the Spanish."[93] Hernández identified several types of domesticated and semidomesticated dogs in New Spain that "are bred in these lands by the natives": the *xoloitzcuintli*, *mechoacanenses*, and *techichi*.[94]

Dogs fulfilled different purposes: "In some regions, they raise herds of dogs to pull wheel-less carts that suffice to carry two men, as other dogs to hunt, and as others for war." Hernández's informant, Bernal Perez de Vargas, "a man of great kindness and virtue," reported that in "Cybola, in the north of New Spain and nearly to the northeast of Florida [in modern Georgia and South Carolina] an army ran into many droves of dogs, with each one carrying four Spanish half pecks of *tiaolli* corn."[95] The resonances with the legendary Seven Cities of Cíbola raise doubts about Hernández's reliance on hearsay. However, he clearly believed that strong American dogs provided labor as well as meat. Of the three types of domesticated dogs he named, Hernández had seen only one variety in person in Europe before arriving in the Americas. He had personally observed *xoloitzcuintli*, which were popular in central Mexico before he moved there (fig. 5.4). These dogs, at more than three cubits, were taller than the other varieties. Like Sahagún's team, Hernández reported that rather than hair, they had had soft and smooth fur on their skin, which was spotted tawny and blue. Hernández wrote that "they are all like ours in nature and habits and not very different in shape."[96]

Hernández thought that the other two types of dog had not yet been taken to Europe, as he had not observed them before moving across the Atlantic. He described a dog from Michoacán, aptly called *mechoacanenses*. It was hunchbacked, with a "certain curious and amusing deformity," and "with its head as if coming from its shoulders themselves."[97] Although they remained on the margins of Hernández's personal experience

IO. FABRI LYNCEI EXPOSITIONE.   479

XOLOITZCVINTLI
Lupus Mexicanus.

FIGURE 5.4. *Xoloitzcuintli*, Lupus Mexicanus, in the Mexican Treasury. In *Rerum medicarum Novae Hispaniae* (Romae: Sumptibus B. Deuersini & Z. Masotti, typis V. Mascardi, 1651), 479. QH107 .H54 1651 F. Courtesy of the Department of Special Collections, Stanford University Libraries.

and knowledge, Hernández had heard remarkable things from the region along the Aconquis River, which reputedly produced a variety of hunchbacks—not only dogs, but also bulls and humans. In a section of the natural history of New Spain subsequently titled as "Of the izcuintepotzotli or humped dogs," Hernández noted that

> there is nothing odd in finding races [*razas*] of hunchbacked bulls and men, and also there is a race [*raza*] of hunchbacked dogs called *mechoacanenses* for its place of origin. These are a bit bigger that the Maltese, and like those of our land, of various colors, but almost no neck, but that the head comes directly out of the back; otherwise they have the same habits and character as all dogs, and likewise they treat their owners tenderly and play with them.

As if to cite an Aristotelian final cause, Hernández noted that the essential characteristic of his dogs was their kindness and love of humans. Whether they were hunchbacked, or silent, or hairless, or gray, Hernández's dogs all loved to play.[98] Finally, Hernández briefly described the familiar *techichi*, "a dog like the little dogs of our land, of bad appearance and in the rest like the common and everyday ones." Unlike Sahagún, who slid from the *tlalchichi* to otters, Hernández subsequently turned to the dogs' wilder relatives that resided in the mountains. He pondered whether these came from European dogs that had interbred with Mexican dogs, or were simply indigenous varieties left to develop on their own without human selective breeding interventions. He also described what he called *tepeitzcuintli*, or a "dog of the mountain," that could be domesticated. Its long hair and tail differentiated it from other canines, but it had a head like a dog. While it was only the size of a little dog, it was "extremely bold, since it attacks deer and at times even kills them." When the animal came to be domesticated, "its diet consisted of egg yolks and bread broken up in warm water."[99] This meatless diet mirrors that found in dogs and dog-wolf hybrids buried in Teotihuacan, but this textual evidence alone is not sufficient to conclude that this animal was, indeed, a dog.[100]

Simplification, Loss, and Textual Dogs

Over the next three hundred years, Hernández was seen as the authority on the dogs of Mexico. By the late sixteenth century, Italian naturalists in Rome, Bologna, and Naples were eager to secure access to Hernández's notes. As a result of his poor reception at the Spanish Habsburg court upon his return from Mexico and the project's sprawling quality, Hernández's great natural history of Mexico had not yet been printed; even so, the importance of his observations was well known throughout Europe. Through Hernández's successor as royal physician, Nardo Antonio Recchi (d. 1595), the Accademia dei Lincei acquired copies of extracts of Hernández's manuscripts. Johann Faber, a German doctor and botanist living in Rome, became responsible for transforming Recchi's extracts of Hernández's notes on animals into book-length natural histories about those animals.[101] Perhaps Recchi had become too eager to redact, or was simply not that interested in canid Mexico, but it seems that he and the subsequent Lincei failed to distinguish between the dog of Michoacán and the *techichi*—two of the three breeds Hernández identified. Later naturalists, particularly Georges-Louis Leclerc, Comte de Buffon (1707–1788) and Johann Friedrich Gmelin (1748–1804), followed the Lincei editors in failing to distinguish between Hernández's second and third dogs.[102]

Like Hernández, Faber defined a dog around its gentle nature, regardless of how monstrous its shape might be. Based on Hernández's notes and a drawing, Faber later described the *ytzcuinteporzotli* (to standardize the orthography *itzcuintepotzohtli*, literally a "dog-like hunchback") as the Mexican domesticated dog.[103] He examined the images Hernández's team had drawn to study the creature, in which he saw a female hump-backed dog with a swollen body and bristly tail. Faber recounted how he drew his conclusions from Hernández's notes and how observations were possible if "one looks at the figure in the picture properly."[104] The "monstrous animal" had a strange body. With a bulging snout resembling that of a snarling dog, its head was unusually small compared with the rest of its body. Faber included colors that are not present in the printed edition of the volume: its snout, face, and eyebrows were red, its ears, jaw, and back tawny. Its back was humped like a camel's, and its short tail was white. The sexual organs were particularly notable in the drawing, "swelling out and likewise big, along both sides of the belly, six teats are visible" (fig. 5.5).

Faber turned to the ancients to make the dog comprehensible. He discussed African dogs, Egyptian dogs, and dogs in Plato and Aristotle. He contextualized the remarkable genitalia with a discussion of Hernández's thoughts on dogs' bone-like penis. Faber thought the *itzcuintepotzohtli* was inclined to be very fat and covered in black-spotted skin. While he thought many of these characteristics odd, a dog's essential trait was in its eyes: "The eyes show no wildness but are rather gentle; they seem tame."[105] Using the paper museum left in the wake of Hernández's travels, Faber saw this animal through already collected materials. His observations were based on a scientific practice that had become comfortable drawing conclusions about character, natural history, and taxonomy from looking at images out of context. Indeed, Faber's attention to the eyes of the dog in the drawing is most revealing. As we have seen, breeders and trainers alike associated an animal's inherent character with its physical expression; this dog's eyes, too, were used as a physiognomic tool to understand its internal nature.

To integrate the Mexican dogs into the larger scholarship, Faber elided different categories of dogs in the New World by switching between sources discussing Caribbean and Central American canines. Hernández, he knew, was by no means unique in describing New World canines. "On the island of Hispaniola, America's history-writers record that small exotic dogs have been discovered, which were partly kept as pets of the home to those Indians, partly outdoors for the enjoyment in hunting certain lowly animals and the seizing of rabbits."[106] While they were useful in killing hares, the dogs resembled them in many ways: like the rabbits "who emitted no sound when battered to death," these dogs also remained silent. Like Oviedo, Faber turned to Pliny to explain the dogs' inability to bark. Faber described Oviedo's experiment in taking a mute dog from

YTZCVINTEPORZOTLI
Canis Mexicana.

IO. FABRI LYNCEI DESCRIPTIO.

NIMAL hoc monſtroſo quodam corporis habitu, ſi probè in pictura figuram intueamur, apparet. Caput cumprimis ad reliquũ ſi corpus referas, adeò eſt paruũ vt pars hæc toti ſuo mediocri vix proportione reſpondeat. Oculi feritatem nullam ſpirant, ſed mites potius ac manſueti videntur. Roſtrum quod gibbum eſt & canino ſimile, totum albicat veluti & frons ac ſupercilia. Aures cum ſubiacente verſus rictum parte fului coloris ſunt & pendulæ. Collo eſt breuiſſimo & obeſo valde, ipſique ſtatim thoraci caput annectitur. Dorſum Cameli inſtar gibboſum, poſt collum ſubito ad pectus accline, ſed coxas verſus decliue coloris itidem fului. Cauda breuis adeò, vt primos haud crurum articulos contingat, albetq́. tota; ſub qua vulua & hæc prægrandis conſpicitur. Turgentes quoque magnaſq. per ventrem vtrinq. conſpicuas mammas, quæ ſex numerantur, gerit. Pinguiſſimũ ventrem monſtrat, quẽ pellis plurimis ijſquè nigricãtibus variegata maculis & pedula ornat; Crura ac pedes albicant digitis caninis diuiſi, & vnguibus valde exertis proſtant.

SCHOLIA EIVSDEM.

Hæc fideliſſima eſt ſuisq. coloribus ex ipſo Autographo excerpta delineatio, qua probè perſpecta, facili quemuis aſſequi coniectura exiſtimo, hoc animal *Canem* & quidem domeſticam *Mexicanam* eſſe. Docent hoc præcipue *Caput*, *pedes* ac *mamillæ*.

Libuit

FIGURE 5.5. *Ytzcuinteporzotli,* or the Lincei copy of one of Hernández's drawings. In *Rerum medicarum Novae Hispaniae* (Romae: Sumptibus B. Deuersini & Z. Masotti, typis V. Mascardi, 1651), 466. QH107 .H54 1651 F. Courtesy of the Department of Special Collections, Stanford University Libraries.

Nicaragua to Panama City, repeating that in both locations it remained silent. While these Mexican dogs seemed to have the tame expression of a European canine, Faber thought their silence defined their difference from European dog varieties: to him, Mexican dogs were "not in any way a dog of the kind of our dogs."[107]

By the Enlightenment, American dog varieties had become yet more simplified. In his eagerness to describe similarities rather than differences across continents, Buffon reduced the different breeds or types discussed by Hernández to one (*alco*), despite the marked differences in habitat and behavior that earlier scholars had identified.[108] In his description of the thirty fixed varieties and seventeen climatic variants of dogs, Buffon began by stressing their malleability and role as companions. He warned readers of his *Natural History* that "nature never preserves its purity in beings which have been long under the management of men." Humans had subjected their familiar companions to training and breeding, and therefore dogs' bodies took on many shapes. However, when left to their own devices in the wilds of the Americas, dogs slipped back to their original form of ferocious hounds—they reverted to being the "wild dogs of the Congo." Yet the Americas were not a wilderness. Buffon believed that pre-Columbian dogs had been bred as well: "It is apparent that the original dogs of America before any communication with those of Europe were all of one race, and they approached most to the dogs with slender muzzles, erect ears, and coarse hair, like the shepherd dog."[109] How curious that so many shepherd dogs traversed a landscape with no sheep. Buffon was actually observing Enlightenment-era dogs that epitomized a transformed New Spanish nature.

Buffon thought Hernández's description of the *xoloitzcuintli* was incorrect, since no other author he had access to mentioned it. Thus, Buffon thought the dog had come from Europe. He was convinced, erroneously, that *xoloitzcuintli* was the word for wolf in Nahuatl, while *alco* was the common name for dog. *Alco* had slipped in textually, probably through the popularization of José de Acosta's *Historia natural y moral de las Indias*, which had asserted that

> there were no true dogs in the Indies, only those similar to little dogs that the Indians called *alco*; and they are so friendly with these dogs that they will go without food for days to give it to them, and when they are walking along the roads, they carry them with them on their backs or in their bosoms. And if they are ill the dog must stay by them, without using them for anything, only good friendship and company.[110]

Acosta's particular experience had been mapped onto the whole of the Americas and concretized as a new certainty through many reprintings by Buffon's day. Likewise, by 1806, Hernández's original dogs had been copied and recopied, and transformed into "the American Obesus, the *Ytazcuinterporzotli* of Hernández," which was "a prodigiously fat dog that has yellowish ears, short neck, and arched back covered in yellow hair."[111] Although sixteenth- and seventeenth-century sources had described different varieties of New World dogs, for this Enlightenment naturalist, the profusion of pre-contact American dog breeds had collapsed into a single type.[112]

In Buffon's day, however, even those who ventured to Latin America did not witness the wide array of animals discussed in the sixteenth century. The radically altered social and political environment that had transformed Mesoamerica into New Spain transformed the physical environment as well. Dogs' bodies had evolved in parallel. As New World dogs changed shape, natural historians struggled to keep up with the flexibility of the canine form. Natural history as an approach found it increasingly difficult to incorporate either human intervention or rapid natural change.

CHAPTER 6

Camelids and Christian Nature

Seeing Andalusia in the Andes

In 1532, as Francisco Pizarro and his company traveled from the Peruvian coast to the Andean stronghold of Inca power, the expedition saw what they expected to see: Moors. Amid the Inca temples and the bodies of llamas sacrificed upon altars, one conquistador—Francisco de Jerez (1495–1565?)—also seemed to see a different kind of sacrifice, and a different kind of worship: mosques, complete with sacrificed sheep. Along with the constant threat that the indigenous people would rise up and "kill all of the Christians," in Jerez's estimation, this animal facet of their religion was alien and threatening, as the "sheep" represented "the best of what [the Peruvians] had and held in veneration."[1] In addition to the other "filthy things" the natives did, he complained that they offered up animals along with "their own children, and with the blood they anoint the faces of the idols, and the doors of the mosques."[2] Jerez, like other Spanish invaders, turned to an existing vocabulary drawn from Spanish homelands to describe what they saw in the Americas, characterizing the Andes through a worldview born of borderland conflicts at the former frontier of Iberian Christendom.[3]

As Karoline Cook has shown through the real and imagined passages of Moriscos and Muslims to colonial Spanish America, even in the so-called New World, "Morisco presence requires us to rethink the colonial category of Spaniard (*español*) by troubling its implication of an "Old Christian" who possessed purity of blood and formed part of a unified Catholic society."[4] In Peru, the precedents of Iberian expansionism (the so-called *Reconquista* and the subsequent forcible conversion or expulsion of Moors and Jews from Spain) shaped Spanish interaction with New World nature. For some, the New World offered a malleable clay, ready to be molded by conversion, education, and "good breeding" broadly defined, as we have seen across previous chapters. Others saw an old enemy in new guise, categorically opposed to European values and challenging European systems by their mere existence.

This chapter employs the book's wider method of suggestive juxtaposition in miniature. It uses two types of camelids—New World llamas and alpacas and Old World camels—to further complicate the history of the Columbian Exchange. I argue that the nature of the New World did not simply mix with that of the Old according to preordained ecological niches.[5] Rather, environmental suitability was filtered by human choice and political whim. As discussed in the last chapter, dogs typified hybridization and homogenization as varieties from Europe and the Americas interbred rapidly, bridging the differences that had separated them when they were on opposite sides of the Atlantic. By contrast, camelids exemplified cultural distance and non-exchange; this case study suggests both what might have passed between hemispheres and why such transfers met their limits. Camelids themselves proved to be symbolically freighted animals whose meaning often changed depending on the interpreter. Sometimes they were analogized to indigenous people, who themselves worked as beasts of burden; sometimes they were tied to black Africans or Moors. The Spanish referred to many Andean camelids as sheep, yet their heavy reliance on llamas' bodies for labor transformed them into still another animal: a poorly adapted donkey, whose bones buckled under the weight of the heavy loads they were prevailed upon to carry. They appeared as exotic collectibles in the gardens of aristocrats, on the sacrificial altars of imagined mosques, as alternative livestock in European governments' solutions to abusive *encomenderos* (those authorized by Spanish administrators to extract tribute from indigenous peoples), and as faux-Moorish extravagances.

Polyvalent and overlapping ideas of species, race, and difference converged on camelids. The Spanish invasion in the 1530s disrupted Andean breeding practices, and llama and alpaca herds were severely reduced, both through the unintentional disruption of Andean animal culture and through intentional culling at the hands of the Spanish invaders. Those invaders—who constructed the Peruvian branch of their empire on top of the Inca empire and appropriated so many of its spaces, resources, and institutions—notably chose not to continue the Incan commitment to llama and alpaca breeding. Meanwhile, courts in Spain bred camelids, and the animals garnered interest for their visual and cultural links to Iberia's Moorish past. The resulting cultural history, complete with disconnected actors, divergent scientific opinions, and Spanish politics, gestures at the links between ideas of irrevocable difference regarding both human bodies and animal bodies.

Why did the Spanish see llamas as sheep and camels as Moorish? One answer requires an understanding of Moorish Spain and *limpieza de sangre*. As Christian Iberian kingdoms seized ever more territory from Muslim polities, the herding patterns on the Spanish plateau, so excellent for sheep, shifted. While in North Africa, especially in the Maghreb, the herding economy was run by independent tribes hostile to centralized government interference, the Kingdom of Castile consolidated control over its sheep, wresting power and animals away from local actors.[6] Under the powerful *mesta* system of ranchers and their flocks, 2.8 million sheep provided the kingdom with wool—Castile's primary export, along with iron—as well as the *servicio y montazgo* tax on pasturage that funded the crown.[7] Castile not only sought to create a Christian kingdom, but also depended on the Lamb of God to fund (and clothe) that kingdom.

The chronology of intolerance, conversion, and ultimately expulsion of Jews and Muslims in Iberia shaped this history. In 1492, the last Muslim kingdom, Granada, fell to Christian forces, consolidating rule of Iberia under Catholic kings; Jews were expelled from Iberia or forced to convert to Christianity; and Columbus encountered the New World. Ten years later, in 1502, *mudéjares*, or Muslim subjects of Christian rulers who had been allowed to retain their laws and religion in return for loyalty, were ordered to convert or leave Castile. Thus, the *mudéjares* that remained became Moriscos, which refers to those who had ostensibly converted to Christianity and normative Castilian culture, dress, and customs.[8] Medieval Spain had been notable in Europe for its combination of Christian, Muslim, and Jewish inhabitants living together—until the end of the fifteenth century. *Convivencia*—peaceful cohabitation of different religious groups—predicated on violence made it possible for minority groups to reside alongside the dominant group, with agreed-upon strictures on their role, in both Muslim and Christian realms, as David Nirenberg has shown.[9] As Barbara Fuchs has argued, a self-loathing hybridity is a defining feature of early modern Spanish culture. Moors played a starring role, such that the crown's "project of imagining a unified nation involved regularizing and regulating Moorishness itself, through both a repressive legal apparatus and the Inquisition."[10] Moors' negotiated place after their forced conversions in the sixteenth century revealed the heightened fears of assimilation in a centralizing Christian state.[11] Many Moriscos adopted a double life in silent revolt, abiding by Christian norms in public but reverting to Islamic traditions and Arabic language at home. In the words of Eric Calderwood, "like the term 'Oriental,' the word 'Moor' tells us more about the person who uses it than it does about the thing it supposedly describes. It is thus best understood as a category of the Spanish imagination, rather than as a descriptor of peoples or cultures."[12]

By the end of the sixteenth century, whatever Spanish tolerance of heterogeneity had existed was fading as the nation began to separate into "different, well-defined cultural spheres," leading to a "new segregation of Mediterranean life," in the words of historian Andrew Hess.[13] Mass education attempts started to fray by the middle of the sixteenth century. Old Christians rioted, highlighting what they considered the social errors of the Moriscos, such as veiling, hygiene in public baths, and culinary customs, like avoiding pork, wine, and meat without drained blood. Alterity, it seemed, could not be erased by language, education, or Mass. The Old Christians questioned whether conversion could indeed create a unified Spain.[14] Tensions grew amid the elite as well, as Philip II in the 1560s turned away from the relative tolerance cultivated by Charles V a generation earlier.[15] Morisco uprisings and rebellions began in earnest on Christmas Eve of 1568, bringing a religion that ostensibly no longer existed in Spain to the forefront of Iberian politics and popular fears. A group of mountaineers in the Alpujarras rebelled in a small protest that exploded into the Morisco Rebellion, which lasted from December 24, 1568, through the autumn of 1570. In response, the crown launched Morisco resettlement projects to break up old communities and drain people from former Moorish strongholds.[16] With concerns about the Turkish threat growing in the Mediterranean, King Philip II gave his bureaucracy the task of expelling the Moriscos, preventing their return, and repopulating the land with Old Christians, as María Elena Martínez has shown. This militancy had drastic consequences. *Limpieza de sangre* cases in both Spain and Mexico increased during the last quarter of the sixteenth century. True conversion was believed doubtful, if not impossible. Tensions continued to rise, as exemplified by the expulsion of the Moriscos early in the seventeenth century.[17] This background likewise contextualizes the political challenges that José de Acosta negotiated, as described in chapter 7. For now, I turn to the Andes.

Breeding Andean Camelids

The domestication and breeding of camelids constituted a cornerstone of the animal economy in the pre-contact Andes. As with the intentional work of maize cultivators and turkey and dog breeders in Mesoamerica, the paucity of written documents from before 1536 compared with the European archive, and the obvious bias of subsequent colonial Spanish sources, mean that the detailed choices of breeders in the Andes are difficult to access. However, the animals themselves provide evidence for the work that went into breeding them.

For the purposes of this discussion, I will rely on the term *Andean camelid* to include domesticated llamas (*Lama glama*) and alpacas (*Vicugna pacos*) and wild vicuñas (*Vicugna vicugna*) and guanacos (*Lama guanicoe*). Modern biologists might sympathize with early modern naturalists here, since today they suggest that the lines between these various camelids are fuzzy. Like the Arabian camel or dromedary (*Camelus dromedarius*) and the Bactrian camel (*Camelus bactrianus*), the Andean camelid was domesticated in antiquity. It fulfilled a role in the Andes similar to that played by camels in North Africa, Arabia, and Central Asia.[18] By the sixteenth century, the animals that we today consider domestic Andean camelids and their wild cousins were called *alpaca, camello, carnero, carnero de la tierra, guacay, guanaco, huancayo, llama, oveja, oveja de la tierra, oveja del Peru, oveja silvestre, paco, puna, urco,* and *vicuña*, among other things. Like these overlapping discursive categories, interbreeding revealed the fragile fault lines between variants. Under human influence, many could combine to make hybrids, some of which were fertile.[19]

Archaeologists and geneticists have suggested that the llama was domesticated from the wild guanaco for multifarious uses, and the alpaca from the wild vicuña with an eye to its fleece.[20] Despite creating clear variations between the llama and the smaller alpaca, domestication transformed camelids in similar ways, altering their size and hair, and often decreasing the size of their brains compared with their wild relatives.[21] Ancient domesticated camelid remains have been found in human settlements throughout the Andes, in patterns that suggest their use in a number of human activities, from herding to gathering of fleece. Most evidence suggests that they were first domesticated in the Lake Titicaca basin in modern Bolivia and Peru between 8,000 and 3,000 BCE.[22] By 500 CE, archaeologists contend, breeders privileged wool-producing animals. As herding grew in social importance, camelids became objects of wealth in this pastoral economy, where intentional animal breeding helped replace losses to human consumption, disease, and predation. Some scholars have argued that breeding more animals made accumulation socially sanctioned, and that this dynamic created not only greater inequality but greater tolerance thereof.[23] The more readily llamas and alpacas reproduced in a particular valley, the greater the generative power of that sacred landscape was held to be.[24]

Alpacas were small and light, and were primarily used for their wool, which was silkier and finer than that of llamas.[25] Their breeders privileged coats that would provide the raw material for the extensive textile and vestment trade that hearkened back to the Nasca (ca. 200–600 BCE) and Wari (500–1000 CE), well before the Inca centered their Tawantinsuyu (four-part empire) in Cusco in 1438.[26] Master weavers relied on camelid fibers from the highlands interspersed with cotton from the coasts. The flesh of naturally mummified llamas and alpacas discovered at El Yaral, a small pre-Columbian village off the Osmore River, offers a window into the animals prioritized by ancient Andeans, which were strikingly different from modern camelids. These animals had been sacrificed with a blow between the ears and buried, with their legs bent as if sleeping, beneath the floors of homes erected five hundred years before the rise of Inca power.[27] Studies focusing on skin and fibers drawn from these remains—including four alpacas and six llamas—yield interesting insights into the history of Andean camelid breeding. Five of the six llama mummies appear to belong to an extinct breed that produced very fine fibers, indicating a conscious selection for that trait. By contrast, experts point out that modern llama fleeces lack uniformity, and suggest that this irregularity was introduced around the time of Spanish conquest. In short, zooarchaeologists hypothesize that the increasing hairiness and coarseness of Andean llama coats over time was the result of a decline in controlled breeding.[28]

Just as in Europe, where color preferences shaped selection of horses and other livestock, Huanachan breeders selected for llama populations with yellow or brown wool, and yellow-brown hybrids became particularly fashionable.[29] Their cultivation of these warmer hues over white, black, and gray reveals an ongoing aesthetic preference realized through the animal population. As Pedro de Cieza de León and other contemporary observers noted, khipukamayuqs (knot keepers or organizers) used the stringed khipu to record herds of animals as well as other tribute including silver, gold, and clothing. Although most coastal khipus were made of cotton, in the highlands, where the herds of domesticated camelids dwelled, they were often made of camelid hair, although a low representation of highland samples makes this hypothesis challenging to test.[30] Llama females were kept for reproduction and meat, while males were usually used as beasts of burden after they were castrated at around two years of age. The meat of llamas, like that of alpacas, was preserved as jerky or carqui; their fat was allocated for candles; their droppings, or taquia, were useful as a fuel for smelting silver at Potosí in the colonial period; and their fetuses, called sullu, served as good luck charms and featured in a variety of rituals related to fertility.[31] As the Incas consolidated their control across the Andes after 1438, llamas carried supplies along the winding Andean trails, and were slaughtered and eaten when their portage was no longer valuable. Fifteenth-century silverwork from temples on the Island of the Sun in Lake Titicaca suggests the animals' growing ceremonial role under Inca rule.[32] Caravans of llamas, with up to a thousand animals per herd, traveled across Cotahuasi in the southern Peruvian highlands.[33]

In the upper Mantaro Valley, herders tended to flocks swelling to a half million animals. Their distribution echoed the Incas' furthermost expansion, as royal armies marched their pack trains from southern Colombia to central Chile.[34]

The world of Andean agriculture and animal husbandry that bred llamas to produce fine coats fractured under the pressure of Spanish colonization.[35] The last four hundred years have been marked by increased hybridization among all four species within the Andean camelid genera.[36] Both morphological analysis and DNA sequencing suggest continual interbreeding and hybridization among the four species. Indeed, variants large and small, wild and domestic. seem to have hybridized much more than originally thought: 41 percent of the llamas tested in one study were hybrids (organisms with parents of two different species), and 92 percent were crossbreeds (organisms with parents of two different breeds within a species).[37] More recently, geneticists have used modern domesticated populations to suggest that the alpaca genome shows considerable introgression—meaning the transfer of genetic information from one species to another as a result of hybridization followed by repeated backcrossing—around the time of the conquest, suggesting the loss of traditional management practices following the Spanish invasion.[38]

Llamas as Labor

Following the inland invasions by Francisco Pizarro González and Diego de Almagro beginning in the 1530s, many elements of indigenous Peruvian culture, such as geographically and symbolically salient *huacas*, retained their cultural resonance. In his writings from Peru in the 1570s, Spanish missionary and naturalist José de Acosta emphasized that Andean camelids remained. They were so prominent that they required him to carefully examine the presence of different animals in Europe and the Americas through the Noah's ark narrative (fig. 6.1), as chapter 7 describes. When it came to classification, Acosta was particularly interested in which animals could produce important medical materials, chiefly the bezoar stones that develop in the stomachs of ruminants. Bezoars were highly valuable and believed to be a powerful anti-poison. All sorts of animals could produce bezoars—including goats, iguanas, and even humans— but those produced by ruminants that ate healthy herbs were thought to be the best. Acosta likened domestic llamas and alpacas to sheep and donkeys and bezoar-producing vicuñas to wild goats. In doing so, he adopted a typical early modern classificatory strategy that grouped living things by the products of their innards. For this reason, radically different types of

FIGURE 6.1. The second age of the world, in which Noah carries a llama on the ark, alongside chickens, lions, and horses. In Felipe Guaman Poma de Ayala, *Nueva corónica y buen gobierno*, (ca. 1615). Royal Danish Library, GKS 2232 kvart.

dragon's-blood trees, from the Canary Islands' *Dracaena draco* tree to the flowering *Croton lechleri* plant of South America, were lumped into the same taxonomic category because all produced similar red, blood-colored resin. Bezoar-producing animals were likewise lumped together.[39] In so doing, Acosta embraced the conflation of *ovejas de Indias* with sheep, presuming "*de Indias*" to be a sort of Aristotelian accidental difference within the wider category of sheep present in both the Old and New Worlds.[40] Acosta noted two variants of these livestock: pacos (*carneros lanudos*, or woolly sheep), and others that were less woolly, between a sheep and yearling calf in size, with a long neck similar to that of a camel, which he thought was necessary because of their tall and high bodies.[41] The long-necked Andean "sheep" were good for food, portage, and making clothing, tablecloths, and coverlets out of their fleece. Like other Spaniards, Acosta marveled at the animal's utility, noting how, without saddles or pack bags, barley, or horseshoes, they could be driven by the thousands to transport maize, coca, wine, quicksilver, and even silver bars from Potosí to Arica with only a few Indian guides as their keepers. With their colored wool of white, black-gray, and multicolored *moromoro*, Acosta saw them as the richest, most profitable asset of Peru and its people, explaining that "God provided sheep [*ovejas*] and beasts of burden [*jumentos*] for them in the same animal."[42]

However, the Andean camelids were as heavily hit by the encounter with Europeans as the human population, the gold-drenched Inca temples, and Potosí's silver-filled landscape. Acosta noted there had been a time when herds of *ovejas de la tierra* covered Peru, when "there were men who owned seventy thousand or a hundred thousand head of these lesser livestock [*genado menor*], and even today there are flocks that are just a little less; to have such in Europe would be great wealth, and here it is only moderate."[43] Population numbers crashed under the Spanish as the result of Old World diseases, competition with new domesticated animals, and most of all, the collapse of the social system that had supported and protected these animals. Archaeologist Jane Wheeler has suggested that the populations of Andean domestic livestock were reduced by 80–90 percent in the first hundred years of contact.[44] Some died from hunting, some from illness. Early modern sources corroborated this decline. Girolamo Benzoni reported in his *Historia del Nuovo Mondo* (History of the New World) that a vast number of American sheep, so large that they were the size of a donkey, had perished from a disease like leprosy.[45]

Llamas, alpacas, and their wild cousins were all affected. The camelid populations first declined in the coastal plains and highland valleys, where their pasturelands were reallocated to Europeans' new imports: sheep, goats, cattle, and pigs. Unfortunately for indigenous flora, the more accessible the pasturage, the more competition camelids met from Spaniards' familiar livestock. As the elevation increased, the landscape became less hospitable to European animals, and the Andean livestock declined less. Nonetheless, by 1651, chroniclers reported that llamas and alpacas had all but disappeared from the Lake Titicaca basin, their former heartland.[46] The political changes had repercussions even for the wild vicuñas. Unlike deer in Europe, which were deemed worth protecting, the vicuñas effectively lost their protected status as royal prey under the Spanish. Acosta suggested that the animals "were not understood to multiply much, and hence the Inca Kings had a prohibition on the hunting of vicuñas, save for on festivals and at their orders." *Chaco*-style hunting gathered thousands of men to surround a large area. They would yell until the animals came together in key places. Then, the hunters would take a few hundred animals and allow the females to go so that they could continue to breed. However, Acosta recorded that "some complain that after the arrival of the Spanish too much license has been granted to the *chacos* and hunting of the vicuñas, and they have diminished in number."[47]

In his description of the process by which silver ore was refined, Acosta explained that "all of the metal that they take from the mountains was brought by the sheep of Peru [*carneros del Piru*], who serve as beasts of burden [*jumentos*]"; they were loaded up and driven to and from the Man-Eating Mountain of the Cerro Rico of Potosí.[48] Zooarchaeological analyses drawn from prominent sites—including colonial wineries in Moquegua in southern Peru and Torata Alta, a nearby colonial resettlement village for indigenous people—offer a material window into colonial animal labor. Camelid bones bear the marks of degenerative paleopathologies that Susan D. deFrance has interpreted as reflecting changes in animal health, use, and herd management following Spanish conquest. Of the seventy pathological camelids studied, forty-five displayed evidence of stress or joint disease, most often on the vertebrae or foot, likely indicating that these animals were employed to transport cargo. In contrast to pre-Columbian specimens, animals that lived in the colonial era showed notably worse bone deformities. The animals must have been in pain from the deformed vertebral phalanges in their arching necks and backs. More than those of the goats, horses, and cattle with which they lived (and near whom their remains rested in the passing centuries), their skeletons had been transformed by the lives they lived in the sixteenth century. These animals had been forced to carry heavy burdens even after age made them frail.[49]

Camels or Sheep?

Like indigenous languages cordoned off into European word lists, American flora and fauna were often squeezed into categories that were not created for them. Spanish invaders, in other words, understood what they saw based on what they had seen.[50] Many brought with them blinders that prevented them from experiencing the alterity of New World nature, and the unsettling encounter often hardened rather than softened their systems of categorization. In spite of an array of evidence, from their anatomy to their use, that suggested otherwise, many Spanish writers emphasized that Andean camelids were sheep, goats, or livestock, but certainly not camels. Males became *carneros* (rams) and females *ovejas* (sheep), often followed by some variation of *del Peru* (of Peru) or *de la tierra* (of the land). Modern scholars, such as geographer Daniel Gade, argue that "the sheep analogy that prevailed for more than two hundred years reflected a European epistemology that honored its perception and discounted that of native peoples."[51] He is right, but the story is a bit more complicated. The generality about "Europeans" crumbles upon inspection, since Spaniards had different views from other Europeans, and Spaniards from Spain used different terminology than did creoles born in Peru.

While other European observers engaged with the Spanish idea of Andean camelids as sheep, they debated the accuracy of this terminology. Noted Swiss naturalist Conrad Gesner (1516–1565) compared the parts of a llama with those of various Old World animals—though not sheep—after a specimen came to Antwerp in 1558 (fig. 6.2). He called it "Allocamelus," and described it thus:

> In the land of the giants it is said, there is an animal with head, ears, neck of a mule; body of a camel, tail of a horse. It is composed from the camel and others. In 1558, . . . this wonderful animal was brought . . . until this time, it was never seen, not by Pliny nor mentioned by any other ancient writers. They said it was that Indian from Piras [perhaps Peru] a region six thousand miles from Antwerp. The animal stood six feet tall and five feet long, with a neck like a swan, pure white. Its body was red and scarlet, its feet like an ostrich.[52]

Despite Gesner's near-monstrous analogies, this is one of those occasions where armchair travel trumped firsthand experience. The distance from Spanish Peru empowered Gesner: disentangled from its context, he pieced this animal back together. Incan names discarded by the Spanish might possit attingere. In posteriori vero parte demissus est instar cerui. Collum habet extensum, caput equinum, licet minus, pedes & caudam vt ceruus: pellem vero sic omni colorum genere diuersimode variatam, vt homo frustra tentet artificio naturalem eius pulchritudinem imitari. Hoc animal nostris temporibus à Soldano Babyloniorum transmissum est imperatori Friderico Romanorum augusto, Hæc Isidorus: ex quo etiam Albertus Magnus de-

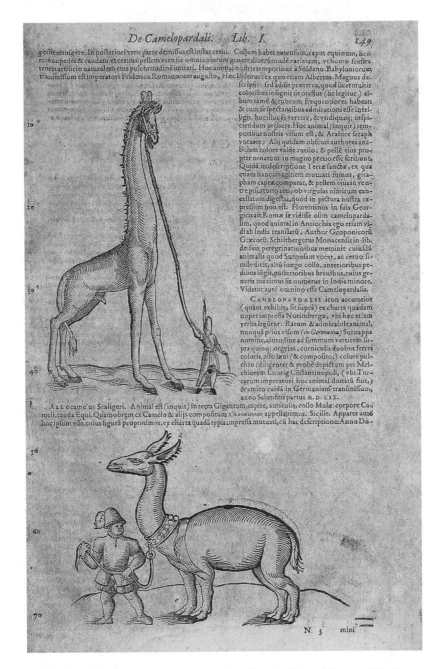

scripsit: sed addit præterea, quod licet multis coloribus insignis sit oraslus (sic legitur) album tamen & rubeum, frequentiores habeat: & cum se spectantibus admirationi esse intelligit, huc illuc se vertere, & vndiquaq; inspiciendum præbere. Hoc animal (inquit) temporibus nostris visum est, & Arabice seraph vocatur. Alij quidam obscuri authores anabulam colore valde rutilo, & pellem eius propter ornatum in magno pretio esse scribunt. Quidam in descriptione Terræ sanctæ, ex qua etiam hanc imaginem mutuati sumus, gitapham capræ comparat, & pellem eius in ventre piscatorio reti, ob virgulas nimirum cancellatum digestas, quod in pictura nostra expressum non est. Florentinus in suis Georgicis ait Romæ se vidisse olim camelopardalim, quod animal in Antiochia ego etiam vidi ab India translatu, Author Geoponicoru Græcoru. Schiltbergerus Monacensis in lib. de suis peregrinationibus meminit cuiusdam animalis quod Surnosam vocat, ac certo similem dicit, alto longo collo, anterioribus pedibus logis, posterioribus breuibus, cuius generis maximus sit numerus in India minore. Videtur aute omnino esse Camelopardalis.

CAMELOPARDALIS icon accuratior (quam exhibita sit supra) ex charta quadam nuper impressa Norimbergæ, vbi hæc etiam verba leguntur: Rarum & admirabile animal, nunquã prius visum (*in Germania*) Surnappa nomine, altitudine ad summum verticem supra quinq; orgyias, corniculis duobus ferrei coloris, pilo læui (& composito,) colore pulchro: diligenter & probe depictum per Melchiorem Luorig Cõstantinopoli, (vbi Turcarum imperatori hoc animal donatu fuit,) & amico cuidã in Germaniam transmissum, anno Salutiferi partus M.D.LIX.

ALLocameus Scaligeri. Animal est (inquit) in tetra Gigantum, capite, auriculis, collo Mulæ: corpore Cameli, cauda Equi. Quamobrem ex Camelo & alijs compositum Ἀλλοκάμηλον appellauimus. Sicille. Apparet aute hoc ipsum esse, cuius figurã proponimus, ex charta quadã typis impressa mutuati, cũ hac descriptione: Anno Do-

FIGURE 6.2. Variations on camelids with handlers. In Conrad Gesner, *Historia animalium*, liber primus, *De Quadrupedibus viuiparis* (Francofurti: In Bibliopolio Cambieriano, 1602), 149. Biodiversity Heritage Library, Smithsonian Libraries. Public domain.

not have found their way to Europe, but neither did the Spanish title of *oveja* gain favor in Gesner's circles. As with animal terms for race in Latin America, this seems to be a case of separate spheres of intellectual influence in the Spanish empire and in other areas of Europe. This separation seems to have been helped by the fact that the Spanish Index of Prohibited Books followed the Venetian index in expurgating Conrad Gesner, among other medical authors, in 1554.[53]

The Florentine merchant Francesco Carletti arrived at Lima's port of Callao in January 1595 and left by June.[54] It is not at all clear that he ever ventured inland, but he reported having seen camelids, although he was more interested in Peru's silver and coca. Camelids featured in his narrative travelogue after he mentioned Spanish jennet horses, as well as mules and asses, that had come to roam wild along the Pacific coast. In Chile, Carletti wrote, European horses roamed free, uncounted. Whoever wanted them simply walked into the countryside to catch them and needed only pay the price of training them.[55] However, llamas and alpacas were in increasingly short supply.

> There are still other flocks of native beasts of this land, which the Spaniards very improperly call *carneros*, that is rams, and the Indians call *pacchi*. But when I saw them, they seem to most resemble little camels except the lack of a hump, because they have the neck, head, and feet of those animals, but are all together smaller in body and strength. Their meat is very good to eat, and their wool is used by the Indians for cloth. This animal is very domestic, simple, and peaceful, but so extraordinarily stubborn and headstrong that they will move along only in their own way and at their own wish. Feeling weary, or having other humor, they throw themselves down to lie on the ground even if they have burdens on their backs, and it is impossible to make them get up again even if one wants to kill them such is their obstinacy. And for that reason, a custom has arisen in that region of saying to a stubborn person: "You are a llama."[56]

Carletti, attentive to the categories used by both the Peruvians and the Spaniards, rejected both on the grounds of the llamas' shape and behavior. Acosta's *Natural and Moral History of the Indies* included an expanded variation of this anecdote, which perhaps provided the basis for Carletti's. Acosta noted that Indian handlers would stop and sit next to the sulky llama, petting and caressing it until the sulky mood abated and it stood up, occasionally waiting for two to three hours until the fit had passed.[57] While Carletti's travels make him anything but representative, his writings provide an important non-Iberian perspective on these animals.

Llamas eventually did make their way to Europe, as we have seen. The Allocamelus from the land of the giants gained a place in European menageries and roving exhibitions, alongside other exotic animals such as elephants, lions, leopards, and antelopes. Ferdinand II of Tyrol (r. 1564–1595) once had a llama in his menagerie. Its likeness appeared in the *sgraffito* of Orpheus in the courtyard of Ambras Castle.[58] In Prague, Holy Roman Emperor Rudolf II (1552–1612) had a llama, and possibly even an alpaca, in his collections. Similarly, in May 1572, the duke of Bavaria received a shipment carrying both feathers from Mexico and exotic animals, including a llama (called an Indian sheep) as well as long-tailed monkeys.[59] Only the Andean camelid's wonder transferred; its utility did not. No projects appear to have been undertaken to breed herds of them for the Alps or Pyrenees. Philip II did try to import vicuñas for the royal gardens of Aranjuez and the Pardo, but they never made it to Spain.[60]

Camels in the Canaries and Peru

Iberians had once known quite a bit about camels. Remains of both Bactrian camels and dromedaries appear at sites in the former Roman Empire, including Iberia, dating from the first to the fifth century CE.[61] Under Muslim rule, Iberian nature more closely resembled that of North Africa, with similar deer, antelopes, and zebras.[62] Paintings and pottery created during the Umayyad caliphate and the Ta'ifa monarchies in the twelfth century regularly featured camels as a symbol of aristocracy and power.[63]

Whether brought by Europeans or earlier African migration, herds of camels were mentioned in reports from the Canary Islands by the fifteenth century. Documents from sixteenth-century settlements occasionally lumped them in with other "livestock," or *ganados*, but most often they were differentiated for purposes of riding. As settlers looked westward to the Caribbean islands and American continents, camels, often from the Canaries, occasionally crossed the ocean with them.[64] In the winter of 1508, a merchant of Gomera named Alonso de Valladolid sought royal permission from the Casa de Contratación to export six camels from the Canaries to the Caribbean island of Hispaniola, as Felipe Fernández-Armesto has shown.[65] While it is unclear whether this voyage ever took place, camel imports to the New World were constantly frustrated and thwarted by skeptical settlers more comfortable with the goats, pigs, rabbits, and sheep with which they were familiar. But some dromedaries did make it to the Americas, though their numbers were few. Acosta mentioned them, writing that, along with many mules and a few donkeys, "I saw a few camels in Peru, which had been brought from the Canaries and had multiplied there, but only for a short period."[66]

A camel scandal arose in June 1552. That year, according to archaeologist Duccio Bonavia, a *cédula*, or warrant, arrived from the king of Spain. The crown deemed that it was time for a new beast of burden in Peru and granted permission to import and breed camels to a member of the Caritate family. In his Anales of Peru for 1552, Don Fernando de Montesinos wrote that Cebrían de Caritate "had permission to take camels to Piru for ten years, and no other person could introduce them in said time." The reason given for importing them was that "they are so necessary for the service of the land, because there no longer are any personal services in it, nor will there be."[67] In sum, Europeans had heard of the seeming lack of domesticated animal labor in Peru and aspired to fill the lacuna currently occupied by human laborers, euphemistically called "personal services." Camels could be swapped in for human porters. The seventeenth-century historian Antonio de Herrera offered a similar account: he noted that "one Baltasar de Cariate presented a royal cédula to the viceroy that granted him alone permission to take camels to Peru for ten years," between 1552 and 1562, "in attendance to the fact that there no longer were any personal services, nor would there be, because it seemed that to carry cargo camels were a great relief."[68] The viceroy, Don Antonio de Mendoza, chief administrator of Peru in the king's absence, seems to have been ill at the time of this pronouncement and died the following month. Uncertainties about precisely who was in charge meant that there was some debate about whether this cédula had been enacted. Some thought it had been read out loud and therefore executed. Herrera, however, concluded that it had not been read, as the Audiencia was waiting for the king's reply. Regardless of the specifics, it does not seem that Caritate was ever allowed to fulfill his dream of a Peruvian camel monopoly.[69] Madrid's encouragement of Peruvian camel breeding by monopoly may well have undermined its very aim.[70]

Bernabé Cobo (1582–1657), a Jesuit missionary and scholar of Spanish American nature, reflected on animals as well as plants and materia medica.[71] Having resided in Peru from 1599 to 1630 and again from 1650 to 1657, Cobo recounted the success of Old World fauna in South America in his *General History of the Indies*, which he completed in manuscript in 1653. Cobo focused on camels as beasts of burden. He began by situating them alongside guinea hens as rare fauna with African origins, which he considered uncommon additions to the Andean landscape. He recounted an oft-cited tale of Captain Juan de la Reinaga, who was "one of the first settlers of this land, and had the camels brought to this kingdom of Peru from the Canaries, which are islands adjacent to Africa, shortly after it was pacified and peopled." In Cobo's telling, the camels originally flourished. They "here formed a lineage [*hicieron casta*] and multiplied much, even so they did not spread over the land or go beyond the jurisdiction of this archbishopric of Lima." While owners of such beasts, presumably including Reinaga, "tamed some to use them, most went wild and untamed in the sierras that extend from this city to the valley of Ica and are vulgarly known as Las Lomas." This wild population flourished due to preference for horses and mules for transportation. With the animals of chivalric tradition already at hand, camels were "not highly valued." Despite this semiferal existence, Cobo recounted that "they lasted for many years and multiplied a great deal." In Cobo's telling, they ended as they began, with African consumption. Cobo recounted that their population lasted only sixty years in Peru, as the last female died in 1615—just as Cobo started writing his history. This female camel (called a cow) had been brought to Lima with one other female by a search party, sent to look for the remaining members of the camel herd so that they would "not die out and their lineage be lost." While the females were brought to Lima, "where they lived for some years," their kindred were killed off. Cobo read their deaths as the result of both cultural disinterest and hunger. "On one hand," he wrote, African camels had been lost because they "lacked the protection and industry of man, as nobody looked after them, and on the other hand because the maroon blacks killed them to live off of them."[72]

Bonavia has connected Cobo's Captain Reinaga to a Don Juan de Reinaga Salazar. He seems to have purchased one male camel and six females from the Canary Islands between 1554 and 1560. In Garcilaso de la Vega's account, he was "a man of noble birth from Bilbao," who Garcilaso purported to have known.[73] After Reinaga served as captain of the infantry in the 1554 war against Francisco Hernández Girón, it appears that he had been paid a sum of 7,000 pesos for "seven she-camels and one male." The man who paid that handsome sum to the commander seems to have been a comrade-in-arms named Don Pedro Portocarrero.[74] Portocarrero was a showman. He had paraded the royal standard of Castile as Gonzalo Pizarro entered the city of Lima. Later in life, he and his wife seem to have showcased their camels, as recorded by chronicler Reginaldo de Lizárraga. When the viceroy aimed to walk across the desert, Portocarrero met him in the arid Huarmey Valley with a camel train in tow.[75] Using sources that are uncited, but which ring true of the pranks of early modern noblemen, later historians have suggested that the party of camels were ridden by men of African descent dressed as Moors.[76] The extensive scholarly literature on Moorishness in Iberia and the Spanish empire suggests how we can interpret this account: Lizárraga wrote that the viceroy dressed up his camel riders, requiring the African slaves to "play the Moor," to employ Barbara Fuchs's term.[77] The phrase "playing the Moor" seeks to unpack a popular element of early modern Spanish ceremonies—donning a turban and dressing in blackface. Through Moorish games, Spaniards impersonated Moors, at once appropriating their practices and consolidating Catholic Spanish victory over their culture; to a foreign audience, the tradition of donning turbans, fake beards, and sometimes even blackface served to render Spain more exotic.[78]

Camels at the Spanish Court

Like the Andean camelids, dromedaries were increasingly relegated to the category of exotic court animals, as were the lions and ostriches that also came from North Africa.[79] Safely contained within the grounds of royal Spanish gardens like Aranjuez, outside of Madrid, a prestigious and useful animal in the Islamic world acquired a new status that undermined its utility in exchange for an increase in its perceived luxury. Court collecting transformed a beast of burden into an exotic. By 1592, Aranjuez experts were breeding camels, like horses, in the fields, as well as lions and leopards for Spanish enclosures. Aranjuez rapidly developed into a supplier of quality camels for European courts.[80] On May 22, 1628, an order was sent to Gaspar Ruiz de Escaray at Aranjuez: Pope Urban VIII's Roman court required camels. He was to send to Rome four camels, two male and two female.[81]

Aranjuez itself developed an institutional permanence that included secure employment for generations. Petitions from family members requesting compensation from the royal household included a passage concerning the "diligence and care" with which the expert had tended to the gardens and stock. Gardener Pedro Tovilla, for instance, worked with royal plants both in Aranjuez and in the gardens of the monastery of San Lorenzo el Real at El Escorial.[82] His career, executed with "care and toil," spanned twenty-four years, from approximately 1584 to 1608. Some gardeners developed varied specialties over decades of employment. One of Tovilla's contemporaries, Francesco Moreno, cared for both the camels of Aranjuez and the trees of its gardens.[83] Like the kennel stewards and stable masters of Mantua, these experts were valued members of the court whose positions were comfortable enough to encourage them to spend decades in the king's service, herding Spanish camels and trimming Aranjuez's many exotic trees. In 1613, Francisco de Corral, a former guard at Aranjuez, requested compensation from the crown for an injury he sustained while working at the garden. Fifteen years before, in 1598, he had been riding through the grounds of Aranjuez on horseback when a camel startled his mount. He fell and broke his arm, which made it impossible for him to continue serving the king. He asked for the necessary restitutions for his wife and children.[84] The "widows of Aranjuez," a named group in administrative documents, likewise lobbied the court for pensions to support their families once their husbands had died after a career of working there.[85]

The fashionable camel herds of Aranjuez flourished well into the seventeenth century as the camel took on artificial scarcity. In the late seventeenth century, Spain retained large numbers of camels in the *presidios* (fortified bases) of North Africa. The Spanish government traced the sale of camels in its North African *presidios*, along with the sale of mares, foals, and oxen raised in the forts.[86] While many camels remained in North Africa, a mere hundred kilometers away from Medina Sedonia, the new duke Juan Carlos Pérez de Guzmán wanted camels from Aranjuez. In the fall of 1668, King Charles II of Spain forwarded the duke's request for six camels to be sent to Medina Sidonia, deep in Andalusia. Several noblemen—including the marquises of Fresno, El Carpio, and Salinas, and Don Germo Carmago and Don Antonio de Alonsas—advocated rejecting the duke's request on the grounds that the number of camels in Andalusia might grow to be dangerously large, lowering the value of those in Aranjuez.[87] Generations of camel keeping at Aranjuez had given the king an exoticized and valuable herd, the scarcity of which was protected. Camels at court were divorced from prosaic labor and turned into display pieces. Meanwhile, elsewhere in the empire, they retained their utility.

Christian Nature

Royal libraries came to hold notes on American nature in published book and manuscript form, such as Hernández's writings on the flora and fauna of New Spain at the Escorial. Other collections included elaborate featherpieces akin to those in Moteuczoma's collections, and bezoar stones from Andean camelids.[88] Herds of dromedaries and llamas were collections of sorts, too. As Paula Findlen has argued, "Strategies for collecting were not only designed to fulfill the humanistic desire for *prisca scientiae*: museums also conveyed political and religious messages." More broadly, "while sixteenth century naturalists faced the dilemma of incorporating the artifacts of the Americas into their cosmos, the seventeenth century Jesuits"—and, I would argue, Catholic collectors more generally—"attempted to develop a moral, religious, and philosophical framework that connected all the different regions of the world."[89] Marvels required a wider system to explain how universal Christian nature developed in the first place.

For Acosta, the specific puzzle of camelids was a gateway into wider questions of migration and population. To conclude this chapter and transition to the next, then, I turn to three interpretations of the Noah's ark narrative to lay the groundwork for Acosta's treatment of this puzzle in chapter 7. As Lydia Barnett has shown, "inspired by new methods for studying nature in the Scientific Revolution, pious natural philosophers reinterpreted the biblical story of Noah's Flood as a global catastrophe that did just as much damage to the natural world as to the human."[90]

Early in his career, before he became an inquisitor, a cardinal, and Philip II's favorite for the papal tiara, Giulio Antonio Santori (1532–1602) provided his own reading of the story of Noah's ark. His vision tellingly centered on the Church as the ark, rather than on the mechanics of global population: he argued that "this Church is represented by the means of Noah's ark: it is unique and only those aboard it were saved."[91] Santori will return in the narratives of the next two chapters.

The second interpretation suggests how Spain's image of the ark had to be updated to include New World nature. Inside the halls of the Royal Palace of Madrid, the values of Christian nature—perfect, selected, preserved—were abundantly displayed. Visitors might have paused before a tapestry made by Willem de Pannemaker in 1563–65 (fig. 6.3), perhaps studying it carefully.[92] The Roman collector Cassiano dal Pozzo described the experience:

> One passes on through a door decorated with silk cloths and gold tapestries depicting the stories of the Flood and Noah's ark. That is to say, they were about the corruption of the epoch, the making of the Ark, the loading of the animals, the moment of departure, and Noah's final call. They had a beautiful frieze on top with a great variety of birds depicted from nature and on the sides the countries and various land animals, and at the bottom the fish and birds of the water, also depicted as in nature. The arms of His Majesty are in the corners.[93]

Noah (complete with a flowing beard) and his family wore Renaissance takes on classical garments as they constructed their ark and summoned the antediluvian creatures aboard.[94] Two scenes appear in the tapestry. The mid-ground shows the first scene. There, Noah opens the doors of the ark for camels, alongside unicorns, rhinos, and American turkeys, to march themselves onto the wooden ship. In the other frame of the moving scene, the ark appears in the distance, raised upon the floodwaters,

FIGURE 6.3. The departure of Noah's ark. Michiel Coxcie and Willem de Pannemaker, 1562–65. Alfonso XIII, 1930, Palacio Real de Madrid, 10007058.

surrounded by desperate people left behind, searching for higher ground. Llamas are not depicted, but other New World creatures found their way onto the ark, including even a pair of white-nosed *coati*, with their striped tails held high.

To conclude this chapter, I turn to the third interpretation: Felipe Guaman Poma de Ayala's 1615–16 *El primer nueva corónica y buen gobierno*, the title of which tacitly critiqued nonindigenous chronicles. Frank Salomon has argued that "in Guaman Poma's eyes, Spain had faced the world-historical opportunity to unite halves of humanity that did not know each other, and thereby hasten the world-historical drama to its redemptive goal. Spain had failed in this, producing instead injustice and a betrayal of faith."[95] Camelids appear prominently in the text and its illustrations. They are shepherded by youths, sung to in feasts, sacrificed, and included among idols and huacas. An administrator of royal mines punishes native lords as one native man rides a llama. Another royal overlord beats his native porter, who accompanies a llama. A native administrator confiscates an elderly Andean's llama.[96] At the start of his project, Guaman Poma centers "The Ages of the World," which follow the Christian narrative. He begins with Adam and Eve and the children that they engendered (*engendró*), who cling to Eve's breast in the titular illustration. He follows with the making of the lineages (*linaje*) leading to Noah. He then recalls that "the world was full of men who did not care and they did not know the Creator and Maker of men. And here, God commanded that the world be punished with the waters of the Flood."[97] The accompanying text then tells of the flood over forty days and nights. To the sons of Noah, Guaman Poma wrote, "God commanded to come out of this land, to pour out and multiply throughout the world. Of these said sons of Noah, one of them brought God to the Indians; others say it came out of Adam himself. Said Indians multiplied, and God, that knows everything and is powerful, left these Indian people apart."[98]

Guaman Poma's illustration on the page before added further information about the details of the flood. It is labeled on the bottom: "By command of God he filled the waters and the world is punished." Noah is at the center of the image, labeled, on his knees with his hands raised in prayer. Above him the rain, also labeled, pours down; below him, water rises. Around him is an ark, also labeled. On the ark, with their bodies appearing out of eight windows, are eight different animals. These animals include a cow, a horse, a lion, a ram, a donkey, and a chicken—and, most important for this chapter, a camelid. Through this illustration, Guaman Poma emphasized that llamas and alpacas were "perfect" animals, made by God, and all the more precious as they served humanity.

PART IV

Difference in European Thought

CHAPTER 7

Thinking Through Conversion, Lineage, and Population

JOSÉ DE ACOSTA

Philosophizing American Nature in Europe

Living animals inspired animal philosophizing. This chapter transitions from material animals, like the camels of Aranjuez and the barkless dogs of the Caribbean, to ideas about animals in late sixteenth-century European philosophy. With a focus on population thinking and physiognomy, respectively, this and the final chapter shift to considering the ways in which animals provided models for understanding heredity and difference. Central to both is late sixteenth-century Rome, where two widely read European scholars, José de Acosta and Giovanni Battista della Porta, navigated imperial and religious politics. Rome, capital of the Counter-Reformation, had such a strong Spanish community and influence that Thomas Dandelet has termed it "Spanish Rome"; it functioned as a key site for negotiating the relationship between universalizing programs and Spanish distinctiveness.[1] As Sabina Brevaglieri has shown, "in the last fifteen years, the relationship between the pope's city and science has been a lively site of scholarship capable of undermining the consolidated myth of the Counter-Reformation Church's direct opposition of modernity. A plural, polycentric, and cosmopolitan city, Rome appears, on the contrary, to be a place of intensive production of knowledge concerning the bodies of man and nature."[2]

In many ways, Acosta, a Jesuit missionary embroiled in Spanish and Catholic high politics, contrasted strikingly with Porta, a Neapolitan magus and aristocrat regularly in trouble with the Inquisition. For all the pathos of Acosta's career, he remained on the side of the Church and the censors, even acting as a censor of apocalyptic texts himself during his time in Peru.[3] Even so, their parallel intellectual projects defined distinct facets of European ideas of animality and inheritance. Acosta and Porta both thought through animals as a means of reckoning with the limits of natural difference, weighed material evidence drawn from experience, published extensively, and survived the intellectual and political influence of the Inquisition and the fraught relations between the Italian and Spanish worlds. Their published works took on lives beyond the environment that produced them as part of a European scientific approach to nature's universal characteristics and particularities.

Acosta's popular published works—particularly his *Historia natural y moral de las Indias*, published in Seville in 1590—offered a nuanced explanation of how American nature fit into a universal Christian world. As previous chapters have shown, Europeans thought about heredity using the noble lineages and family histories of humans and their nonhuman animals. The unprecedented scale of the project to comprehend the Americas, however, required explaining how large groups, whether of peoples or animals, came to be. Acosta's unshakable belief in the convertibility of all humans and the role of Noah's ark in preserving all perfect animals required him to jump through analytical hoops. His discussion of animal populations, particularly those of domesticates like dogs and Andean camelids, shows the unresolved tensions around the categories of species, breed, and lineage; he weighted the essence of creatures against their accidental features on a continental scale rather than a courtly one. For Acosta, at the heart of each perfect animal type rested the essential version of that animal that had landed on Mount Ararat, just as all humans shared Adam as their progenitor. When Acosta returned to Europe, he published on nature and conversion and entered high politics, representing the Jesuits in Philip II's court in Rome. Acosta's writings, in tandem with the choices he made in advocating for policy, demonstrate his commitment to the possibility of universal conversion.

José de Acosta in Rome

In 1592, José de Acosta knelt at the feet of Pope Clement VIII (r. 1592–1605). Acosta had left Europe in 1571 to sermonize in the Andes among mines and camelids, and became a naturalist; he had returned to Europe in 1587 and subsequently became entangled in Spanish Rome and the Catholic facet of Spain's imperial project. His work in Rome was the latest stop in his career as a go-between. Following Acosta's positive reception as an expert on Americana at the Spanish court, the superior general of the Society of Jesus, the Neapolitan Claudio Acquaviva (1543–1615), had appointed him to be the Society's representative to Philip II.[4] One of Acquaviva's original tasks for Acosta had been to prevent a visitation of the Society in Spain by a non-Jesuit, a political feat that Acosta had achieved in part by carrying out many of the *visitas* (evaluative visitations) across

Andalusia and the south of Spain himself.[5] By 1592, Acosta found himself imperfectly serving three masters—the Society of Jesus, the Spanish crown, and the papacy—although all ostensibly represented the same religious interests.

The papal audience belonged to the period that Acosta's recent biographer, Claudio Burgaleta, has cast as the nadir of Acosta's life: a dance of politics and principles on the stage of Rome.[6] Acosta's new goal, encouraged by Philip II, was to persuade the pope to call the Fifth General Congregation of the Society of Jesus in order to address the concerns of the *memorialistas*. The *memorialistas*, a group of about thirty Jesuits from the provinces of Castile and Toledo, had complained about the Society's affairs through anonymous memoranda addressed to the Spanish king and Inquisition, which raised concerns about the Society's centralization under Acquaviva and its imperfect adherence to Spanish policies. If Pope Clement were to call the congregation, it would be the first held within a superior general's lifetime, thus offering a serious rebuke to Acquaviva's leadership. This responsibility placed Acosta in an impossible position that would only get worse during the congregation itself, which did indeed take place from November 1593 to January 1594. On one hand, assuaging the *memorialistas'* concerns was essential to preventing Philip II's interference in Society affairs, which was Acosta's original task. On the other hand, calling the congregation went over Acquaviva's head and led many Jesuits to see Acosta as a Spanish stooge rather than a loyal Society member.[7]

No wonder Acosta was anxious to keep the superior general from knowing too much about his movements as he walked the fine line between secrecy and obedience to his many superiors that December in Rome. Letters of introduction led to audiences, which led, in turn, to more audiences, until Acosta at last entered the pope's chambers. Clement was alone, which was all to the good for Acosta's secret negotiations and all the more challenging linguistically. In his diary, Acosta captured the linguistic tensions and overlaps at the intersection of Italian and Spanish worlds.[8] After kissing Clement's feet with devotion, he said in halting Italian, "Most Blessed Father, I understand Tuscan well, but I don't know how to speak it expeditiously." Acosta recalled that the pope smiled at this and responded, "Speak your Spanish language, which I understand well enough." With that, Acosta dove in to "the things of his religion" that "the Catholic King had commanded he report to his holiness."[9] Sharing mutually intelligible languages but unable to speak in each other's native language fluently, Acosta and Pope Clement exemplified the reciprocal reliance of Spanish and Italian worlds, but also their imperfect overlap. Following the death of Pope Sixtus V, whose reign had been marked by campaigns of centralization and alliance with Spain, Philip II had been eager to influence the next papal appointment. However, the papacies of Urban VII, Gregory XIV, and Innocent IX were notoriously short, lasting about twelve days, ten months, and one month respectively. In the end, in late January 1592, Ippolito Aldobrandini had taken the papacy and the regnal name of Clement VIII, backed by a French coalition in spite of strong support for the Spanish favorite, the inquisitor Giulio Antonio Santori, whose family came from the Kingdom of Naples.[10] In the process of taking the reins of the Church, it was all the more important for Clement to prove that he was neither soft on heresy nor inattentive to Spanish interests. As with the *memorialista* affair more broadly, Acosta was negotiating both the extent to which the Spanish empire needed to rely on Italian cooperation and the universalizing aspirations of Rome.

As recalled in his diary, Acosta began his papal audience with a personal narrative substantiating his expertise. He had served the Society for forty years and noted that he had been raised on the doctrine of founding father Ignatius of Loyola.[11] Indeed, in 1540, the same year the Society of Jesus was founded, Acosta had been born in Medina del Campo. Along with three of his brothers, he partook in Jesuit education, which brought him from Salamanca to Coimbra to Alcalá de Henares. He was ordained at twenty-seven and promptly requested to be sent to the Indies; he was dispatched on the third Jesuit expedition to the Americas in 1571. As he told Clement, after studies that focused on scholastic theology and preaching, he had traveled to the "West Indies with the desire to help the conversion of those peoples, and in that I occupied myself for sixteen years taking all six of the provincial offices."[12] These American appointments brought him to the Caribbean, Central America, and most extensively, South America, where he oversaw the college in Lima, founded new colleges in the Andean highlands, and served as official theologian to the Third Provincial Council of Lima.[13] Acosta's publishing career had started in Peru with *Doctrina Christiana*, a trilingual collection of catechetical books written during the Third Provincial Council in Spanish, Quechua, and Aymara; the text continued to be widely used in religious instruction in Peru for centuries.[14]

During his time in the Americas, Acosta developed expertise based on experience. First-person observations animated his descriptions of the North and South American continents. When Acosta passed the equator and the sun was at its zenith, he went outside to warm himself in its rays, laughing at how Aristotelian meteorology predicted violent heat in the Torrid Zone.[15] Acosta had seen for himself a volcano in Mexico spitting ash and smoke, debated the role of *huacas* in conversion projects in

Chuquisaca, examined rocks shot with gold from the mines of Zamora, described pearl-wearing practices, and observed a few plantains of the Indies growing in the royal Spanish gardens of Seville.[16] Acosta valued both his own experience and that of others. He prefaced his analyses by noting his qualifications as an on-the-ground expert with access to other on-the-ground experts in indigenous affairs, writing that "I resorted to experienced men who were very knowledgeable in these matters, and from their conversation and abundant written works I was able to extract material . . . with the experience of many years and my diligence in inquiring and discussing and conferring with learned and expert persons."[17] These observations underpinned Acosta's writings, where they served as the basis for his authority.

New World Nature, Education, and Taxonomy

In 1587, Acosta returned from the Americas to Spain, laden with manuscripts and memories of his New World experiences.[18] In the following five years, he both plunged into ecclesiastical politics and published at a remarkable tempo, drawing on writings he had been compiling since the 1570s. *De procuranda Indorum salute* had been composed in Latin between 1575 and 1577, three or four years after Acosta had arrived in Peru. The book first appeared in 1588 in Salamanca, published with *De natura Novi Orbis*.[19] The first two books of *De natura Novi Orbis* became the first two books of the *Historia natural y moral de las Indias*, first published in Seville in 1590. *De procuranda Indorum salute* was originally meant to be read after Acosta's introduction to New World nature, *De natura Novi Orbis*, in a structure that showcased how those preaching in indigenous communities needed to understand the natural world as foundational to the differences they encountered. By 1590, *Historia natural y moral* had received all of the required permissions for publication. These included permissions granted by King Philip II, which allowed for printing under ecclesiastical privilege through his realms for ten years; permission from Acquaviva with the sanction of the Society of Jesus; and the approval of Augustine Friar Luis de León on behalf of the Inquisition. In these same years, Acosta also published two religious works in Rome — *De Christo revelato* and *De temporibus nouissimis* — both printed by Giacomo Tornieri in 1590 and intended to aid in the writing of sermons on the deeds of Jesus and the apocalypse.[20] A study of the divine connected both sets of projects, as Acosta wrote that "all natural history is pleasant in itself, and, for he who has a somewhat higher consideration, it is also profitable to praise the author of all nature."[21] To know nature was to seek to know God.

While his standing in the Society of Jesus suffered a formidable blow following the congregation in Rome, Acosta's posthumous reputation as a naturalist bloomed in the centuries after his death. His writings on American nature would be printed and reprinted, and his texts were translated into other European languages, including Italian, Latin, French, Dutch, German, and English.[22]

At once committed to the potential for conversion and insistent on the need to reckon with American natural diversity, Acosta oscillated between nurture and nature, but ultimately landed on the former. As an educator and missionary, Acosta was concerned with how individuals and populations might improve.[23] Human improvement could be cultivated. In *De procuranda Indorum salute*, botanical metaphors describe the conversion efforts in the Indies. Better priestly strategy could yield a greater "harvest of souls," "with patience and effort, the Lord's field yields abundant fruit," and the "great fruit" of conversion led to "hope for fruit in the future."[24] The Jesuits took a relatively flexible approach to diversity in their global Catholic mission, which relied on a vision of the potential to convert all humans by living among them, knowing their customs, and bringing them to Christianity using their own languages. Conversion, Acosta argued, had proven successful in the past, such that while members of the early Church faced challenges, subsequent generations born into Christianity and educated by Christian parents forgot that the communities with the strongest faith had emerged from generations of labor.[25] Christianity had everywhere been a process of education and socialization, and the same efforts would be required to convert the New World. Anthony Pagden has read Acosta as fundamentally an educator, committed to human instruction that could shape the forms of the human mind.[26] The Americas could provide lessons for educating other non-Catholics across Europe, Africa, and China.[27] Drawing on the Jesuit trope of the "Other Indies" that painted remote areas of Europe as inhabited by people bemoaned as backward, Acosta pointed out that uneducated villagers could be taken to schools, courts, or famous cities, where they excelled. Education and new customs could be transformative, thereby proving that nature was not the innate deciding feature of behavior.[28] As Anthony Grafton has argued, "The point of the new history that Acosta wrote was to lay down the foundations for what would genuinely be a new world: a Christian New World, one in which the energy, skill, and practical intelligence of the Indians were turned to new purposes."[29]

Acosta's optimism about the possibilities for conversion was paired with an insistence on understanding the extent of both human and natural difference. He deployed a tripartite hierarchy of "barbarians," classifying peoples by the extent of their attainments in language, religion, and state-craft and elaborating appropriate conversion strategies for each group. Acosta classified the Chinese, with their many books and academies, as being not so distant from right reason and Christianity; the Mexica and Inca, with their empires but without written laws, as subject to conversion through languages of symbols and substituting their ceremonies with Christian ones; and the lowest "savages," who he saw as without laws and without states, "who are close to beasts and in whom there is hardly any human feeling," as requiring conversion by force.[30] Acosta's universalism came with a sting. He saw both humans and the natural world as participating in a hierarchy, which unsurprisingly put Amerindians near the bottom, but still within the field of legitimate study. Acosta likened Indians to rocks in his defense of his larger project, writing that their histories deserved recording, just as "in natural things we see that authors write of not only noble animals, famous plants, and precious stones, but also low animals, common herbs, and rocks, and very ordinary things because in them there are also properties worthy of consideration."[31] Similar questions animated Acosta's discussion of the boundaries between animal kinds and of those between human kinds. As Pagden noted, "Acosta the natural historian was well aware that to understand and to classify men one had to treat their cultural differences with as much care as one would the differences between separate species of plants."[32] While Acosta believed nature to have been created by God, he was very interested in puzzling out exactly how its present state had come into being.

A Great Chain of Being in Peru

Beneath Acosta's vision of nature rested the notion that the universe comprised a Great Chain of Being.[33] Arthur Lovejoy has characterized this idea, maintained from the medieval period to the late eighteenth century, as a chain

> composed of an immense, or—by the strict but seldom rigorously applied logic of the principle of continuity—of an infinite, number of links ranging in hierarchical order from the meagerest kind of existents, which barely escape nonexistence, through "every possible" grade up to the *ens perfectissimum*—or, in a somewhat more orthodox version, to the highest possible kind of creature, between which and the Absolute Being the disparity was assumed to be infinite—every one of them differing from that immediately above and that immediately below it by the "least possible" degree of difference.[34]

Historia natural y moral de las Indias constructed a scaffold rising from the nature of the earth, climbing up the links of the Great Chain of Being, and ending with human morals and histories. For example, in moving from book 3 to book 4, Acosta shifted from what he saw as elements and single entities—water of the rivers, earth of the landmasses, fire of the volcanoes, and air of the winds—to mixtures and compounds, noting that metals, plants, and animals mixed the elements, but fell into a clear hierarchy from inanimacy to greater perfection.[35] Among the infinite links of the chain, however, Acosta held that perfect and imperfect animals fell into distinct categories, defined by their reproduction. As Acosta expounded, "Animals exceed plants in that they have a more perfect being, and they also need nourishment that is more perfect" than simply the food that nature provided them at birth, which he presumed to supply plant sustenance; for this reason, "the same plants are nourishment for animals, and the plants and animals are nourishment for men. The lower order always serves to sustain the superior one, with the less perfect subordinating itself to the more perfect."[36]

Natural things could problematically extend up or down the Great Chain of Being. The human pursuit of gold and silver, never far from the mind of an expert on Spanish Peru, could be condemned both on grounds of avarice and as an unnatural pursuit of something lower on the Great Chain of Being.[37] Order could also be challenged in the other direction. Metals grew in the mountains like roots of plants, and monkeys could be trained to perform antics that looked like human intelligence.[38] Failure to recognize categories could risk moral crimes. Acosta "almost felt scruples" as he wondered whether the manatee he ate in Santo Domingo on a Friday really was a fish. The chops from the shoulder of the cow-sized animal resembled veal in taste and color, a likeness buttressed by his description of the manatees' use of teats to care for their young.[39] Still, these doubts were not enough to change Acosta's classification. He grouped the manatee with other fishes in his description of water in book 3. But even more pressing than the moral problems embedded in dietary rules was the imperative to classify New World nature according to a European system that emerged from Noah's ark.

The Noah's Ark Problem

Much of Acosta's writing about nature in the Americas sought to answer the question where did the people and animals that populated the Americas come from? While they had created so much worthwhile knowledge, ancient authorities had proven deeply unreliable when it came to

predicting the existence of the New World, let alone the kinds of beings that lived there. Acosta began his critique in *Historia natural y moral
de las Indias* by wryly noting that "the Ancients were so far from thinking that this New World was peopled that many of them refused to believe that there was any land in these parts" and that the heavens even
existed there.[40] Acosta dismissed stories of monstrous humanoids, like
the antipodes thought to live in the South of the globe; imagination, he
warned, had the ability to exceed the realities of nature, spinning off stories
from passages of ancient texts to invent all sorts of beings that did not, in
truth, exist.[41]

The Jesuits taught that nature came from one God and, as Acosta put it,
"all men come from one man."[42] Acosta's commitment to Christian scripture required him to see a universal nature, regardless of the analytical
gymnastics required. He was uncommonly plain about this:

> The reason that we are forced to admit that the men of the Indies were from
> Europe or Asia is so as not to contradict sacred scripture, which clearly
> teaches that all men descend from Adam; and thus we cannot give any
> other origin to the men of the Indies. For the same divine scripture also
> tells us that all the beasts and animals on earth perished except those that
> were preserved for the propagation of their kind [*propagacion de su gen-
> ero*] on Noah's ark.[43]

If one held to universalism, there could not be a truly separate Old World
and New World, but only one world.

How did people and animals come to the Americas? Had they traveled
by boat, without compass or lodestone? Probably not, especially in the
cases of the many "useless" animals, such as lions, living in the American
landscape.[44] Did they spring from American soil as a unique population?
Biblically impossible, concluded Acosta, despite indigenous peoples' stories to the contrary.[45] Had they sprung from Atlantis, or descended from
Jews? No and no. Despite the evidentiary challenges of relying only on
ancient texts and in-person observations of living populations to guide his
deductions, Acosta at last concluded that humans and animals must have
crossed by land.[46] The continents were not so far apart, and their points
of connection must have served as a bridge of sorts. Acosta argued that
"the lands of the Indies and Europe and Asia and Africa have a connection
among themselves or at least come very close together at some point."[47]
Thus Acosta concluded that "the Indian lineage has proceeded and multiplied for the most part from savage and fugitive men."[48]

Likewise, animal populations needed to be "reduced to the propagation of all of the said animals that emerged from the ark in the mountains of Ararat where it set down: in this way just as for men, it is also necessary to look for the way that the beasts came, for where they had passed from the Old world to the New."[49] This meant that Acosta needed to explain the origins of perfect creatures, which were defined by their reproduction, or the "natural order of breeding," wherein clean animals like lions, tigers, and wolves could reproduce themselves, but imperfect creatures like frogs, mice, and wasps could be engendered from the earth.[50] Acosta's reliance on this mechanism helped him explain why megafauna seemed to be missing from New World islands but present on the larger landmasses: they must have bred on land and walked, while the imperfect creatures were engendered from the earth or water itself. For this reason, wild animals on islands included fewer birds. Certainly, there were parrots, who could fly long distances, and other birds, but Acosta asserted that there were not as many, and that islands lacked partridges, which were common on the mainland. Likewise, the islands had none of Peru's guanacos, vicuñas, or "sheep of the indies."[51] Still, the presence of some animal kinds in both Europe and the Indies meant that the New World could not be completely divided or separated from the Old World.[52]

To explain the origins of animal populations in the Americas, Acosta supplemented the category of perfect animals with three further divisions: animals brought by the Spaniards, animals that had not been brought to the Indies but were of the same species (*misma especie*) as those in Europe, and animals present in the Indies but not found in Spain.[53] Asses, cattle, cats, dogs, goats, horses, and sheep had all been brought by the Spaniards, Acosta claimed. Sheep had multiplied so abundantly across the available pasturage that people did not even bother owning them.[54] Goats were used for their milk, their leather for shoes, and the fatty tallow from the females that proved cheaper than oil. Spanish horses had become widely available and were regularly used for transportation, racing, and showing; Acosta argued that "there are some breeds [*raças*] that are as good as the best in Castille."[55] To prove that these animals had not been in the Americas before the Europeans' arrival, he turned to both living witnesses and the lack of indigenous names and use of loanwords for these animals.[56]

Spanish domesticates brought to the Americas had the potential to go wild, and thereby to change.[57] Among cattle, there were domestic herds in the Charcas district and other regions of Peru and New Spain used for meat, butter, calves, and plowing, as in Spain. However, there were also those that had changed, that had taken to the mountains, where they roamed. Like the sheep, there were so many that no one bothered to own them or brand them properly, and whoever hunted or killed them first became their owner. In Hispaniola, these cattle were so plentiful that hunters left their flayed corpses to rot, transporting their hides alone back to Spain. Acosta marveled at the scene on the sandy riverbank in Seville where a 1587 fleet unloaded 35,444 hides from Santo Domingo and 64,350 hides valued at 96,532 pesos from New Spain.[58] Acosta also believed that there were no true dogs in the Indies, as he saw European dogs as having likewise slipped during the decades of Spanish settlement, "increasing in number and size such that they became a plague of Hispaniola, eating livestock and roaming in herds." Just as those in Spain who killed wolves won a prize, similar awards were given for hunting wild dogs.[59] Although Acosta does not expand in this direction, the common designation of "de la tierra" or "de las Indias" could be read as an Aristotelian accident, a slight variation from the essence of such creatures.

The second category of perfect animals comprised species found in both Europe and the Americas. In one of Acosta's notable redundancies between book 1 and book 4 of *Historia natural y moral*, and one of the imperfect integrations of *De natura Novi Orbis*, he turns to this argument twice, first in his cosmographic description of the New World and again in his description of the creatures populating it. This second category of perfect animals included bears, bees, boars, deer, foxes, lions, stags, and tigers, as well as birds, like doves, eagles, falcons, hens, herons, parrots, partridges, and pigeons, that were present in both regions, and which might have flown or traveled by water.[60] These animals were both like and different from Old World varieties. The "lions" of Peru were "large and savage," although they lacked the "reddish color of the famous lions of Africa."[61] American tigers were spotted, fiercer, and able to leap farther; the bears were similar, the bees smaller and able to produce less honey, and the deer hornless like European roe deer.[62] Acosta concluded that God would not have simply produced them from the land (that would be against the order of breeding), and that it would have been improbable for them all to swim (they would have needed to have swum for days). Men might have brought them for hunting or for their menageries, as Acosta recognized that princes and lords all over the world had a desire to cage beasts for their own prestige, but why would they have bothered taking "stinking" or "useless" species such as the foxlike *añas* aboard their ships for such long voyages?[63] Acosta subsequently concluded that the animals, like the Indians themselves, provided further evidence for his land-bridge thesis, writing that "it follows that for some part where one hemisphere is near another they penetrated and little by little populated the New World."[64]

The third category of perfect animals, those found only in the Indies, proved the most challenging to explain. If the Creator had produced distinct animals like alpacas, armadillos, chinchillas, condors, guinea pigs, guanacos, hummingbirds, llamas, macaws, peccaries, vicuñas, vizcachas, and thousands of others in the Americas alone, then there would have been no need for an ark, which would be against scripture.[65] Rather, Acosta's insistence on the ark required him to develop a means of explaining animals' agentive migration to different areas, disappearance from other regions, and ability to have variants within broader categories of relation.

Kind, Species, Lineage, Race

The terms *species* and *lineage* broadly indicate different gradations of separation between one living thing and the next. The 1589 Latin edition of *De natura Novi Orbis* organized groups of people through the category of *genus*. For instance, Acosta's chapter 22 was titled "How Indian kinds [*genus*] did not come by way of Atlantis as some believe."[66] *Genus* took on a capacious meaning of "kinds" in the centuries that preceded Linnean taxonomy, which would use *genus* (pl. *genera*) as the category wider than species in a move to linguistic precision in science. In Acosta's world, *genero* and related terms linked ideas of generation and reproduction, preserved a sense of perpetuation; this idea meant that they maintained the potential for either improvement or degeneration, according to the medieval Latin tradition preserved in Giovanni Balbi's *Catholicon*, which was printed in Latin in 1506.[67] In a telling change, the 1590 Castilian version of *Historia natural y moral de las Indias* translates that same passage using *lineage* (*linaje*).[68] Acosta's Latin does employ the term *stirpes* when speaking about the wrath of God against the disobedient stock whose offspring led others to sin.[69] Modern English translations further the conflation of different levels of relatedness.[70]

   In Acosta's 1590 Spanish edition of the text, he considered the muddied divisions between kinds through the example of camelids such as vicuñas. Indeed, the question of likeness became even trickier when it came to bezoar stones, mineral deposits traditionally taken from the stomachs of some goats, which were also extracted from the bellies of vicuñas. Andean camelids were far better at producing this highly desired materia medica than Castilian sheep or goats, which produced no stones at all. In this way, the vicuñas were more like goats of the East Indies.[71] When seeking to clarify the nature of vicuñas, the 1590 Spanish edition held that "nor are they the goats of the East Indies whence comes the bezoar stones, or if they are of that kind [*genero*] they must be a different species [*especies*], just as in the lineage [*linaje*] of dogs the mastiff is a different species [*especies*] from the greyhound."[72] Like early modern Latin, vernacular Castilian in Acosta's era did not draw a tight, scientific distinction around these terms; *genero* and *especies* were structurally ambiguous. The broadest category of difference was "kind" (*genero*), which divided into related but not identical creatures; the human race could be *genero humana*, and one might note different kinds of livestock both as and within species.[73] Acosta considered these categories species (*especies*) or, more broadly, types. Covarrubias's 1611 dictionary would describe *especies* as akin to *genus* in Latin and noted that this term was used in the vernacular, writing that "there are certain species of animals; in his species (*especie*); to change species; return a thing to its type (*especies*); fruit in species."[74] From there, Acosta turned to the category closest to the Italian *razza*, that being the lineage (*linaje*) of dogs, which he presumed the reader to understand through his allusion to well-known dog breeds, mastiffs and greyhounds. Covarrubias defines *linaje* as the "straight line of inheritance" that "descended from fathers to sons to grandchildren, etc."[75] Acosta used the well-known world of human-orchestrated dog breeding as a means of differentiating livestock, East Indian bezoar goats, and Peruvian vicuñas.

Variations among translations of this passage in subsequent editions reveal the subtle vagaries of and debates around variants within types and kinds. The 1596 Italian edition maintained the same sentence structure, employing the terms *genere* and *specie* for larger and smaller categories, respectively.[76] However, this version dropped the category of "lineage," using *genere dei cani* instead of *linaje de perros*. Florio's 1598 Italian-English dictionary defined *génere* as "a kind, manner, or sort of any thing."[77] More direct translations were available, such as *legnaggio*, *schiatta*, or *stirpe*.[78] Perhaps the Spanish discourse focused more on discussions of family lines, especially given the fixation on *limpieza de sangre*. By contrast, the 1598 French text used the phrase *race des chiens* for *linaje de perros* to preserve the Castilian emphasis on selection.[79] A Dutch version published the same year differentiated between *soorte* and *specien*, while the English version from 1604 cut the redundancies and used the term *race*, writing, "For if they be of that kind, it were a divers one, as in the race of dogges, the mastie is divers from the greyhound."[80]

Aristotelian accidental difference provided Acosta with a supplemental tool for understanding how unique animals found only in the Indies could fit into the Noahic narrative of universal nature. Once again revisiting the Noahic argument and reasserting his belief in a land-based connection across continents, Acosta argued that the animals that came out of the ark followed natural instinct and divine providence to make their way to different regions. Those who ventured away from their intended regions "were not conserved, as has happened many times before."[81] However, Acosta supplemented his migration and extinction thesis with a turn toward the Aristotelian distinction between essence and accident. An accident, in this sense, is a property that an entity has contingently, without which the entity could still maintain its identity. Accident did not alter essence. Acosta continued:

> We must also consider whether these animals differ specifically and essentially from all others or whether their difference is accidental; this could be caused by various accidents, as in the lineage [*linaje*] of men some are white and others black, some giants and others dwarfs. Thus, for example, in the lineage [*linaje*] of monkeys, some have tails and others not, and in the family of sheep some are bare and others have fleece; some are large and strong and long-necked like those of Peru, and others are small and weak with short necks like those of Castile.[82]

He ended the chapter on a middle ground between the ark-migration argument and the Aristotelian taxonomic one, noting the absolute requirement to describe New World nature with attention to difference rather than amalgamating all new animal kinds into those familiar in Europe:

> But what is most certainly true is that anyone who, by trying to establish only accidental differences attempts to resolve the propagation of Indies animals and reduce them to those of Europe must take great care because he could be very wrong about the matter. For if we are to judge the species of animals by their properties those of the Indies, they are so diverse, that to reduce them to species [*especies*] known in Europe would be like calling an egg a chestnut.[83]

In the end, Acosta started to imagine that the ark might have carried progenitors who would yield animal families rather than what we would see as animal species or subspecies, and those could give rise to accidentally different creatures as time passed. Even so, there were some animals whose ancestors had to have been on the ark but were not familiar in the Old World.

Like the language of Genesis itself, Acosta's explorations of the population thinking required to explain the human and animal inhabitants of the Americas relied on the terminology of multiplication, procreation, and generation.[84] For him, there remained a *linaje* of Jews and of Incas, as well as a *linaje* of dogs and of monkeys. Breeding need not be bodily, but could also be geographic, born of the earth itself as naturally as "innumerable vicuñas breed [*se crían*] in the Sierra."[85] While he would have rejected the idea that land directly engendered beings, Acosta's descriptions of the properties of the land of Peru focus on the direct correlation between geographic fertility and living bounty, as the "heathy land is the most populated and richest in the Andes, with an abundance of livestock that breed well."[86] By this reasoning, New Spain's superior pasturage thus bred the largest number of horses, cattle, and sheep, while the earth's excessive fertility prevented seeds from growing to maturity on the islands.[87]

Throughout his writings on the Americas, Acosta considered populations on a continental scale, which had been made possible by the passage of time and the migration of peoples and animals.[88] Acosta came to these ideas neither from secular, mathematical theorizations of political economy nor bureaucratic collection of tax revenues, but instead through an effort to reconcile a literalist reading of scripture with the reality of a populated New World. The boundary conditions of the Noahic Flood pushed Acosta to come up with migration theories to explain how perfect animals could have come out of the ark and populated a whole set of continents.

Conversion Meets Its Limits

The final section of this chapter traces how, during the time of Acosta's return to Rome and partially successful political machinations, Jewish and Moorish lineages were being defined as beyond the purview of conversion. During Acosta's time in Peru, the situation for New Christians and their descendants in Spain was deteriorating. Although his writings have been characterized as picking up after the utopian phase of evangelization had ended, Acosta's nature studies and sermonic texts alike sought to support the argument that all humans were ultimately perfectible and capable of salvation.[89] In a remark likely intended to respond to Diego Durán without validating him with an explicit citation, Acosta wrote that "ignorant folk believe that the Indians proceed from the lineage [*linaje*] of Jews."[90] Jews had long been used as an example of the persistence of traits in human character despite relocation.[91] This explanation, linked to customs and clothing, seemed specious to Acosta, who responded by pointing out divergences in the practice of circumcision and reliance on money. Most of all, however, he argued that Jews were defined by the memory of their migration: was it likely that, given that the Jews were so adamant about "preserving their language and antiquity, and so much that all the parts of the world where they live today they differ from the rest, but that only in the Indies they have forgotten their lineage, law, ceremonies, Messiah, and finally all of their Jewishness?"[92]

The issue of conversion and bloodlines continued to preoccupy Acosta after he had returned to Europe and published his American writings. Following Acosta's audience, Pope Clement VIII did indeed order the Fifth General Congregation of the Society of Jesus, the first to be held within a Jesuit superior general's lifetime; it convened from November 3, 1593, to January 18, 1594. The event itself served as a rebuke of Superior General Acquaviva's imperious leadership. Over the previous year, Acquaviva had worked hard to discredit Acosta before the Spanish court. He dispatched Fr. Alonso Sánchez as his envoy to Philip II, where Sánchez repeated allegations that Acosta descended from *converso* stock, raising doubts about his loyalty to the Jesuits' cause and the crown. Meanwhile, Acquaviva himself repeated the charges that Acosta was a New Christian to the pope.[93] Although recent scholars have found no evidence to support these rumors, their circulation alone tainted Acosta's reception while he was in Rome, though not enough for him to either lose his position in the Society or become unwelcome in the Spanish court. Still, his time in Rome must have been fraught, even for a man used to living surrounded by different cultural priorities, as Acosta was exiled from a professed house, alienated from his fellow Jesuits, and treated with increasing doubt by Philip II.[94]

As superior general, Acquaviva presided over the meeting, and Acosta himself attended as King Philip's representative.[95] Among many matters taken up by the congregation was Decree Twenty-One, which renewed the rule that "in order to show deference to the Holy Inquisition, Ours in Spain are to refrain from using the privileges of reading banned books, absolving from heresy, and of claiming exemption from dignities and offices outside the society."[96] In deference to Philip II's search for "faith and peace in his kingdoms," members of the Society were to refrain from exercising their reading privileges unless the Holy Office accorded them permission. All lingering privileges were nullified, such that Society members would be required to obtain them from the Apostolic See afresh.[97] The Society established a further restriction separating itself from the Spanish Holy Offices by decreeing that "under penalty of excommunication Ours are forbidden the ambition of becoming officials of the Inquisition in Spain."[98]

The Fifth Congregation also brought serious changes in the Jesuits' treatment of New Christians.[99] The Spanish concerns about purity of blood entered directly into its new decrees. Decree Fifty-Two put forward that "those who are of the Hebrew or Saracen races are not to be admitted into the Society. If anyone like this has been admitted by error, let him be dismissed after awaiting the Superior General's reply."[100] On December 23, 1593, Acosta joined Francisco Arias in opposing the other sixty-two delegates who voted to adopt this decree. Burgaleta points out that members of the congregation were both eager to please Philip II and the Inquisition and convinced that the *memorialistas* raising trouble in Spain were New Christians. Nonetheless, in passing this decree, they had succumbed to pressure that had been successfully resisted since the Society's foundation, overriding a principle that had been key to Ignatius of Loyola's intentions: he had centered the Society's educational program around the presumption that education, not blood, determined one's faith.[101]

The final language of the decree articulated a clear *limpieza de sangre* argument intended to counter the universalism of the conversion project. The decree began with a caveat about the Society's aims, suggesting that "the ministries of our society are exercised with greater fruit in the general quest for the salvation of souls in proportion to the distance Ours are from those human situations that can prove offensive to others." To reach out to those distant from Christian ways of living was a key goal of a Society that sent representatives to found colleges across the globe from China to the Andean lakes. The decree continued that

> those, however, who are descended from parents who are recent Christians have routinely been in the habit of inflicting a great deal of hindrance and harm on the Society (as has become clear from our daily experience). For this reason, many have earnestly requested a decree on the authority of this present congregation that no one will hereafter be admitted to this Society who is descended of Hebrew or Saracen stock [*Hebraeorum aut Saracenorum genere descendat*].

The decree continued that "those who had been admitted by mistake" should be immediately dismissed from the Society, thereby retrospectively enacting this requirement.[102]

The argument for exclusion embedded in the decree waffled between two mutually incongruous ideas of inheritance. To have been born to New Christian parents was not itself harmful, but the habits of such parents could be dangerous. Perhaps a New Christian raised in a different family would be able to join the Society? No. The congregation then turned to a sweeping use of *stock* as a means of eliding one generation into the next and tying habits to lineage. Some differences, the decree contended, simply ran too deep for the conversion project to alter, continuing that "for even though the Society, for the sake of the common good, wishes to become all things to all men in order to gain for Christ all those it can; however it is not necessary that it recruit its ministers from any and all human kinds [*hominum genere*]."[103] Continuing in this anti-Semitic vein, the decree insisted that there were enough workers who were "very acceptable to other nations" who could be better employed by the Church.[104]

Three formal appeals requested reconsideration. One father asked that the superior general be given the right to grant dispensation to remain in the Society in the case of those of Hebrew or Saracen descent who had entered the Society in good faith.[105] A second asked that the superior general would have the "prerogative of dispensing in certain very rare cases so that a person who is of either of those bloods [*sanguine*] but is of very noble extraction [*genere tamen nobilissimus*] could be admitted if he so requests."[106] A third asked that these impediments be categorized as essential as well as indispensable impediments. After debate, the congregation voted to reject the first two appeals and issued the decree over again.[107] Its language concluded that "doubt has arisen about the force and efficacy of this decree," and "after many arguments had been adduced on either side," the congregation had concluded that not even the superior general could grant dispensation from this decree, "hence it is hereafter to be kept entire and inviolate in the Society."[108] This was one of the cases in which the Spanish pathology, based on Spain's recent conversion experience and its desire to police social boundaries and pander to anti-Semitism, was written onto global Catholic missions based in Rome.

Despite tensions around this exclusionary decree, the Sixth General Congregation in 1608 preserved this requirement, extending it from parents to the "fifth degree of family lineage [*quinque generationibus*]."[109] The decree clarified that "no account would be taken of rumors," instead requiring an intensive collection of documents and witnesses who could speak to the family.[110] Likewise, the later congregation took pains to specify exactly who was considered suspicious. It stipulated that they took "for granted the exclusion of those who in Spain are called Moriscos and of whoever else has been regarded as Christians in appearance only [*qui christiani quasi specie tenus haberentur*]," as well as "those who are held infamous by reason of Hebrew or Saracen descent [*generationis*]." The exclusion was directed at those "who are otherwise of respectable families or are popularly regarded as belonging to the nobility or are of good name." Only if five generations inclusive proved to be free of such background would there be no impediment to joining the Society. The language of the decree hints that both nefarious scuttlebutt about New Christians and an exact definition of Hebrew and Saracen remained persistent concerns. The 1608 decree noted that "uncertain rumors or reports spread among Ours which could be injurious to superiors or other outstanding workers" needed to be avoided. Not only should "the inquiry not be pursued further" if one had five generations of Christians in one's family, but members of the Society "should not engage in conversation about these matters" once the issue had been decided. Likewise, the Sixth Congregation clarified that lineage did not have the same meaning across different geographic contexts. The "Tartars and other Mohammedans, whether in Poland or elsewhere," were to be considered specially, and for them, the "Reverend Father General is entrusted the power to dispense with the aforesaid decrees of relationship, since among them the concept of infamy and indecency does not obtain."[111] Concerns about purity of blood were thereby limited to the particular politics of the Spanish world; indeed, the Spanish crown would expel the Moriscos in the following year.

For all his disagreement with the congregation, Acosta remained a Jesuit and ended his career interpreting the Psalms as a rector at a Jesuit college in a Spanish university town.[112] The debate, however, on nature and nurture, blood and inheritance, animals and taxonomy remained far from resolved outside of the narrow ambit of Jesuit rules. The sweeping challenges of global conversion had pushed Acosta to think broadly in terms of populations. Meanwhile, another widely read scholar of nature, Giovanni Battista della Porta, faced with a different context, looked to individuals and their animal attributes to think about the predictive nature of external appearance.

CHAPTER 8

Seeing Inside from the Outside

GIOVANNI BATTISTA DELLA PORTA

From Natural Magic to Physiognomy

On April 9, 1592, Cardinal Giulio Antonio Santori (active 1570–1602), head of the Congregation of the Inquisition, sent Giovanni Battista della Porta a clear message: Do not publish your "book of physiognomy [*fision-omia*] in the common language, nor any other books, without the express permission of the Holy Tribunal of Rome."[1] Having just been passed over for the papacy in favor of Ippolito Aldobrandini (Clement VIII), despite being the Spanish favorite and the precedent that the position of inquis-itor often served as a stepping-stone to the papacy, Santori was not the sort of Catholic official with whom one would want to trifle.[2] In the days following Santori's message to Porta, a bevy of prohibitions sought to fur-ther restrict the circulation of Porta's ideas: the Inquisition of Venice pro-hibited the vernacular publication of *Fisionomia*, and the printer Barezzi was barred from publishing it as well. By April 20, the Roman Inquisition had gathered together a number of cardinals to reopen their conversation about Porta; they sentenced him to *purgatio canonica*, as there were lin-gering concerns about his possession of books on necromancy.[3] Closer to home, the archbishop of Naples, Annibale di Capua, repeated Santori's warning: Keep the book unpublished to avoid a fine of 500 golden duc-ats and censure from the ecclesiastics.[4] His youthful forays into natural magic had raised hackles, but Porta's new monograph—which claimed to teach the reader how to read the book of nature—arguably posed an even greater threat to religious orthodoxy.

This chapter follows another approach to natural difference that was circulating in Southern Europe at the same time José de Acosta was pub-lishing his writings on American nature and conversion. By beginning in 1592 in Rome, I aim to show some of the parallel institutions that shaped Porta's publication efforts, which included patronage from Church of-ficials, the role of the Inquisition, and imperfect overlaps between the

Spanish and Italian worlds. This juxtaposition highlights the multiplicity of approaches to the tangle of issues related to breeding, population, and taxonomy in European thought without imposing one discourse on them all. While Acosta thought across wider populations to construct a universal natural history, Porta focused on individuation, the challenge of explaining why particular people behaved the way they did, and how that related to their physicality.

If the first half of Porta's career focused on natural magic and nature's malleability, then the second part (1580–1615) included a focus on physiognomic vision. By 1558, the first version of his magnum opus, *Magia naturalis* (Natural Magic) had appeared, as discussed in chapter 1; it was eventually issued in twenty Latin editions and was translated into Dutch, English, French, German, and Italian.[5] Porta's commitment to natural magic did not immediately provoke concerns from either the Spanish or the Roman Inquisition. Sometime in the 1560s, he established the Accademia Secretorum Naturae, an academy that likely brought together Porta's teachers, friends, and colleagues who were experts in the classics, medicine, and alchemy to distill further secrets from nature. However, as Porta neared forty in 1574, Cardinal Scipione Rebiba, the man who had introduced the Inquisition to Naples, wrote to the archbishop of Naples to ascertain Porta's whereabouts so that he could be arrested and sent to Rome. That same year, Neapolitan authorities closed the academy.

Rather than turning away from controversial topics at the boundary between the natural and the preternatural that risked raising Church hackles, Porta forged ahead, with a particular focus on the art of reading patterns in nature. In his early fifties, the magus unleashed a stream of new natural treatises, including *On Human Physiognomy* (1586), a revised version of *Magia naturalis* (1588), *Phytognomonica* (1588), and *Villae* (1592).[6] Porta now had a new bastion of social protectors in Rome. He sought patronage from the influential Cardinal Luigi d'Este, to whom Porta dedicated *De humana physiognomonia* in 1586; the duke of Monteleone wrote to the secretary of the Congregation of the Index on Porta's behalf in 1593; and Porta would later allow himself to be collected by the young Prince Federico Cesi (1585–1630).[7]

The power of physiognomy to suggest general likelihoods in nature and its propensity for linkages—between plant, animal, and human; between medicine, philosophy, and natural observation—captured Renaissance readers' imaginations and bore unintended consequences. Bodies reflected their souls, Porta contended, therefore one could read the essence of something from the exterior alone.[8] Porta's relatively late vision of physiognomy, which returned to the ancient treatise *Physiognomics*, played on interests that had been growing throughout medieval Europe. It wove together two epistemological threads: humanistic knowledge (found in publications and libraries, and curtailed by the Inquisition) and artisanal knowledge (cultivated as the magus researched plants and animals in villas, menageries, and personal collections). With greater exposure to natural diversity came a more nuanced vision of how external characteristics revealed internal truths in plants, animals, and humans, which in turn undermined the divide across these types of beings.[9] In embracing this vision, Porta participated in a medieval tradition that had Greek and Arabic textual precedents, which had been recodified in handwritten manuscripts throughout the medieval period and then in printed volumes held in the libraries of physicians, surgeons, theologians, and lawyers. The core of this chapter provides a close reading of sections of *On Human Physiognomy*, its frontispiece, and the rendering of the leopard to consider what it meant "to see like a lynx." Through Porta's growing alliance with Cesi, the magus joined a new scientific academy, the Accademia dei Lincei (Academy of the Lynx-Eyed), as a senior member of the society that would eventually bring the Mexican Treasury, that compilation of New World nature and medicine mentioned in chapter 5, to publication. The Lynx-Eyed took up their name from one of Porta's own mantras, *aspicit et inspicit*—"he that sees the outside sees the inside"—a phrase that captured Porta's observational practice and analytical method, as exemplified in his physiognomic publications in the late sixteenth century.[10]

Bolstered by his reputation following the publication of *Natural Magic*, Porta brought physiognomy to the center of European intellectual discourse. More than other theorists of physiognomy, it was Porta whose intellectual framework and reputation spread beyond Italy and into the Iberian world. The definition of *physiognomy* in Covarrubias's 1611 dictionary of the Castilian language included a direct mention of Porta. First, Covarrubias defined physiognomy as "a certain conjectural art by which the conditions and qualities of the man are pointed out through consideration of his body, and particularly the signs of his face and head." After citing ancient authors' descriptions of the craft, Covarrubias continued, "In our times the one who has best dealt with this matter was Iuan Bautista Porta Napolitano, who wrote four books on human physiognomy with great acuity. This author made comparisons between the figure of man and the animals, applying their qualities and conditions to them: it is a pleasure to read."[11] Porta's ideas spread beyond the learned readers of dictionaries to become popular knowledge in Iberian discourse. Ruth Hill has traced their lingering presence in ideas concerning race and caste that developed in the Spanish Americas.[12] Porta's mastery as a magus would be remembered in the Spanish theater, too, as he was woven into the plot of Pedro Calderón de la Barca's *El astrologo fingido* (1618). Diego, the play's eponymous fake astrologer, tells of arriving in Naples to find Porta, whose world was filled with planets, signs, and astrolabes, all obedient to his commanding knowledge. The fictional character stayed with Porta, learning about the stars and nature's secrets for years until he surpassed the master in controlling the secrets of nature, with ever more power over the world of signs.[13] Through the success of Porta's publications, knowledge created across Roman villas and Southern Italian stables, laboratories, and menageries became commonplace in the wider Spanish empire.

Mirrors of the Soul

Deep, textual awareness undergirded Porta's study of a theory of "mirrors of the soul," which was already prominent in Aristotle's time (384–322 BCE) as a means of creating logic and clear boundaries within the often messy human and nonhuman natural world. By the Renaissance, the discipline's foundational text, *Physiognomics*, had found its way into humanists' libraries by way of a false attribution to Aristotle.[14] This ancient treatise was based on the idea that the mind and body were not separate, but deeply intertwined. As *Physiognomics* put it, physiognomy traced how the "fundamental connection of body and soul and their very extensive interaction may be found in the normal products of nature."[15] Because the body and soul were so inextricably linked, each type of body contained certain characteristics. The treatise linked physiognomic theory to artisanal practice, explaining that, by reading these signs in the natural world, "experts on animals are always able to judge character by bodily form: it is thus that a horseman chooses his horse or a sportsman his dogs."[16] Rather than suggesting that they were independent of one another, the ancient text argued that souls and bodies were "affected sympathetically by one another: on the one hand, an alteration of the state of the soul produces an alteration in the form of the body, and inversely an alteration in bodily form produces an alteration in the state of the soul."[17] Even as discourses of the soul became deeply intertwined with Christian doctrine, this competing theory insisted that souls resided inside nonhuman animal bodies as well as human forms. In short, the ancient theory of physiognomy sought to predict behavior by looking at the exterior casing that grew around the interior *anima*. The rhetorical figure of *ethymeme* prevailed across physiognomic treatises, linking the animal likeness to the human figure: the lion resembles the man; the lion is brave and virtuous; thus, the man is brave and virtuous.[18]

Pseudo-Aristotle contended that female domesticated animals all shared similar characteristics: they were tamer, gentler, less powerful, and more manageable than males. He then connected these animals in behavior and physical appearance to women, writing that "everyone can see this is so in women and in domesticated animals, and according to the unanimous evidence of herdsmen and hunters it is no less true of the beasts in the field."[19] Untroubled by the problem of the male leopard, Pseudo-Aristotle saw inherently female behaviors as inseparable from the leopard's body and proceeded to offer a second gender theory difficult to reconcile with the ostensibly common characteristics of all females, stating that leopards as a species (apparently male and female) exhibited characteristics inherent in human females:

> The leopard, on the other hand, of all animals accounted brave, approximates more closely to the feminine type, save in its legs, which it uses to perform any feat of strength. For its face is small, its mouth large, its eyes small and white, set in a hollow, but rather flat in themselves: its forehead is too long and tends to be curved rather than flat near the ears; its neck too long and thin: its chest narrow and its back long: haunches and thighs fleshy: flanks and abdomen rather flat: its color spotted; its whole body ill-articulated and ill-proportioned. Such is its bodily aspect, and in soul it is mean and thievish, in a word, a beast of low cunning.[20]

All leopards shared similar characteristics, like the shape of their heads and spines and general musculature. Slinking and cunning, sneaky and ready to pounce, to look at a leopard was to look at some of the fundamental characteristics of women.

Like so much classical knowledge, physiognomic thinking proliferated in Greek and Latin, and was considered anew by medieval Arabic scholars engaging with multiple classical heritages, as Martin Porter has shown. The Pseudo-Aristotelian text became the Arabic tradition of the *Secretum secretorum* (Arabic: كتاب سر الأسرار or "Book of Secrets"). In translated fragments, these ancient ideas reemerged in High Medieval Europe, where Mediterranean courtly learning fostered new physiognomic inquiries by Michael Scot (1175–ca. 1232) and John of Seville (1133–1153). From Pamplona to Padua, physiognomy slipped into medical learning, both in books explicitly devoted to the subject and via excerpts in surgical treatises and medical texts. In Padua, Italian philosopher, astrologer, and professor of medicine Pietro d'Abano (1257–1316) turned to physiognomy as a means of inquiring into the occult sciences and nature's secrets.[21] Also in Padua, physician and professor of natural philosophy Alessandro Achillini (1461/3–1512) argued that the face portrayed an image of the soul. Prominent readers of physiognomic theories included medical professionals, especially surgeons and physicians, in epicenters of scholarship from Toledo to Bologna, Padua to Valladolid.

Those who ventured into questions of physiognomy inevitably found themselves thinking about free will and determinism. The wide-ranging author and theologian Miguel de Medina (1489–1578), who represented the University of Alcalá at the Council of Trent, warned about the limits of prophesy through physiognomy and chiromancy, or the art of judging character by looking at the hands. He argued that these techniques allowed prophesies about the universal at the level of a kind (*genere*) and, at most, a species or subtype. When speaking of generalities and large groups, physiognomy and chiromancy could be tools to predict likely outcomes.[22] However, Medina argued that it would be unwise, indeed problematic, to use such ideas to predict the choices one individual was destined to make. Likewise, Juan de Horzco y Covarrubias (1540–1610) tried to disentangle physiognomy's implications for determinism and free will through a new focus on shared matter in living bodies and the predictable nature of that matter. Recalling the biblical assertion of dust to dust,[23] Horzco y Covarrubias argued that man has a body formed of elemental clay, and in that clay are found diverse temperaments. No wonder, then, that nature gives signs predictive of behavior and character, some of which are visible to the trained observer.[24] Medical scholars, including the physician Guglielmo Grataroli (1516–1568), who had studied in Padua, puzzled out physiognomy and its relationship to free will.[25] Astrologers taught that movements of the heavens could be linked to the reading of human bodies. The stars, too, informed physiognomic interpretation, complete with the seven different natures of the planets and their links to conditions of the body.[26]

Artists also drew on these connections between human nature and animal character. Physiognomic likenesses feature prominently in Titian and his workshop's *Allegory of Prudence*, completed between 1550 and 1565, which depicted the three ages of man as linked to animals. Under a Latin inscription urging one to act prudently in the present are three faces of a man. Beneath the youthful visage in profile appears a dog, likewise in profile. Beneath the mature man appears a maned lion in the bloom of health. Finally, beneath the aged man is a grizzled hound. In the hierarchy of past, present, and future, the lion of the present portended the greatest power, which could slip away to a lesser carnivore with age and mistaken choices. Recently, art historians have connected the interpretation of dogs in this work to Titian's 1529 portraits of Federico II Gonzaga with a loyal Maltese and of Charles V in 1533 with a large, collared hunting dog.[27] To physiognomists, a decisive criterion (the soul and reason) separated humanity from nonhuman animals, but animals in nature were laden with symbolic indicators with implications for the humans that could be likened to them. The Gonzaga had their loyal lapdogs, the emperor his hounds of war.

Publishing Physiognomy, Banning Physiognomy

On a prima facie basis, physiognomy seemed useful but dangerously predictive, especially given that free will and determinism were especially sensitive subjects after the Council of Trent (1545–1563). Porta's own career straddled this shift in ecclesiastical history and the Church's approach to magic, as Neil Tarrant and Michaela Valente have shown.[28] Porta's early magnum opus *Natural Magic* was at first resoundingly successful, circulating in many vernacular languages and Latin editions, although it was published just as censorship was becoming a priority of the Catholic Church.[29] However, the book walked a fine line in defining natural magic. As he would do again later in the case of physiognomy, Porta argued that magic could be either bad or good: wicked and revealed through the power of demons, or natural, a "knowledge of natural things and a perfect philosophy" which "all revere."[30] Meanwhile, indexes of prohibited books were published, proclaimed from pulpits, and dispatched to booksellers in France (1544); Louvain, Spain, Portugal, and Venice (1554); Florence (1555); and Rome (1559). The Roman index wholly forbade the circulation of texts by anonymous authors, heretical authors like Martin Luther, and certain books by otherwise acceptable authors; Porta's early work fell into this last category.[31] His scholarship and reputation piqued the interest of inquisitors following the Decretal of the Holy Office in 1577–78. Porta was arrested, and his trial opened in October 1577. Although the exact motivation for the investigation is not known, Tarrant has argued that the issue likely centered on "the orthodoxy of his natural magic activities to determine whether they entailed contact with demons."[32] In November 1578, Porta's trial concluded with a sentence of *purgatio canonica*, meaning that the magus had a reputation as a heretic, but it was not possible to prove the crime from facts, witness depositions, or confessions. By 1581, *Natural Magic* was included on an index of books whose sale was prohibited in Rome and, by 1583, on the Spanish index.[33] Even as he ventured into physiognomy, Porta and his allies continued to contend that their methods were natural, not preternatural. A 1586 manuscript copy of *On Human Physiognomy* written in vernacular Italian, now held in the Biblioteca Nazionale di Napoli, is titled "Book of *Natural* Physiognomy" (italics added for emphasis) in hopes of clarifying the method, just as Porta had done with the title of *Natural Magic* nearly three decades before.[34]

Throughout the sixteenth and seventeenth centuries, agents of the Catholic Church, by way of councils, inquisitions, and pastoral reforms, worked as "a network of actors with distinct goals and motivations for controlling thought and behavior," as Hannah Marcus has pointed out.[35] To some degree, there were territorial and jurisdictional overlaps of the reach of the Spanish, Roman, and Neapolitan Inquisitions, but policies are often best explained by the proclivities of individual inquisitors like Santori.[36] On one hand, written precedent was often preserved; for example, some sections of the officially adopted 1607 Roman expurgations reproduced the Spanish 1584 index verbatim. On the other hand, indexes from Antwerp, Spain, and Rome differed, reflecting local concerns perpetuated by particular censors.[37]

Spanish authorities focused on forbidding prediction, as Folke Gernert has demonstrated. The Quiroga Index of 1583 listed physiognomic texts among those banned, and forbade Porta's *Natural Magic* in particular. Alongside the expurgation of medical authors such as Conrad Gesner, Leonhart Fuchs, and Paracelsus, the ninth rule of the Quiroga Index forbade texts that provided rules for knowing the future by looking at the stars or through aspects of the human form, whether lines on the hand or the face.[38] The censors were less concerned about the inaccuracy of such predictions than about the ethics of knowing the future; the future might be known, but ought not to be, realizing it was up to human free will.[39] Censorship—by officials forbidding publication or by readers wielding black ink and knives—was an imperfect, highly political business that sought to be more comprehensive in its elimination of dangerous ideas than it ultimately succeeded in being.[40] Further inconsistencies emerged in exactly what ought to be censored in physiognomic books. In the case of a censored 1615 copy of *On Human Physiognomy* once owned by the Capuchin Order of Lucignano in the province of Arezzo, someone took the trouble to cross out all of the genitalia, including those of the statue of Hercules from the prince of Parma's collection in Rome, rather than worrying about the passages that might infringe on freedom of action.[41]

Whether such texts were published or not, physiognomic thinking slipped out. In the case of Porta's work, scholars like Giovanni Antonio Magini, a mathematics professor in Bologna, wrote to the magus to ask whether it would be possible to see manuscripts—thus presuming that printing such materials under inquisitorial sanction would prove difficult, yet unafraid to ask for the content nevertheless.[42] More broadly, texts on physiognomic ideas were an important part of the libraries of many learned people even as inquisitions across the Catholic world tried to prohibit their circulation, as Gernert has demonstrated.[43] To predict the character of one who walked through their doors—whether he or she was heathy, ill, or charged with a crime—was a useful skill. In 1557, the Jesuit and bishop of Calahorra Juan Bernal Díaz de Luco owned a chiromancy manual by Ioanni ab Indagine. In 1573, the library of the Marquess of Astorga contained a French translation of Tricasio's chiromancy manual as well as a Spanish physiognomy manual published in Zaragoza in 1567. Three years later, a scholar of the liberal arts in Salamanca, Bartolomé Barrientos, was accused of having books on chiromancy and other secrets. The inventory of the famous humanist Benito Arias Montano in 1598 likewise included a book of physiognomy, as well as a text on chiromancy by Diondro. In 1618, Francisco Martínez Polo, a professor of medicine in Valladolid, owned a whole series of physiognomic books, including Porta's book of physiognomy and his book of secrets (presumably *Natural Magic*). Banned or not, physiognomic thinking had stolen into libraries, even in the form of books with the forbidden method inscribed in the title.[44]

Given the extent to which personal politics determined whether one could circumvent the negative attentions of the Church, Porta's meeting with Cardinal Luigi d'Este (1538–1586) opened a world of possibility. Luigi was the second of five children of Ercole II d'Este, duke of Ferrara, and Renée of France. (It was to Renée that Isabella d'Este hoped to send the offspring of her "race of little dwarves," as discussed in chapter 3.) Indeed, Porta's intellectual interests included subjects that had long fascinated the Este, including both breeding and physiognomy. In the previous century, Michele Savonarola (1384–1464) had written his own physiognomic text as the Este court doctor after holding a professorship at Padua.[45] In an environment fixated on competitive performance and experimentation, two threads in the tapestry of race—selective breeding and physiognomy—were woven together. Like many influential patrons-to-be, Luigi d'Este had received works dedicated to him from a young age, from books on ancient Roman hunting practices to epic poems, laying the groundwork for his future role as an influencer of both intellectual court cultures and the Roman Church. Known for his integrity and precocity, and protected by his powerful family, Luigi became a cardinal under Pope Julius III (r. 1550–1555) in 1554, and took up the full responsibilities of his office at twenty-five.[46] Later in his career, Luigi divided his time between Rome and the Villa d'Este in Tivoli, a pleasure palace replete with fountains, gardens, and grottos. It was into this relationship of villas and ostentatious performance that in November 1579, Teodosio Panizza, the cardinal's doctor, drew Porta with a letter of invitation to join the Este household in Rome.

Porta accepted the invitation, establishing a patronage relationship that allowed him to focus on his research and investigations.[47]

The magus was in residence at Villa d'Este in Tivoli from January to July of 1580.[48] Remaining correspondence suggests that he was eager to set to work. While the majordomo of the villa complained about his need for complete and total silence and his strange working habits, Panizza reported that Porta remained obsessively focused on physiognomy. According to a letter that Panizza wrote to the cardinal on January 21, 1580, shortly after Porta's arrival at Tivoli, the magus appeared to be far along with the work that eventually reached publication six years later. Panizza wrote that Porta was "concerned with the Book of Physiognomy and he found a painter that would serve in coloring in the book the types of the eyes, that such a book he would want to be transcribed in a beautiful hand to make it as easy as possible to read, but he could not do this for not having the required money."[49] Illustrations were key to Porta's method, and he needed patronage to secure expert assistance.[50] The publishing process would prove to be much longer and more complex than he had hoped, given his blooming relationship with such a senior Church official. Long after Porta left Tivoli, the unpublished book of physiognomy remained central to their correspondence. On May 14, 1583, Porta wrote to Cardinal Luigi d'Este announcing that "the book on Physiognomy is completed though it wants more time to have the license to print it."[51] Following this, the Inquisition held the text for three long years. The imprimatur was granted on June 18, 1586, by Onorio de Porta, the vicar general of the Neapolitan diocese, and Paolo Regio, the bishop of Vico Equense, the site of one of Porta's family villas. When it was finally sanctioned, Porta dedicated the text to Luigi d'Este. He then turned to a local printer in Vico Equense, Giuseppe Cacchi, to create the final text.[52] While his correspondence indicates that he had been willing to circulate nature's secrets in the form of manuscripts to friends and patrons, Porta's physiognomic theories were about to populate library shelves far beyond Southern Italy.[53]

Learning from the Villa, Museum, and Menagerie

Over the last thirty years, scholars have revealed the intimate relationship between sites of knowledge production and the experimental and observational insights that distinguished the new sciences of the early modern era. As we have seen in the case of the Villa d'Este outside of Rome, Porta sought out research spaces like Bacon's vision of Salomon's house, including his own family villas in Southern Italy. Building on the scholar-

ship on the magus's scientific performances in the stillroom, I turn instead
to other venues of experiential knowledge-making where he encountered
living nature as he considered questions of embodiment and shared attributes of plant, nonhuman animal, and human life.[54] Across these experimental spaces, Porta acquired observational knowledge about animals and
plants, which in turn fueled his physiognomic project and overall *venatio*,
or hunt for the secrets of nature. He consorted with stable masters, gardeners, and other artisans and craftsmen who worked with their hands. As
he did so, the stables, menageries, villas, and museums became "trading
zones" wherein Porta learned not only from the collections themselves,
but from the experts working at those sites.[55] Historians have shown the
presence of craft workers' expertise at the roots of alchemy, perspective,
instrument-making, and more. Porta, likewise, sought out expertise in
complex practical tasks that required handiwork, and those handworkers,
in turn, influenced his scientific practice directly and indirectly, as chapter 1 discussed in relation to his notes from breeders. Because of this cross-
pollination, as Pamela Long has argued, binaries between artisan and
scholar, handworker and theorist, and even art and nature had little meaning, given the fluid overlaps that undergirded Porta's experimentation and
theoretical work.[56]

Through the 1580s, both the villa as an idea and the specific villas in
which he spent his time played key roles in Porta's studies of nature. The
villa, a civil outpost in a rural environment, provided both space and inspiration for the study and improvement of nature. There, the gentleman
might experience the beauty of human interactions with the natural world.
Like the Romans, away from the *negotium* (business and work travails) of
the urban environment, Renaissance husbandmen found a refuge of *otium*
(pleasure or retirement from the world of business) there.[57] In the mid-
sixteenth century, Alberto Lollio (1508–1568), an author who worked for
thirty years in the Este capital of Ferrara, wrote that the villa was the perfect place to observe natural phenomena, including everything from the
"splendor of comets and the milk of the heavens" to "the variety of fruits,
the sentiments of animals, and the nature of fish."[58] For Porta, his family's
Villa delle Pradelle in the hills of Naples was both a place of respite and a
site for experimentation and observation that influenced his writings on
natural magic, plants, and animals. When he sought the pleasure of the
countryside along the Bay of Naples, he traveled out to Vico Equense, a
short distance from the city, where the hillsides grew olives, lemons, figs,
and even the occasional pear and plum.[59] Here, space provided room for
experimentation. There are archaeological hints of subterranean meeting rooms used in the early modern period beneath the Vico Equense

property; perhaps it was there that the Accademia Secretorum Naturae—Porta's scientific society to which one could earn entry only with the discovery of a previously unobserved natural fact—met and compared their understandings of Renaissance nature, nearing the status of magi with every discovery.[60] In such pleasant surroundings, Porta penned his comedies and studies of nature and later invited close friends like Cesi, founder of the Accademia dei Lincei, to visit him.[61]

Characteristically, Porta did not just retreat to the villa, but also wrote an entire book—*Villae*—about his experience with plants there.[62] In 1583 in Naples, he had published *Suae Villae*, which included an extensive discussion of orchards (*Pomarium*), followed by a book on olives.[63] In these texts, Porta described the plants of Campania, first as detailed by ancient scholars—naturalists such as Aristotle, Pliny, and Aelian; medicinal experts like Theophrastus and Dioscorides; and agriculturalists like Cato, Columella, Palladius, and Varro—then observed the varieties found in Naples. Porta noted that "by continuous inspection of all kinds of fig trees of our Campania, and by debate, reading old books, and comparison, we tried to apply common names, and we hope that our attempts not be rejected."[64] He laid the groundwork for *Phytognomonica* by tracing the peculiar shapes of plants in relation to their nutritive or medicinal properties. These included "ugly good pears," "badly dressed pears," pears that look like a woman's leg, and *pero paccone bastardo*.[65] From there, he started to sort the plants according to analogies between their forms and those of animals. One plum, called *pruna cogliopecore*, for instance, "surpass[es] all in size, with a golden color, and soft texture ripens in July and August," but he deemed it "tasteless and imitative of the shape of the testicles of he-goats."[66] Porta would next draw an analogy between the external features of the plants and their hidden qualities. Whether human faces or medicinal plants, features decoded meaning and utility.

Signs provided the key to knowing nature's inclinations before they were completely borne out. For example, in *On Human Physiognomy*, Porta turned toward Virgil's description of the proper bovine choice for breeding. Like the ancient theories of Pseudo-Aristotle, Virgil's *Georgics* taught readers good agriculture. Porta wrote that "Virgil still teaches signs by which the plowman ought to know the successful ox" before quoting Virgil's ode to the ugly-headed cow.[67] For good horses, he also turned to Virgil. In his chapter "Signs of the body through which one can know the inclinations of the other animals," Porta quoted from Virgil's description of the preferred horses, "the first yon high-bred colt afield, his lofty step, his limbs' elastic tread: Dauntless he leads the herd, still first to try the threatening flood, or brave the unknown bridge, by no vain noise affrighted;

lofty-necked, with clean-cut head, short belly, and stout back; his sprightly breast exuberant with brawn. Chestnut and grey are good; the worst-hued white and sorrel."[68] It is worth noting that despite its ubiquitous use in records of horse and mule breeding in the Neapolitan archives, Porta used *razza* only on occasion. He wrote of matching good pigs to *far razza*, or breed one's own line of swine.[69] To *far razza* required the same skills as to grow a garden: "These same signs match the vines with long hops and good seeds, so good breed [*razza*] comes from good country."[70] *Razza* was indicated by signs, and those signs, the technology of breeding, and appropriate location were all important.

Just as his experience of the villa influenced Porta's approach to physiognomy, social and family dynamics shaped his knowledge of nature. As sons of Nardo Antonio della Porta, a Neapolitan noble, Giovanni Battista della Porta and his two brothers—Giovan Vincenzo, the eldest, and Giovan Ferrante, the youngest—learned, like other young men of their class, to appreciate music and dance, and to ride horses.[71] Porta's early education had focused on music and mathematics, laying the groundwork for his later theorizations on cosmic harmony across nature, as Nicola Baldoni has shown.[72] In addition to their studies of the classics with tutors, such as the influential instructor Giovan Abioso da Bagnola, the young Portas each developed an interest in collecting human and natural works of art. Giovan Vincenzo collected busts, marbles, and coins and cultivated an expertise in antiquity, eventually manifested in his own museum; Giovan Ferrante collected crystals and geological wonders.[73] After years of observing his family members' collections, Porta relied on them as evidence for his ideas.[74]

For the illustrations of his writings on natural physiognomy, meant to show the correlation between animals and humans, Porta turned to art from his relatives' collections. Giovan Vincenzo's museum offered particularly essential material to observe and about which to theorize physiognomically. With sketches of owls, chameleons, leopards, lions, dogs, and wolves in hand, Porta could compare animals drawn from life with the busts, medals, coins, and marble statues his brother collected. He recorded these comparisons in his vernacular manuscript *Libro di fisionomia naturale*: "Here is shown the portrait of the head of the dog near the head of [Cato], which is found [in] the Museum of Giovan Vincenzo della Porta, learned brother and diligent investigator and conservator of antique medals" (fig. 8.1).[75] The image of Socrates, drawn from his brother's museum, was also supposed to go into the book; it was a "statue to demonstrate his deformed head and curly hair."[76] Porta used these images to ensure that his readers understood the link between head shape, animal, and underlying character. He encouraged the reader to "recognize in the

FIGURE 8.1. A canine-human comparison drawn from the museum of Giovan Vincenzo della Porta. Printed in Giovan Battista della Porta, *De humana physiognomonia*, libri 4 (Vici Aequensis: apud Josephum Cacchium, 1586), 30. Wellcome Collection. Public Domain Mark.

table the effigy of Caligula Emperor of the large brow, as we have it in the portrait of the medal of Giovan Vincenzo della Porta, brother" and think about what that pronounced brow might mean for the character of others with a similar head shape.[77] But Giovanni Battista did not confine his use of his family's collections to those of his brothers. An "owl's head" was intended to appear alongside the "great head of Emperor Vitellius, as we have depicted from the marble statue in the study of Hadrian Sparafora, my learned uncle, and most studious conservator of ancient memories."[78] A figure modeled on a marble statue by the famous ancient sculptor Praxiteles (fourth century BCE) also came from Porta's uncle's museum; Uncle Sparafora was "most curious about antiquity."[79] Unconcerned by the extent to which the artistic representations accurately captured the ancient emperors, Porta took for granted that their likenesses, as well as their famous character attributes, were faithful.

   Both well-known ancient faces and famous animals—who might have hopped out of the pages of bestiaries—illustrated the signs that Porta wanted to highlight. To conduct his analysis, he tried to find a live specimen of the animal or relied on a portrait drawn from life. Live specimens could be quite close at hand. For a chameleon, he noted in the manuscript that the accompanying image would be "the true figure of the chameleon, which I keep alive at home, rendered from life."[80] Porta's arguments about male and female human physiognomy relied on animals that came to Naples. Like Pseudo-Aristotle, he explained male and female physiognomy through lions and leopards. Porta had found a Florentine portrait of a leopard insufficient for his study of physiognomy. His quandary and its resolution appear in the manuscript of the *Book of Natural Physiognomy*, in the space where the leopard's picture would appear in the print edition: "The graphic we have been sent by courier from Florence with the portrait of the leopard was not satisfactory. We have made a new portrait from life here in Naples created in 1584."[81] Spotted predators—whether leopards or cheetahs, conflated under variations of the term *leopardus* or *gato pardo*—were an important part of any good menagerie, although Renaissance Italians did not clearly distinguish between their species.[82] Incidentally, by 1586, at least two years after the Neapolitan *liopardo* provided its visage for Porta's text, Renaissance Italy was so teeming with leopards that it was possible for some Renaissance lords to have too many.[83] In his manuscript, Porta recorded that the leopard portrait was necessary because "here we find the resemblance between this [creature] and woman."[84] As he believed Pseudo-Aristotle's treatise to be the work of true Aristotle, his corresponding text relied on its analysis and organization, often adopting full passages verbatim, just as in the leopard's relationship to the human woman.[85] In the middle of his discussion of the leopard, Porta inserted his debate with acclaimed naturalist Conrad Gesner (1516–1565) about the proper interpretation of the Greek text, which mentioned the leopard's rounded ears and long face.[86] Even as Porta insisted that the leopard be drawn from life, his humanistic text was more focused on careful translation of Greek than it was on an examination of the leopard itself.

As the eagle was to man and the partridge to woman, the viper to woman and the snake to man, Porta saw the leopard as the encapsulation of the human female, just as brave men resembled lions in other long-standing classical corollaries. Porta also required a lion to provide the visual basis for brave, masculine features. This time he had the privilege of comparing a few. His manuscript indicates personal experience: "We have depicted in the shape below, the form of the lion, which we have drawn from life from several brought here to Naples so that comfortably I might look at its parts with that of the man."[87] If the human male was supposed to have large eyes, a large body, a thick neck, and a great spine, then so too should the lion.[88] Leonine features led to a magnanimous character, a desire for victory, strength, and ease in conversation. Both censors and playful commentators on the printed version of the book crossed out some of the private parts in the subsequent vivid depictions of the male human body.[89]

Empowered by his status as an aristocratic practitioner—a Neapolitan noble rather than one of the university-educated, upwardly mobile elite— Porta intended his theorization of the physiognomic body to transcend the disciplinary boundaries around medicine, agriculture, husbandry, and magic rather than to further inculcate them.[90] To show this boundary crossing in action, the next section will follow Porta's approach to physiognomic vision, teaching the reader how to see like a lynx.

Seeing Like a Lynx

Whiskers outstretched, nose in the air, the lynx gazed at the side of a rocky mountain. Its image was featured in the foreground of an insert in the frontispiece of Porta's 1588 *Phytognomonica*. The idea that one could understand the interior of nature by looking at the exterior carefully enough fueled *Phytognomonica*, an analysis of plants' features that was supposed to help in the identification of medicines. Intended as Porta's magnum opus, this capstone required a mantra and mascot. The lynx emblem represented something essential to Porta's intellectual outlook.[91] He coined his personal motto, *aspicit et inspicit* ("he observes the exterior and the interior"), with the legendarily sharp eyes of the lynx in mind. Speaking in the third person, Porta wrote:

> When the Author found the "Physiognomy of Plants," which had not ever been thought of by anyone, it seemed to him that he had penetrated into the secrets of plants. He made this lynx the *impresa* for his book because everyone writes that its sight passes through mountains. Beneath it he wrote the motto "*Aspicit et inspicit*," he sees the outside of that inside. Laugh at those pedants who say that they do not believe that the lynx sees all secrets, as if humans had to see what the lynx did not to believe a foolish story. But the truth was written by many authors.[92]

The *impresa* connected the discerning sight of sly Lynceus of Greek mythology with the short-tailed Eurasian feline. In spite of Porta's dislike of Gesner, he was clearly influenced by Gesner's ideas, or at the very least, unlike those unfortunate pedants, he too was a believer in the power of the lynx's sight. Porta copied Gesner's lynx exactly and dropped it into the frontispiece of *Phytognomonica*. The same lynx in the same posture was depicted in the Lincei's Mexican Treasury, as well as in the research of Galileo, including his famous treatise on sunspots.[93] The *impresa*'s endurance symbolized an appealing approach to the discovery of secrets in nature: an expert could use external observations to classify nature and determine characteristics.

Two ironies make the choice of the lynx particularly revealing. The first was that the Lincei, a society devoted to the empirical observation of nature, had neither observed real lynxes nor wrestled with all of their classical connotations. Indeed, despite its Renaissance fame, many ancients had not been so enthralled with the lynx's legendary vision. Aristotle had not focused on it. Pliny mentioned that the lynx was "of all quadrupeds possessed of the most piercing sight."[94] However, he then went on to devote much more attention to the "Five Remedies Derived from the Lynx"; on the Isle of Carpathus, people reduced the lynx's nails and hide to ash and then drank these substances, expecting them to reduce the "abominable desires in men"; similarly, "if sprinkled upon women, all libidinous thoughts will be restrained."[95] The second irony derived from the choice to celebrate and emulate the lynx's vision. This act suggested that one could choose to adopt certain animal characteristics, indeed, even choose the animal one wished to resemble. Porta's philosophy, however, is ambivalent on the issue. On one hand, he personally chose the motto of the lynx, as did his fellow Linceans. On the other, he argued strongly that physiognomy was based on inalterable, intrinsic characteristics. Porta's emblematic view offered limited space for trainability because he insisted on the essential animal nature of each and every human. A leonine person would be unable to transcend their physical features and act like a vulpine person.

It should come as no surprise that a major contribution to sixteenth-century science required a bestial emblem. Renaissance authors used animals literally, but also allegorically, since so much of their understanding of the natural world came through the symbolic landscape of the ancient past. Natural history developed amid pervasive interest in adages and epigrams, emblems and devices.[96] Within the covers of a book, early modern naturalists built on bestiaries, zoos of animal anecdotes both real and imagined, which translated nature into familiar, moral terms.[97] The creatures described existed not as individuals or species, but as ideal-types. The natural world's constituent parts, including animals, could be read symbolically as words within the book of nature. During the sixteenth century, however, greater exposure to natural diversity through the travel narratives and exotic specimens brought back to Europe required natural historians to nuance their reliance on classical and Christian stories.[98]

To see through exteriors like the lynx, one need only know how to read nature with an eye to animal-human connections; human character would then suddenly become clear. By analytically dividing animals into their physical features and understanding those features in relation to the animal's essential character, Porta proposed using animal parts as guides to explain variation in human temperament as a function of external appearance. He mused, "If animals' parts are similar to those of men which reveal man's customs as similar to animals', what else is there to say but that the animals' temperament produced that part similar to man's?"[99] Porta's take on observation was both essentializing (he understood animals and humans not as individuals, but as emblematic ideals or character types) and democratizing (anyone who could read Porta's De humana physiognomonia might understand both their own bodies and the bodies of others through the animal parts that constituted them). Readers came away with a new understanding of how human faces and bodies contained a mirror of the diversity of the animal world.

The frontispiece to On Human Physiognomy taught this basic technique (fig. 8.2). Porta stares unflinchingly out at the reader from his place in the center of a roundel. On his right, he is surrounded by men's faces with elongated noses, slanting eyes, weak chins, and overt cheekbones. Such unattractive visages drew on the emerging trope of the Renaissance grotesque, made famous by Leonardo da Vinci's "Study of five grotesque heads" (ca. 1493). In pursuit of an understanding of beauty, artists and theorists had become fascinated by ugliness and physical anomalies in the human form. The left side of the frontispiece, in contrast, resembles a Renaissance hunting lodge. It includes animal heads in profile: a rabbit, boar, bird, stag, cat, and ram. The fused head of a conjoined twin—half glowering human, half an ass with lowered ears—appears at the bottom of the frame, head emerging from a donkey's body. This hybrid creature resembled Pan, both animal and human, from Ovid's Metamorphoses. The frontispiece represented the point of the book as a whole: to see the human in the animal, and the animal in the human. Porta trained his readers' eyes to look across these images, comparing each human face with the beast opposite. At the center, the magus solicits an examination of his own savvy, vulpine features, which he intends us to read according to the physiognomic characteristics he outlined in his book. In short, one could really judge the book of nature by its cover.[100] Porta's physiognomic system also functioned as a mnemonic device for recalling personalities and characters at a single glance. In this way, he catalogued human characters' idiosyncrasies in their features, preserving an artificial memory. Many subsequent printings of Porta's work reused this same frontispiece, centering the magus encircled by monstrous and animal portraits, long after he had died. To subsequent generations of readers still interested in his methods, he remained a good and virtuous man, despite being encircled by beastly natures.[101]

FIGURE 8.2. Seeing the inside from the outside: Giovan Battista della Porta, *De humana physiognomonia*, libri 4 (Vici Aequensis: apud Josephum Cacchium, 1586), frontispiece. Wellcome Collection. Public Domain Mark.

On Human Diversity

This vision of nature was based on coherences. While Porta's early work on natural magic had focused on the rivers of nature, interconnected and continually flowing across perceived barriers, his physiognomic turn sought to look within different types to understand character in a neo-Aristotelian sense. Unlike many of his contemporaries, including naturalists such as Ulisse Aldrovandi and Conrad Gesner, he was not primarily invested in collecting and sorting taxonomic specimens; his relatives could do that work.[102] Rather, Porta found natural similarities, affiliations, and coherences more interesting than the rules that separated different types of nature. In a characteristic merging of disciplinary categories, Porta provided the clearest Renaissance answer to the question of why differences in body and character exist in plants, animals, and humans. He did not doubt, however, the top of the hierarchy: humans. Man in particular was the apex of all animals, for we take on all of the characteristics of nature, from the strength of the lion to the slyness of the fox. It was in that multitude that Porta saw God and the signs God had left to improve human life.

Human diversity, then, could be read through animal forms. Porta continued his argument, contending that he believed that people from different places were astonishingly varied in both appearance and character. In *On Human Physiognomy*, he imagined a cascading relationship between climate and custom:

> This temperament makes the Italians different from the Spanish, the Spanish from the Germans and the Turks. Air, sky, land, food, water, age, and time make temperament and customs. This does not only pertain to such distant places, but nearby ones, too. How different are the customs of the Neapolitans from the Calabrians, and the Calabrians from the Puglians, and these from those of Abruzzo?[103]

Even in the center of the Mediterranean world, where urban culture had flourished since the Ionians decided to call their settlement a "New City" (*Neapolis*), such a variety of peoples required explanation. Porta did not call them different races, but his vision of social groups and their physical features moved actively in that direction.[104] Cultures, aesthetics, or priorities alone could not explain their differences. Civilized nature had a certain aesthetic. Eliding savage, wild, and uncontrolled, Porta wrote that "one sees, therefore, in the aspect of the wild animal a brutish, horrid, hairy, rustic and, finally, more deformed variant than the urban."[105] Country mice differed conclusively from their city brethren.[106] Porta thought the same pattern held for rural Abruzzo compared with sophisticated Naples. Brutish people and animals not only came from brutish places— they had been born brutish largely because of those places. As Porta concluded that different "air, sky, land, food, water, age, and time" made features that in turn yielded different temperament and customs, he raised a major problem that would only grow as the century came to a close: What happened to city mice if they moved to the country? Would they change their form to match their new environment? If that new environment bronzed their skin with sun and wind, would their offspring, like the bipedal dogs, be born darker or "deformed"?[107] This passage, like much of Porta's published writing, leaves his final position unresolved. Although it suggests a kind of climatic determinism, ascribing natural traits to national groups, Porta did not push his conclusions to the aggressive extent that those in later centuries would. On the contrary, rather than taking simple premises and pushing them to rigid conclusions, the magus focused on fruitful associations and suggestive connections.

The Perils and Possibilities of a Conjectural Science

Physiognomy offered not simply a way of seeing human character, but a much more powerful analytical tool. Porta understood it as a method that could transcend a debate between determinism and free will; it was a science of conjecture that could predict likely features of the human and nonhuman natural world. Furthermore, Porta believed one could extend the traditional practice of physiognomy to plants, as he did in *Phytognomonica*. In the 1588 volume's dedication to "the illustrious Marino Bobali, son of Andrea the Ragusan," Porta wrote with confidence in his power to see nature, for the eyes of the lynx can be neither occluded nor shut. Nothing, even the innards of mountains, could be invisible.[108] *Phytognomonica* united physiognomy with the doctrine of signatures, an ancient idea central to the texts of Dioscorides and Galen, which held that herbs with a shape and features similar to a body part could be used to treat that body part. Porta, ever fascinated with links, saw this as just the beginning. Could not the nature of metals, herbs, gems, and stones be understood by looking at their external features? And could not their physical attributes, in turn, be linked to their inherent characteristics? How did the stars, tied to their own celestial physiognomy, likewise shape the forms of living things? And what happened when beings moved from one region to another?[109] With great eagerness, if analytical irregularity, Porta tried to establish a climatic link between regions and the creatures that resided there. Different "air, sky, land, food, water, age, and time" made the Calabrians different from the Neapolitans, and mountain elephants different from valley ones.[110] No wonder that urban creatures differed from rural ones: they were fueled by elements of the regions from whence they came.[111] Like Nonio, the breeding doubter in Mantua with whom chapter 1 concluded, Porta believed plants, animals, and humans who moved to be subject to change as they inhabited different climates.

To look at nature physiognomically presumed knowledge without contradiction and left the magus in awe of his own power. Porta wrote to Cardinal Luigi d'Este on June 27, 1586, that his patron ought to be pleased: Porta had been extremely productive despite the summer heat, which he was eager to escape. With characteristic boldness, he wrote:

I will bring with me the book on the physiognomy of herbs, which I have awaited all of the times that I came to Rome. With this method, I am scrutinizing all of the natural secrets and providing most subtle speculations and new inventions that our humanity has never seen before, of which I am the least. [In the book, *Phytognomonica*] are more than two thousand secrets of medicine and other beautiful things. I will also bring with me the book that I had started more than thirty years ago [when I was writing *Natural Magic*], in which I put all of the chosen and proven secrets in all sciences, that is the most subtle things in which learning has grown tired, as in the perspective to make a mirror that burns a mile away, another with which it is possible to reason with a friend a thousand miles away by means of the night moon, and to make glasses that can depict a man some miles away, and other marvelous things. On agriculture, there are near three hundred rare secrets: to seed a bushel and come back with a thirty-fold yield, to preserve one uncorrupted grain for a hundred years, to make a fruit inside another fruit such that when one is opened in half there is another fruit inside, to make fruit and flowers no longer ever seen, to double the production on the vines, and similar things. On [household] economy, [there are secrets about how] to make a hundred types of bread without flour and to make it grow double without mixing it with something else, as that of the Philosopher's Stone [*lapis filosoforum*], a hundred sorts of recipes for wine, olives, squash, apples, and other sorts of things, and other sorts of sciences that I have done myself and those of my friends and patrons. I have deliberated not to make them visible to humanity and for jealously that the others would not know them and for doubt of that wickedness they teach. Still, I am getting annoyed with life all together, I will give the book to Your Most Illustrious Lordship, as I am sure that you will not use it to work evil, and still having to give it to someone I do not know of anyone in the world who is more worthy than you.[112]

Amid his descriptions of gazing up at the stars and his endless requests for lenses for his *occhialo*, the device that would become his iteration of the telescope, Porta renewed his focus on secrets of the agricultural arts. The same fascination with tips and tricks that would help him develop invisible ink and the camera obscura had implications beyond illusions alone. Whether grown on the villa or in the fields along the bay, plants and animals obeyed the same cycles of nature, which the magus aimed to control. *Phytognomonica* joined the shelf of other books of physiognomic secrets, whose teachings could be put to either good or evil use, to be shared among a select group of friends and patrons (fig. 8.3).

Nonetheless, Porta was aware of the danger of seeming to preempt choice and exclude free will in his physiognomic theories. He sought to get in front of the accusations to come in his 1586 dedication to Luigi d'Este in *De humana physiognomonia*. In it, Porta sought to limit the uses of physiognomy:

> This is a conjectural science, which [however] does not always reach the desired end. Its signs can only indicate natural propensities, not the actions of our free will, or whatever [*vel quae ex . . .*] comes out of a habit of vice or effort [*studium*]. Indeed, it is good and bad actions, subject to our power, that constitute virtue & vice; but not propensities, that are not subject to our will.[113]

In this dedication, if not elsewhere, Porta attempted a synthesis to avoid allegations of determinism. He argued that people have particular talents and characteristics and that they have the freedom to operate within those confines. Crucially, the paratext treated humans as moral actors with full scope to exert effort, to indulge their own passions, and to realize their virtuous and vicious qualities through their own choices. Humans' moral independence, however, does not give them unlimited scope for action. Rather, their bodies develop in tandem with the environment around them, from the stars above them to the plants they eat. For Porta, the powers of self-fashioning met their limits in corporeality.

A Lynx-Eyed Academy

After reading Porta's physiognomy books, young scholars like the Roman noble Federico Cesi, prince and founder of the Accademia dei Lincei (1603–1630), were taken by the lynx's legendary sight. They wanted to see nature just as Porta's lynx had. Porta told Cesi that he had "penetrated the secret of plants with my thought. I made the lynx, whose sight

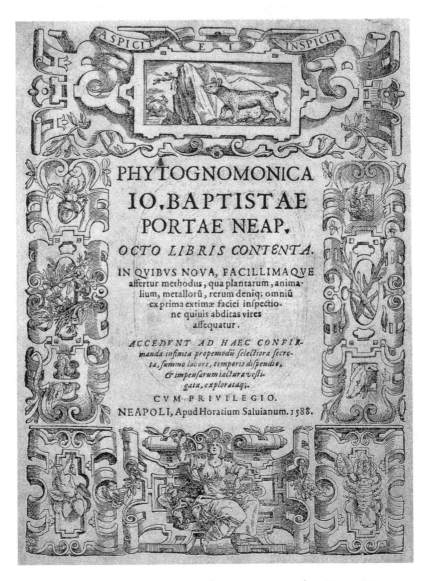

FIGURE 8.3. Giovanni Battista della Porta, *Phytognomonica* (Naples: apud Horatium Salvianum, 1588), frontispiece. Wellcome Collection. Public Domain Mark.

passes through a mountain according to all writers, the *impresa* for this book."[114] Cesi was so impressed by the image of an animal that could penetrate nature simply by looking that he adopted Porta's lynx as the symbol for the Accademia dei Lincei, Europe's first scientific society.[115] The Roman academy, whose name literally translated as the Academy of the Lynx-Eyed, made Porta its senior member in 1610 and took the feline's vision as its emblem for its own penetrating investigations of nature for the next two decades. When Galileo Galilei joined the academy as its other distinguished member in 1611, he subsequently marked his books on heavenly bodies and physical mechanics with the lynx.[116]

Porta proved just as committed to eking out support from Cesi, his new patron, as he was under Cardinal Luigi d'Este. His letters to the young prince remained cloying and self-promoting even as he continued to write prolifically. He dedicated his works *De distillatione* (1608), *Elementorum curvilineorum* (1610), and *De aeris transmutionibus* (1610) to Cesi. Over the same years, Porta and Galileo, the younger and soon to be more famous member of the Lincei, worked with, and recognized the utility of, the telescope.[117] Despite these common threads of empirical observation, Porta remained focused on using his Roman connections to work around the ban against all predictive sciences. In 1608, Porta wrote to Cesi in the hopes of publishing the next installation of his physiognomy series—this time a book on chirognomy. Porta's book consolidated tips and tricks on popular methods of reading the future, using only the powers of lynx-eyed observation, of course.[118] As "work on Chironomia has been coming along well," Porta sought to nudge his Lincean colleague into helping him in the publication process, writing, "I hope to God, and in the favor of Your Excellency, that soon a good license [to publish] would come, which would cheer me up, as well as all of my friends who are awaiting it with great desire."[119] Porta was happy, of course, to receive the license to publish one of his comedies, but, forgoing all irony, he nonetheless begged, "Oh when might be that happy day when it would please God to have a license for *Chiromantia!*"[120]

In Porta's final years, the research agenda he had envisioned around physiognomy and direct likenesses fell on the deaf ears of a scientific community that was moving in a different direction. Mechanism and experimentation in the realm of physics took on the central position that had once been reserved for naturalists and natural historians, who in turn looked like increasingly antiquated collectors.[121] Porta's pursuit of physiognomy never gained the mainstream attention that Galileo's sunspots did. Rather, its influence proved more subtle. The Academy of the Lynx-

Eyed continued to see like lynxes, and in their scientific publications they plowed forward, ordering the natural world with great taxonomic confidence. Their hubris is epitomized in the Mexican Treasury, the compendium they created from Francisco Hernández's notes on the natural history of New Spain that sought to identify the new flora and fauna of Mesoamerica, as discussed in chapter 5.

While publishing this massive tome was a colossal accomplishment for European naturalists who had never been to the New World, the text compiled in Rome in the mid-seventeenth century often disregarded the natural mutability and the indigenous knowledge that had revealed such flora, fauna, and minerals in the first place. By putting all the different plants and animals that they had managed to pull from Hernández's notes side by side, the Lincei compared and contrasted the specimens described, just as Porta had done with the specimens he had drawn from the Neapolitan menagerie. Interweaving New World nature into memories of Dioscorides and Theophrastus, the Lincei integrated new nature into old categories. The Nahuatl names were shortened and butchered. For example, in search of a balsam tree of the Indies to replace the declining balsam populations of Europe, they located the Mexican balsam *hoitziloxitl*; a later early modern Spanish reader more competent in Nahuatl corrected its spelling to *huitziloxochitl*. Similarly, in the Mexican Treasury, the plant later known as an amaranth lost its name, which had meant "the flower that hummingbirds came to" (*huitzil*, for hummingbird; *xochitl* for flower).[122] Likewise, in elaborating Nardo Antonio Recchi's notes on maize, or "*Tlaolli, seu Maizio*," the Lincei transformed growing corn into dried kernels alienated from growing stalks, to which the term *tlalloli* refers in modern Nahuatl. The processes that made the plant were lost in the bastardization of its name, frozen in time in that lynx-eyed commentary by observers so committed to images of the plant's exterior that they bothered neither to grow the plant nor to view it for themselves nor to understand the meanings embedded in the indigenous names.

Cesi himself aimed to simplify nature through visual order. Although he was more interested in seeing like a lynx by observing the inner workings of nature through the transmutation of air and curves rather than by embracing physiognomy, Cesi's own intellectual project found itself playing on links, trees of life, and taxonomies that take seriously both the insides and outsides of things. However, for Cesi, order and separations were the goal, rather than slippage across nature or its manipulation. In his tables in the Mexican Treasury, Cesi drew a distinction between infinite and finite things. Within the category of finite things, there were spiritual (spirit, intellect, etc.) and corporeal things, the latter of which have substance. These corporeal things could include mixed and simple sub-stances. Simple substances included parts of the terrestrial sphere, comets and meteors, and the natural heavens. In the terrestrial sphere, there were the elements of land, water, air, and fire, while in the celestial, there were the stars and the heavenly ether. Most important to this story, however, was his vision of the plants, brutes (animals), and humans. Man was ratio-nal and intellectual, tied to abstraction, ideas, reason, and morals. Brutes lacked rationality, and at one end bled into plants, which in turn bled into inanimate substances.[123] Cesi went on to map out the differences between plants, and even between fields of knowledge (indeed, in his table of the liberal arts, he even taxonomized Porta himself, alongside Gesner, as an expert in botany). Just as a plant could be sectioned off into roots, flowers, and seeds, so could knowledge, beings, and everything else.

Leopards had been frozen in Porta's drawings, which had in turn be-come likenesses that could be compared with yet other drawings of hu-man characters disassociated from life. Similarly, the paper museum of esteemed naturalist and collector Cassiano dal Pozzo (1588–1657) and other members of the Lincei would lose the life behind the images of na-ture in favor of legible categories. While chapter 2 discussed the creation of brands to show what breed one was supposed to see, branding books and bureaucratic categories were, among other things, technologies that attempted to correct the fact that *razza*-making had not managed to yield animal populations obviously different from one another. Those signs had been a means of making difference more visible. This chapter has told a different story in parallel. As the Mexican Treasury shared details of New World nature with European naturalists, it also normalized a new use of visual and analytical technologies that fixed that nature. The rise of physi-ognomy through Porta's work had supported that visual project. By seeing like lynxes, collecting images, and linking visual features, the Lincei sought to show that external features offered a reliable means of organizing and seeing patterns in nature. Surely, they did not reveal herbs' medical prop-erties so directly as Porta had hoped, but the analytical move remained the same: on the level of genus and species, surface appearances often did reveal internal differences in habit, custom, growth, and life.

Epilogue

In this book's epigraph, from Plato's *Republic*, Socrates pushed Glaucon to think about embedded hierarchies in nature and the potential power of controlled breeding for shaping human society by drawing on his personal experience breeding hunting dogs and fighting birds. Socrates's questions encouraged Glaucon to think about how his choices in their breeding kept these animals noble, hinting that if Glaucon no longer intentionally bred from the best, his stock would decline in quality. When asked to consider whether the same principles could pertain to horses and other animals, Glaucon responded that, indeed, it would be strange if they were any different. Through this tactic—highlighting likenesses in breeding across the diverse types of animal stock (birds, horses, and dogs)—Socrates prepared the wedge for his key argument. The presence of reproductive likenesses and the power of selection to improve each variety allowed Socrates to corner Glaucon: if all of these domesticated animals can be improved using the same principles of selection, then ought not excellent humans to be matched to yield the best human herd? In the dialogue's framing, it would have been impossible for Glaucon to answer with anything but an affirmative. Yet without Socrates's pointed questions, it seems unlikely that Glaucon would have taken his work as a husbandman to this most extreme conclusion about governance, hierarchy, and human selective breeding.

The choice to begin and end this book with Plato is not an arbitrary one. When Renaissance Italian humanists rehabilitated Plato's *Republic* through its translation into Latin over the course of the fifteenth century, they perceived the text as cohering with many of their own values. For instance, the 1402 translation of the *Republic* had propagandistic valences, as the humanist Uberto Decembrio bemoaned the systematic alienation of the best, most knowledgeable, and most noble men as well as Florentines' proclivity to fall under the allure of false liberty. Decembrio wrote that Plato had worried about the political health of his homeland; the *Republic* might be the doctor's hand that corrected its degeneration into a savage state.[1] The *Republic*'s vison of a just state culminated in the utopian city-state of Kallipolis ruled by philosopher-kings. There, the right to rule could come to those best suited for the task without falling into the traps of other regimes. If only their rulers were learned enough and their inhabitants assigned to their rightful positions, new republics might offer an alternative to the weaknesses of oligarchies, democracies, and monarchies. The rise of Platonic thought might even be seen to temporally match a rise in eugenic animal projects among Italian elites.

Without Socrates to push them, early modern animal breeders tasked with the care of flocks, herds, and kennels only rarely articulated the implications of their work for humans, at least in the documents that have survived. If pushed, however, they probably would have emphasized that all beings—whether dog or horse or otherwise—come from breeding. Indeed, we have seen how some early modern patrons, animal experts, and authors, such as Isabella d'Este and Tommaso Campanella, casually, actively, and intentionally played with the boundary between human and animal, describing human generation with the same language and principles that animated animal breeding.

Talking with Glaucon seemed like a productive exercise. The modern breeders of livestock and small animals with whom I spoke while researching this book generally avoided conversations about humans. Often, when I raised the topic they shied away, or changed the subject, or just looked nervous, a tacit recognition of the ethical ground at the edge of their work, the slippery slope between the breeding that they did, which no one thought twice about, and the broader possibilities and dangers of eugenics. However, as our conversations became more comfortable and intimate, connections emerged, particularly when they turned to their own choices regarding family planning. One woman who had inherited her father's stable of horses mentioned that, unlike the horses she had chosen to breed, she knew too well her own physical frailties to impress them on another generation, no matter the pressure to have children who could preserve the family farm. Many comments resembled those made by the modern horse breeders Rebecca Cassidy studied, one of whom noted that "there is no difference between a woman and a mare, except that a mare is more agreeable. The mare is a self-contained foaling unit and nursery, and that is all a woman would be if she didn't talk too much."[2] By contrast, in the human-centric world of fertility treatment, artificial insemination, and anonymous sperm donation, the discursive avoidance of the mechanics and animality of reproduction is striking. To liken human maternity to animal maternity risks undermining human personhood at a moment of vulnerability; thinking too carefully about the eugenic principles behind such selection would sully our vision of children, our insistence that they emerge unblemished by human desires as innocent tabulae rasae.

It would be unfair to make too much of these comments from breeders and reproductive specialists about the human condition, not least because I am an early modern historian and not an anthropologist. It seems, however, that eugenic principles are most often explored on nonhuman animal bodies; human eugenics is often an epiphenomenon of the desire and technologies first used to shape nonhuman animals.[3] Today's breeders, like the early modern practitioners whose accounts have featured in this book, emphasize the practicalities of husbandry and animal life rather than abstract boundaries of personhood. Thinking broadly with animals risks emphasizing a search for vague likenesses and differences; working with real animals requires dealing with specificity around the needs of particular species, breeds, and even individuals. Thus, animal specialists come in many guises for the simple reason that nature takes so many diverse forms.

This book is designed to show how people in the early modern era lived in an animal world, and how their understanding of animals shaped their vision of human and nonhuman natural diversity. I have aimed to understand how *razza* and related concepts evolved across animal-human boundaries by studying the categories used by early modern thinkers, rather than by either anachronistically imposing those of my own time or by presuming race to be a solely modern invention. An attention to breeding has a historical relationship to the modern definition of race. But, for breeders then and now, animal bodies are the focus, and connections to human difference often quite irrelevant. Alas, origin stories are both tempting to historians and fearsomely difficult to prove. The evidence here—of breeders' reports, dictionaries, and natural histories produced under European intellectual influence—can at most be suggestive of how and under what conditions linkages between human and animal were evoked. What seems to be the case, however, is that a hubristic belief in the power of humans to shape the natural world reemerges regularly over the centuries, only to be dispelled anew by doubt, lie dormant, and resurge.

Acknowledgments

Vital support from institutions and people made this book possible, beginning with Stanford University's Department of History. There, Paula Findlen's incessant curiosity and erudition guided me from the beginning; I am most indebted to her mentorship. Jessica Riskin pushed me to hone my argument, and her History of Evolution course laid the foundations for integrating Renaissance projects into an improved understanding of inheritance. Laura Stokes fueled my creativity and offered feedback at crucial junctures. Lisa Surwillo encouraged me to interweave history and literature in pursuit of a wider cast of historical actors. I am thankful for the Lane Research Travel Grant in History and Philosophy of Science and the Stanford Graduate Fellowship in the Digital Humanities, as well as the Program in Feminist, Gender, and Sexuality Studies. For their mentorship, I thank Keith Baker, David Como, Rowan Dorin, Zephyr Frank, Estelle Freedman, Fiona Griffiths, Londa Schiebinger, Kären Wigen, and Ali Yaycioglu.

I am thankful for the support and scholarly community that national funding agencies made possible. The Council on Library and Information Resources Fellowship for International Research in Original Sources, the Fulbright Foundation, and the Mellon Foundation Dissertation Fellowship generously funded this transatlantic project. The Mellon Summer Institute in Italian Paleography gave me the chance to acquire skills necessary for this research, and the Renaissance Society of America Huntington Grant provided access to key sources, archival and botanical.

In the United States, I am grateful to the staff at the Green Library Special Collections, Stanford University; the Olin Library Special Collections, Cornell University; the Getty Research Institute Library; the Widener and Houghton Libraries, Harvard University; the Huntington Archive and Library; the John Carter Brown Library; the Benson Library, University of Texas at Austin; the Beineke Lesser Antilles Collection,

Hamilton College; the Library of Congress; the National Museum of the American Indian; and the Smithsonian Institution. In Italy, I thank the staff of the Archivio di Stato di Mantova and Biblioteca Teresiana, Mantua, especially Laura Melli, Giancarlo Malacarne, and Lisa Valli; Archivio di Stato di Milano and Biblioteca Ambrosiana, Milan; the Biblioteca Nazionale Marciana and Palazzo Giustian Recanati alle Zattere, Venice; Archivo di Stato di Firenze, Florence; Archivo di Stato di Napoli, Biblioteca Nazionale di Napoli, and Società Napoletana di Storia Patria, Naples; and Archivio Accademia dei Lincei, Rome. In Mexico City, I thank the Archivo General de la Nación and Museo Nacional de Antropología e Historia. In Lima, I thank the Archivo General de la Nación and Biblioteca Nacional del Perú. In Spain, I thank the Biblioteca Nacional de España, Biblioteca Francisco de Zabálburu, El Escorial, Instituto Valencia de Don Juan, and Real Jardín Botánico, Madrid; the Biblioteca de Catalunya, CSIC-Milà i Fontanals, and Jardí Botànic, Barcelona; Archivo General de Simancas; Archivo General de las Indias, Seville; and the Monasterio de La Vid. In Austria, I thank the Österreichisches Staatsarchiv, Vienna. In the UK, I thank the Warburg Institute and the Wellcome Library. A network of international mentors made this research possible. I thank José Pardo-Tomás and colleagues at the IMF-CSIC; Giovanni Muto, for teaching me the ropes in the Neapolitan libraries and archives; Annemarie Jordan Gschwend for her indefatigable support and suggestions on how to use the Habsburgs' vast documentary collections; Thomas and Pia Wallnig, who taught me about Schloss Ambras and Vienna's Spanish Riding School, and welcomed me into their home with such generosity. I thank Antonio Dávila Perez, Nieves Baranda, Maddalena Signorini, Giulia Calvi, Sarah Duncan, Jens Amborg, and Byron Hamann for their insights and encouragement. As is often the case, some of the most challenging research for this book was also the most beautiful. I thank equestrian Annie Morris for sharing her expertise on Iberian horses and dressage training in Portugal, as well as equine experts at Cloverfield Farm, Greyrock Farm, and Stanford University's Red Barn.

The Cornell University Presidential Postdoctoral Fellowship and Society of Fellows launched the writing of this book. Jessica Cooper, Tyler Coverdale, Stephen Sansom, and Hannah Leblanc wrote alongside me. Rachel Weil, Judith Byfield, Barry Strauss, and Peter Dear offered integral feedback on shaping earlier drafts into a book. Paula Cohen, Nerissa Russell, Matt Velasco, Susan deFrance, and Mariana Federica Wolfner challenged me to think about inheritance biologically rather than through discourse alone. Conversations about animals, metaphor, and evolution that I had as an undergraduate with Dominick LaCapra, Camile Robcis, and William Provine stayed within me as I moved between the schools for Agriculture and Arts and Sciences. As a visitor to the Max Planck Institute for the History of Science, I was pushed by colleagues each day to consider new connections between animals, race, and intellectual history. I am especially grateful to Dagmar Schäfer, Lisa Onaga, Rebecca Woods, He Bian, and Aleksandar Shopov for support and pointed conversations alike.

My formal training is as an early modern Europeanist. It has been a pleasure and honor to learn from the community of experts in Nahuatl and other indigenous American languages. Sincere thanks to John Sullivan, Justyna Olko, Eduardo de la Cruz Cruz, Carlos Cerecedo Cruz, and Abelardo de la Cruz de la Cruz for including me in the IDIEZ (Instituto de Docencia e Investigación Etnológica de Zacatecas) Nahuatl program. Thanks to Chris Valesey and Joshua Fitzgerald, who read and commented on two chapters. Emphatic thanks to Alanna Radlo-Dzur for her guidance through Nahuatl grammar, period dictionaries, and the use of archaeological evidence in history writing, and her editorial suggestions on chapters 4 and 5, as well as her camaraderie in analyzing the *Tira of Don Martín*, the Florentine Codex, and the writings of Diego Muñoz Camargo. Thanks to Andrew Laird, Joe Campbell, Kevin Terraciano, Helen Ellis, Chris Heaney, and Julia Madajczak for their conversations, critiques, and sharing of their own research.

Sally Hayes, Erica Brittany Feild, and Farah Bazzi challenged me to think more about belonging and difference in the Spanish empire. As part of our Natural Things research group in global natural history, Anna Toledano and Duygu Yıldırım pushed me to expand my definition of naturalist. During a whirlwind visit to the Institute of Advanced Study, Patrick Geary and I discussed the use of physical biological evidence in the writing of deep history, and Silvia Sebastiani carefully reviewed sections of this manuscript and pointed me toward a more nuanced understanding of the philosophy of race; I am very grateful for their input. I have published early versions of the research now dispersed throughout chapters 2 and 3 in "Marketing Nobility: Horsemanship in Renaissance Italy," in *Animals at Court: Europe, c. 1200–1800* (Berlin: De Gruyter, 2019), and I thank Mark Hengerer and Nadir Weber for their editorial insights.

My colleagues at Hamilton College supported the completion of this book. Particular thanks to John Eldevik, Usman Hamid, Kevin Grant, Celeste Day Moore, Lisa Trivedi, Thomas Wilson, Rebecca Wall, Mariam Durrani, and Edward Halley Barnet for their advice and encouragement. Beyond the history department, my thinking benefited from collaborations with scientists Natalie Nannas, Rhea Datta, and Aaron Strong; conversations with animal studies scholar Onno Oerlemans; and discussions about theories of metaphor with Suzanne Keen; Stephanie Bahr, the Medieval and Renaissance Studies Program, and the "Imagining Race in Early Europe: From Antiquity to the Renaissance" lecture series; and the Humanities Center workshop "Boundaries and Transgressions of a Comparative Race Project." My students have likewise invigorated this research with their questions and discoveries; thanks to Kate Biedermann, Elizabeth Atherton, Thomas Anderson, Antton de Arbeloa, Erica Ivins, Ben Kaplan, Emma Tomlins, Liam Garcia-Quish, Philip Chivily, Isabelle Crownhart, Edsel Llaurador, Kayla Self, and Ali Zildjian.

This book took its present form following a book manuscript workshop that took place during the height of the coronavirus pandemic of 2020. Daniel Lord Smail, Justin Smith, Marcy Norton, and Jorge Cañizares-Esguerra generously offered their feedback on the text even as panic gripped our countries and universities. I offer them my deepest thanks for extending their acumen and kindness to developing this project especially during at a moment of such profound uncertainty.

I have presented versions of this research at Cambridge University, Cornell University, Binghamton University, the University of Toronto, Harvard University, Princeton University, the University of Vienna, the American Academy in Rome, the Smithsonian Institution, the University of Verona, the University of Stockholm, the University of Minnesota, and Stanford University, and through working groups including Cambridge's History and Philosophy of Science "Generation to Reproduction" seminar, the Central New York Early Modern Colloquium, Stanford's Center for Medieval and Early Modern Studies' "Race in the Archives" (Primary Source Symposium), and the Biblioteca Hertziana's project "Visualizing Science in Media Revolutions." Panels at conferences of the Renaissance Society of America, Scientiae, the History of Science Society, and the Northeastern Nahuatl Association prompted me to hone different sections of this project. I am grateful for the questions and suggestions I received at each venue.

Chapter 7, on Acosta, seeks to answer a question Yair Mintzker asked: How does this book revise or build on Anthony Padgen's description of human difference in *The Fall of Natural Man*? It took shape in a paper presented at the "Global Conversion" panel at the April 2021 Renaissance Studies Association conference, and I am indebted to my fellow panelists and chair Emily Michelson, as well as to Claudio Burgaleta S.J. for his help in navigating the Jesuit archive. I thank Bradford Bouley for his expertise on sixteenth-century Rome, his invitation to the History of Science Colloquium at UCSB, and his suggestion to explore *La città del sole* back in 2014. The chapter also benefits from conversations with Florencia Pierri concerning her Princeton dissertation, "A World of Wonder: Exotic Animals in Early Modern Europe." Finally, I thank Lydia Pyne for her inspiration on questions of species and kinds beyond a European context.

As a Fellow at Villa I Tatti: The Harvard University Center for Italian Renaissance Studies, I completed the final edits to this manuscript. During that time, I benefited from lunchtime conversations about the nature of the Renaissance, case studies as evidence, and the motivating nature of aesthetics. My thanks to Katy Park, Marty Brody, Alina Payne, Deborah Parker, Dario Gamboni, Enrico Piergiacomi, Amy Chang, and Alexander Bevilacqua for their suggestions and comments during this period.

The editors and staff at the University of Chicago Press have ushered this book into existence. Karen Merikangas Darling saw its core story; I have been so grateful to her vision. Four anonymous readers offered extensive reports—I am exceptionally grateful for their rigorous feedback. Thanks to Tristan Bates, Erin DeWitt, and Deirdre Kennedy, and the rest of the Chicago team, including Norma Sims Roche and June Sawyers, as well as Bill Nelson, who developed the maps for this book.

To the friends and colleagues on whom I tested these ideas in their earliest forms, thank you. Hannah Marcus, Zeb Tortorici, Alexander Statman, and Rebecca Gruskin helped shape this project through their suggestions and support over the years. Michelle Kahn pushed me to make the early modern muddle into comprehensible history. Diana Garvin shared in many adventures, conversations about writing, and discussions of "fiamma e fama" from Rome to Oregon, the Alps to the Rockies. To Kathleen Breitman, who offered inspiration, perspective, and companionship in Italy, Greece, Spain, France, Germany, California, Miami, New York, London, and many other places where we pretended to belong along the way, I am especially grateful. I thank my family, especially Lauren Cooley and Mary Bronner, whose love of animals, medicine, and scientific processes always pushes me beyond my comfort zone in historical sources. My mother, Anne Cooley, and godmother, Kathy Czerwiak, made bold choices in building our family, cherished more-than-human relationships at home and at Greyrock Farm, and taught me the richness of language study; they are fundamental to the shape of this book.

Most of all, I thank Brian Brege. Writing one's first book requires one to become the kind of person who could write in the first place. Brian held my hand along that winding path and guided each step. He read these words time and again, entertained each idea, opened Italy to my imagination, and envisioned history worth writing. He built a life of love and thought, and we welcomed Artemis Brege, who came into the world as I finished these pages.

Abbreviations

A&D Fray Bernardino de Sahagún, *General History of the Things of New Spain: The Florentine Codex*, 12 books, trans. and ed. C. E. Dibble and A. J. O. Anderson (Santa Fe, NM: School of American Research/ University of Utah, 2012)

ASF MDP Archivio di Stato di Firenze, Mediceo del Principato, Florence, Italy

ASMN AG Archivio di Stato di Mantova, Archivio Gonzaga, Mantua, Italy

ASN DS Archivio di Stato di Napoli, Naples, Italy, Dipendenza della Sommaria

AGS CSR Archivo General de Simancas, Simancas, Spain, Casa y Sitios Reales

BAV Biblioteca Apostolica Vaticana, Vatican City

FC Fray Bernardino de Sahagún, *General History of the Things of New Spain: The Florentine Codex*. MS Med. Palat. 220, Biblioteca Medicea Laurenziana, Florence

HHSTA Haus-, Hof- und Staatsarchiv, Österreichisches Staatsarchiv, Vienna, Austria

HNMI José de Acosta, *Historia natural y moral de las Indias [...]* (Sevilla: Iuan de Leon, 1590)

MAP Medici Archive Project

ÖSA Österreichisches Staatsarchiv, Vienna, Austria

RG René Acuña, *Relaciones Geográficas novohispanas del siglo XVI*, 10 vols. México: Universidad Nacional Autónoma de México– Instituto de Investigaciones Antropológicas, 1982–88

SNSP Società Napoletana di Storia Patria, Naples, Italy

Notes

Introduction

1. On Campanella's trial, see Thomas F. Mayer, *The Roman Inquisition: On the Stage of Italy, c. 1590–1640* (Philadelphia: University of Pennsylvania Press, 2014), 46–63. On the life and works of Tommaso Campanella, see Germana Ernst, "Tommaso Campanella," in *Stanford Encyclopedia of Philosophy*, Fall 2010 ed., https://plato.stanford .edu/entries/campanella/; Bernardino Bonansea, *Tommaso Campanella: Renaissance Pioneer of Modern Thought* (Washington, DC: Catholic University of America Press, 1969); Luigi Amabile, *Fra Tommaso Campanella, la sua congiura, i suoi processi e la sua pazzia*, 3 vols. (Naples: Morano, 1882); Luigi Amabile, *Fra Tommaso Campanella ne' castelli di Napoli, in Roma ed in Parigi*, 2 vols. (Naples: Morano, 1887); Germana Ernst, *Religione, ragione e natura: Ricerche su Tommaso Campanella e il tardo Rinascimento* (Milan: Franco Angeli, 1991). On the relationship between Campanella's writing and Francis Bacon's ideas, see Francis Bacon and Tommaso Campanella, *The New Atlantis & The City of the Sun* (Mineola, NY: Dover, 2003); Margherita Palumbo, *La città del sole: Bibliografia delle edizioni (1623–2002)* (Pisa: Istituti Editoriali e Poligrafici Internazionali, 2004). On Campanella's library, see Kristine Louise Haugen, "Campanella and the Disciplines from Obscurity to Concealment," in *For the Sake of Learning: Essays in Honor of Anthony Grafton*, vol. 2, ed. Ann Blair and Anja-Silvia Goeing (Leiden: Brill, 2016), 602–20.

2. See Michael Subialka, "Transforming Plato: 'La città del sole,' the 'Republic,' and Socrates as Natural Philosopher," *Bruniana & Campanelliana* 17, no. 2 (2011): 417–33. Campanella hearkened back to Thomas More's *Utopia*, written nearly nine decades before; "utopia" itself, coined by More and his eponymous text (Latin 1516, English 1533), carries an inherent pun, meaning both "no place" (o + topos) and "happy place, fortunate place" (eu + topos). Building on an existing set of classical and medieval traditions about fantastical, distant societies, Renaissance utopian writers added a Neoplatonic, socially critical element to compose a new society. In homage to traditions established in Plato's *Republic* and his description of Atlantis in *Timaeus*, this utopian literature reported on fictional ancient societies in the amorphously defined, far-off Indies contacted through overseas empire. Examples include Thomas More, *Utopia* (Louvain: Thierry Martens, 1516); Albrecht Dürer, *Etliche Unterricht, zur Befestigung der Städte, Schlösser und Flecken* (1527); Francesco Patrizi, *La città felice* (Venice: Giovan. Griffio, 1553).

3. Tommaso Campanella, *La città del sole*, manoscritto anonimo, Biblioteca comunale di Trento, 1602, 18v: "Una parte del territorio, quanto basta s'ara, l'altra serve

per pascoli delle bestie. Hor questa nobil arte di far cavalli, buoi, pecore cani, et ogni sorte d'animal domestico, è un sommo preggio appresso loro come fu in tempo antico d'Abramo, e cõ modi li fan venire al coito, chi possano ben generare innanzi à cavalli pinti, o bovi o pecore, e nõ lasciano andar in campagna li stalloni con giumente, ma li donano a tempo oportuno innãzi alle stalle di campagna." See also Tommaso Campanella, *La città del sole: Dialogo Poetico/The City of the Sun: A Poetical Dialogue*, trans. Daniel J. Donno (Berkeley: University of California Press, 1981). See also William Eamon, "Natural Magic and Utopia in the Cinquecento: Campanella, the della Porta Circle, and the Revolt of Calabria," *Memorie Domenicane* 14 (1995): 369–402; and Lorenzo Bianchi, "La Magia Naturale a Napoli tra Della Porta e Campanella," in *La magia naturale tra Medioevo e prima età moderna*, ed. Lorenzo Bianchi and A. Sannino (Firenze: Sismel, 2018), 203–28.

4. Campanella, *La città del sole*, manoscritto anonimo, 4r: "Il Amore ha cura della generatione, d'unir li maschi alle femine in modo che faccin buona razza; e se rideno di noi ch'attendemo alla razza delli cani e cavalli, e trascuramo la ñra. Tien cura della educatione, delle medicine, speciarie, del seminare e racogliere i frutti, delle biade, delle mensi, e d'ogni altra cosa p̃tinente al vitto, vestito e coito, et ha molti maestri, e maestre dedicate a q̃st'arti." The treatise was first published in Latin as an appendix to Campanella's *Politica* (Frankfurt, 1623).

5. Dániel Margócsy, "Horses, Curiosities, and the Culture of Collection at the Early Modern Germanic Courts," *Renaissance Quarterly* 74, no. 4 (2021): 1210–59, esp. 1239.

6. Campanella, *La città del sole*, 4r. For a previous etymology suggesting origins from *generatio*, see Leo Spitzer, "Ratio > Race," *American Journal of Philology* 62 (1941): 129–43. For its connections to the French *haras*, which, in turn, had its origins in Latin *haracium*, see Gianfranco Contini, "I più antichi esempi di razza," in *Studi di Filologia Italiana* 17 (1959): 319–27; Contini, "Tombeau de Leo Spitzer," in his *Varianti e altra linguistica: Una raccolta di saggi (1938–1968)* (Turin: Einaudi, 1970), 651–60; Rosario Coluccia, "L'etimologia di razza: questione aperta o chiusa?," *Studi di Filologia Italiana* 30 (1972): 325–30.

7. On the "perfect" horse, see Sarah Duncan, *Privileged Horses: The Italian Renaissance Court Stable* (Wimbledon: Stephen Morris, 2020), 28–35.

8. Margócsy, "Horses."

9. On Philip II's horses, see Kathryn Renton, "Defining 'Race' in the Spanish Horse: The Breeding Program of King Philip II," in *Horse Breeds and Human Society: Purity, Identity and the Making of the Modern Horse*, ed. Kristen Guest and Monica Mattfield (New York: Routledge, 2020), 13–26; Renton, "Breeding Techniques and Court Influence: Charting a 'Decline' of the Spanish Horse in the Early Modern Period," *Court Historian* 24, no. 3 (2019): 221–34.

10. An undated manuscript is preserved in the Biblioteca Nacional de España: Tomasso Campanella, *Compendio della Monarchia di Spagna*, S. VII, MSS/1416.

11. Germana Ernst, *Il carcere, il politico, il profeta: Saggi su Tommaso Campanella* (Pisa: Istituti Editoriali e Poligrafici Internazionali, 2002).

12. Germana Ernst, *Religione, ragione e natura: Ricerche su Tommaso Campanella e il tardo Rinascimento* (Milan: Franco Angeli, 1991).

13. The issues of breeding and heredity are in some ways already part of Spanish history, although not considered systematically. Concerns about interbreeding and inherited conditions started to increase with the beginning of the seventeenth century. Most famously, this came in the form of the hirsute Gonzalvus family and the sickly

Habsburg progeny. Today diagnosed with hypertrichosis, Petrus Gonzalvus (1537–1618) hailed from the Canary Islands. His excessive amounts of facial hair challenged early modern understandings of heredity and the limits of humanity. At least five of his family members were afflicted with this genetic abnormality, which made them famous in the sixteenth-century court circuit of Ambras and Italy. Merry Wiesner-Hanks, "The Wild and Hairy Gonzales Family," *Apparence(s): Histoire et culture du paraître* 5 (2014). Concerns about the feeble nature of the Habsburg line increased over the course of the War of the Spanish Succession as the Iberian line died out. Although anxiety over consanguineous unions is more the result of modern ideas of incest than a prevalent notion from the time, clearly the power and risks of breeding were visible to early moderns in both the ruling classes and their courtly entertainment, animals and humans alike. The Habsburgs' disastrous inbreeding typified a warning about the dangerous consequences of incest. From 1516 to 1700, it has been estimated that over 80 percent of marriages within the Spanish branch of the Habsburg dynasty were consanguineous; that is, they were marriages between close blood relatives. Now scholars can trace an extraordinarily high inbreeding coefficient that foretold problematic health effects for the Habsburgs. G. Alvarez, F. C. Ceballos, and C. Quinteiro, "The Role of Inbreeding in the Extinction of a European Royal Dynasty," *PLoS One* 4, no. 4 (2009): e5174.

14. On counterpoint, I have tried to confirm, as best as an amateur might, that this model is sustainable. Thanks to the musicologists at I Tatti who entertained my questions. See Iain Fenlon and James Haar, *The Italian Madrigal in the Early Sixteenth Century: Sources and Interpretation* (Cambridge: Cambridge University Press, 1988). In his survey of physiognomic texts, Martin Porter has noted that "there is much more evidence in other treatises to suggest that there appears to have been an increasing tendency to use physiognomy not so much in search of a business partner but in search of a marriage partner, or breeding partner, and in some cases, one might say even in the business of breeding." Martin Porter, *Windows of the Soul: The Art of Physiognomy in European Culture, 1470–1760* (Oxford: Clarendon Press, 2005), 249. See also Joseph Ziegler, "Physiognomy, Science and Proto-racism, 1200–1500," in *The Origins of Racism in the West*, ed. Miriam Eliav-Feldon, Benjamin Isaac, and Joseph Ziegler (Cambridge: Cambridge University Press, 2009), 181–99.

15. On embodiment and animal erotics, see Karen Raber, *Animal Bodies, Renaissance Culture* (Philadelphia: University of Pennsylvania Press, 2013); Juliana Schiesari, *Beasts and Beauties: Animals, Gender, and Domestication in the Italian Renaissance* (Toronto: University of Toronto Press, 2010); Schiesari, *Polymorphous Domesticities: Pets, Bodies, and Desire in Four Modern Writers* (Berkeley: University of California Press, 2012). See also Sarah Cockram and Andrew Wells, *Interspecies Interactions: Animals and Humans between the Middle Ages and Modernity* (New York: Routledge, 2017).

16. GMOs are largely prohibited in Europe, but the traditional methods of breeding and selection (which use the same principles enacted more slowly) are permitted and considered natural.

17. Francis Galton, *Inquiries into Human Faculty and Its Development* (London: J. M. Dent and Sons, 1943), 17, cited in David J. Galton, "Greek Theories on Eugenics," *Journal of Medical Ethics* 24, no. 4 (August 1998): 263–67. See also Sarah Wilmot, "Breeding Farm Animals and Humans," in *Reproduction: Antiquity to the Present Day*, ed. Nick Hopwood, Rebecca Flemming, and Lauren Kassell (Cambridge: Cambridge University Press, 2018), 397–414, esp. 404, 412, which contends that "pedigree thinking intensified a two-way traffic with discourses on human breeding."

18. Siddhartha Mukherjee, *The Gene: An Intimate History* (London: Vintage, 2017), 73; Peter J. Bowler, *Evolution: The History of an Idea*, rev. ed. (Berkeley: University of California Press, 2003).

19. Galton, "Greek Theories on Eugenics."

20. On the Iberian relationship with minorities, see David Nirenberg, *Communities of Violence: Persecution of Minorities in the Middle Ages* (Princeton: Princeton University Press, 1996). Indeed, even catastrophic population collapses across the Americas were generally regretted by the Spanish and other Europeans as (at the very least) a loss of human resources. Both the expulsions from the Iberian Peninsula and the losses in the Americas gave credence to tales of Spanish cruelty, fostering the Black Legend. J. N. Hillgarth, *The Mirror of Spain, 1500–1700: The Formation of a Myth* (Ann Arbor: University of Michigan Press, 2000); Margaret Rich Greer, Walter Mignolo, and Maureen Quilligan, eds., *Rereading the Black Legend: The Discourses of Religious and Racial Difference in the Renaissance Empires* (Chicago: University of Chicago Press, 2008).

21. See Archivio di Stato di Mantova (hereafter cited as ASMN), Archivio Gonzaga (hereafter cited as AG), bk. 258, "Osservazioni particolari di Lodovico Nonio sopra l'infermità che sono occorse, ed occorrono nelli Razze di S.A.S. [Sua Altezza Serenissima] alla Roversella."

22. John Florio, *A World of Words* (London: Melch. Bradwood, 1611), s.v. "generáre." I have followed Hopwood in generally avoiding the language of "reproduction," which developed in the eighteenth century. See Nick Hopwood, "The Keywords 'Generation' and 'Reproduction,'" in Hopwood, Flemming, and Kassell, *Reproduction*, 287–304.

23. The history of breeding in the sixteenth century has focused on Northern Europe; many parallels remain, although Northern Europeans were directly influenced by Southern European practices. See Nicholas Russell, *Like Engend'ring Like: Heredity and Animal Breeding in Early Modern England* (Cambridge: Cambridge University Press, 1986), who focuses on Gervase Markham, Thomas Blundeville, Prospero D'Osma, Christopher Clifford, and the later work of William Cavendish. Harriet Ritvo's scholarship on the Victorian era explores the epistemological consequences of animal labor at the so-called "animal estate," outlines animal race as a tool of taxonomy, and shows how theories of heredity linked to favored animals reveal wider ideas about class, gender, and race. See Harriet Ritvo, *Noble Cows and Hybrid Zebras: Essays on Animals and History* (Charlottesville: University of Virginia Press, 2010); *The Platypus and the Mermaid, and Other Figments of the Classifying Imagination* (Cambridge, MA: Harvard University Press, 1997); and *The Animal Estate: The English and Other Creatures in the Victorian Age* (Cambridge, MA: Harvard University Press, 1987); Rebecca Woods, *The Herds Shot Round the World: Native Breeds and the British Empire, 1800–1900* (Chapel Hill: University of North Carolina Press, 2017); Arthur Oncken Lovejoy, *The Great Chain of Being: A Study of the History of an Idea* (Cambridge, MA: Harvard University Press, 1936) and its critique in Michel Foucault, *Les mots et les choses: une archéologie des sciences humaines* (Paris: Gallimard, 1966).

24. For debates around taxa, see J. R. Hendricks, E. E. Saupe, C. E. Myers, E. J. Hermsen, and W. D. Allmon, "The Generification of the Fossil Record," *Paleobiology* 40 (2014): 511–28; G. Giribet, G. Hormiga, and G. D. Edgecombe, "The Meaning of Categorical Ranks in Evolutionary Biology," *Organisms, Diversity & Evolution* 16 (2016): 427–30.

25. My summary in this paragraph relies on Julia D. Sigwart, Mark D. Sutton, and K. D. Bennett, "How Big Is a Genus? Towards a Nomothetic Systematics," *Zoological Journal of the Linnean Society* 183, no. 2 (2018): 237–52, esp. 237–38.

26. Donna Landry, "Habsburg Lipizzaners, English Thoroughbreds and the Paradoxes of Purity," in Guest and Mattfield, *Horse Breeds and Human Society*, 27–49, esp. 27.

27. John Wilkins, *Species: A History of the Idea* (Berkeley: University of California Press, 2009), 10–11, 19, 40, 168.

28. Wilkins, *Species*, x, also 4–5, 192. I would agree with his characterization that "effectively every philosophical issue raised about biological species in the modern period was raised in one form or another, about philosophical, or logical, species during this interregnum" between Aristotle and Linnaeus's first work on classification.

29. Justin Smith, *Nature, Human Nature, and Human Difference: Race in Early Modern Philosophy* (Princeton: Princeton University Press, 2015), 12.

30. Fernand Braudel, *The Mediterranean and the Mediterranean World in the Age of Philip II*, trans. Siân Reynolds, 2nd rev. ed., vol. 1 (Berkeley: University of California Press, 1995), 226. For the impact of New World practices on Europe, see Marcy Norton, *Sacred Gifts, Profane Pleasures: A History of Tobacco and Chocolate in the Atlantic World* (Ithaca: Cornell University Press, 2008).

31. Exciting work in the field of Spanish Atlantic history aims to engage with the question of empire in such a way that the New World territory is taken seriously as part of Spain. See, for instance, Tamar Herzog, *Defining Nations: Immigrants and Citizens in Early Modern Spain and Spanish America* (New Haven: Yale University Press, 2008); Herzog, *Frontiers of Possession: Spain and Portugal in Europe and the Americas* (Cambridge, MA: Harvard University Press, 2015); and Antonio Feros, *Speaking of Spain: The Evolution of Race and Nation in the Hispanic World* (Cambridge, MA: Harvard University Press, 2017). See also Matthew Restall, "The Renaissance World from the West: Spanish America and the 'Real' Renaissance," in *A Companion to the Worlds of the Renaissance*, ed. Guido Ruggiero (Oxford: Blackwell Publishing, 2007), 70–88.

32. María Elena Martínez, *Genealogical Fictions: Limpieza de Sangre, Religion, and Gender in Colonial Mexico* (Stanford: Stanford University Press, 2008), 2.

33. Michael Servetus (1509–1553), a theologian and physician born in Navarre, explored the idea of pulmonary circulation in his anti-trinitarian treatise *Christianismi Restituito* (Vienna, 1553), linking the physicality of blood with ideas about clean blood and predestined individuals.

34. Camilla Townsend, *Fifth Sun: A New History of the Aztecs* (Oxford: Oxford University Press, 2019), 213–15.

35. James Lockhart, *The Nahuas after the Conquest: A Social and Cultural History of the Indians of Central Mexico, Sixteenth Through Eighteenth Centuries* (Stanford: Stanford University Press, 1992).

36. Frances Dolan, *Digging the Past: How and Why to Imagine Seventeenth-Century Agriculture* (Philadelphia: University of Pennsylvania Press, 2020), 4. On Renaissance agriculture, see Mauro Ambrosoli, *The Wild and the Sown: Botany and Agriculture in Western Europe, 1350–1850* (Cambridge: Cambridge University Press, 2009).

37. William Shakespeare, *Othello*, in *The Annotated Shakespeare* (New Haven: Yale University Press, 2005), act I, scene 1, lines 109–12. Thanks to Stephanie Bahr for suggesting this passage.

38. Shakespeare, *Othello*, I.1.113–15.

39. *Oxford English Dictionary Online*, s.v. "jennet, n.1."

40. Here, I draw on a manuscript created by Martín de Villaverde for Don Juan of Austria. The original manuscript is held in the Biblioteca del Monasterio de Santa María de la Vid in Burgos. I use the manuscript's pagination of 116r for this quotation. I have also consulted the facsimile, Martín de Villaverde, *Bestiario de Don Juan de Austria* [Texto impreso]: S. XVI: original conservado en la Biblioteca del Monasterio de Santa María de la Vid (Burgos) (Burgos: Gil de Siloé, 1998) held at the Biblioteca Nacional de España.

41. Don Sebastian de Covarrubias Orozco, *Tesoro de la lengua castellana o española* (Madrid: L. Sanchez, 1611), *honestidad*.

42. *Bestiario*, 116r.

43. David Reich has shown that genetics teaches us something about populations, but some have inappropriately extended that to behavior and culture. Thinking about genomic data requires a middle ground between the two horns of the dilemma: "On the one side there are beliefs about the nature of the differences that are grounded in bigotry that have little basis in reality. On the other side there is the idea that biological differences among populations are so modest that as a matter of social policy they can be ignored and papered over." This desire to find an analytical category as a middle ground between two poles is itself an omnipresent feature of discussions of difference in the history of biology. In his synthesis of aDNA research to address questions of human origins, Reich resisted the idea that humans could be biologically grouped into populations that separated tens of thousands of years ago that correlate to the idea of "races"; rather, genomic data suggests that populations in the recent past were divided along radically different fault lines. Deep genomic differences do not map onto comparatively recently devised racial categories. Social fissures waxed and waned, and the current fracture patterns of discrimination emerged not as an inevitability, but as the specific result of admixture and discriminatory events in the last five hundred years. David Reich, *Who We Are and How We Got Here: Ancient DNA and the New Science of the Human Past* (New York: Vintage Books, 2018), 253.

44. For a summary and use case, see Kenneth Gouwens, "Erasmus, 'Apes of Cicero,' and Conceptual Blending," *Journal of the History of Ideas* 71, no. 4 (2010): 532–45; George Lakoff and Mark Johnson, *Metaphors We Live By* (Chicago: University of Chicago Press, 1980); Gilles Fauconnier and Mark Turner, *The Way We Think: Conceptual Blending and the Mind's Hidden Complexities* (New York: Basic Books, 2002), esp. 39–57. Thanks to Dan Smail and Suzanne Keen for pointing me toward this literature. For animal metaphor and its links to violence, see Dominick LaCapra, *History and Its Limits: Human, Animal, Violence* (Ithaca: Cornell University Press, 2009).

45. Covarrubias, *Tesoro* (1611), *raza*. Kathryn Renton has shown how Covarrubias's definitions of *raza* and *casta* conflated horse race and animal race: Renton, "A Social and Environmental History of the Horse in Spain and Spanish America, 1492–1600" (Ph.D. dissertation, University of California, Los Angeles, 2019), 204–6; on the difference between *raza* and *casta*, see 136–48. See also Arturo Morgado García, "The Animal World Vision in 17th Century Spain: The Covarrubias Bestiary," *Cuadernos de historia moderna* 36 (2011): 67–88.

46. Javier Irigoyen-García, *The Spanish Arcadia: Sheep Herding, Pastoral Discourse, and Ethnicity in Early Modern Spain* (Toronto: University of Toronto Press, 2014), 42.

47. Joshua Bennett, *Being Property Once Myself: Blackness and the End of Man* (Cambridge, MA: Harvard University Press, 2020), 4. Bennett has shown how African American authors have worked to reclaim personhood from the animal metaphors, and the legacy of the history discussed below, which systematized their erasure as the centuries wore on. For a compelling argument concerning race and animality in a multicultural arena, see Claire Jean Kim, *Dangerous Crossings: Race, Species, and Nature in a Multicultural Age* (Cambridge: Cambridge University Press, 2015).

48. Duncan, *Privileged Horses*, 36.

49. Russell, *Like Engend'ring Like*, 13–18, 64; Margócsy, "Horses," 1215, 1222, 1242.

50. Jacob Burckhardt, *The Civilization of the Renaissance in Italy*, trans. S. G. C. Middlemore, revised and edited by Irene Gordon (New York: New American Library, 1960), 157.

51. Burckhardt, *Civilization*, 157.

52. *Mestizaje* plays a key role in colonial Latin American history. See Ben Vinson III, *Before Mestizaje: The Frontiers of Race and Caste in Colonial Mexico* (New York: Cambridge University Press, 2017); and Joanne Rappaport, *The Disappearing Mestizo: Configuring Difference in the Colonial New Kingdom of Granada* (Durham: Duke University Press, 2014).

53. Lovejoy, *Great Chain of Being*, 58–59.

54. José de Acosta, *Historia natural y moral de las Indias [. . .]* (Sevilla: Iuan de Leon, 1590) (hereafter cited as *HNMI*), bk. 1, chap. 20; Acosta cites Genesis 7:2–3.

55. Renton, "Defining 'Race' in the Spanish Horse," 13–27, esp. 13–14.

56. Landry, "Habsburg Lipizzaners," 27–49.

57. Woods, *Herds Shot Round the World*, 98–99; Rebecca Cassidy, *The Sport of Kings: Kinship, Class, and Thoroughbred Breeding in Newmarket* (Cambridge: Cambridge University Press, 2009), 20.

58. English Springer Spaniel Club, "History of the Breed," accessed December 28, 2021, http://www.englishspringer.org/breed-history/; American Kennel Club, "Xoloitzcuintli," accessed December 28, 2021, https://www.akc.org/dog-breeds/xoloitzcuintli/.

59. Emma Spary, "Political, Natural and Bodily Economies," in *Cultures of Natural History*, ed. N. Jardine, J. A. Secord, and E. C. Spary (Cambridge: Cambridge University Press, 1996), 178–96, esp. 184, 186–187, 194. On *casta* paintings, see Rebecca Earle, "The Pleasures of Taxonomy: Casta Paintings, Classification, and Colonialism," *William and Mary Quarterly* 73, no. 3 (July 2016): 427–66; Ilona Katzew, *Casta Paintings: Images of Race in Eighteenth-Century Mexico* (New Haven: Yale University Press, 2004); and Katzew, *Inventing Race: Casta Painting and Eighteenth-Century Mexico* [exhibition brochure for eponymous LACMA exhibition] (Los Angeles: Los Angeles County Museum of Art, 2004); William Max Nelson, "Making Men: Enlightenment Ideas of Racial Engineering," *American Historical Review* 115, no. 5 (December 2010): 1364–94, esp. 1369.

Chapter One

1. Timothy J. Reiss, "Structure and Mind in Two Seventeenth-Century Utopias: Campanella and Bacon," *Yale French Studies* 49 (1973): 82–95. See also Bacon and Campanella, *The New Atlantis & The City of the Sun*; Jürgen Klein and Guido Gigli-

oni, "Francis Bacon," in *Stanford Encyclopedia of Philosophy,* Fall 2020 edition, https://plato.stanford.edu/entries/francis-bacon/; Francis Bacon, *Novum Organum,* ed. Joseph Devey (New York: P. F. Collier & Son, 1902).

2. Francis Bacon with William Rawley, "New Atlantis: A Worke Unfinished," in *Sylva sylvarum: or, A Natural History* (London: W. Lee, 1627), 35–36; Paolo Rossi, *Francesco Bacone: Dalla magia alla scienza* (Turin: Einaudi, 1957).

3. Paula Findlen, "Anatomy Theaters, Botanical Gardens, and Natural History Collections," in *Early Modern Science,* ed. Katharine Park and Lorraine Daston (New York: Cambridge University Press, 2006), 272–88, esp. 272–73.

4. Graeme Barker, *The Agricultural Revolution in Prehistory: Why Did Foragers Become Farmers?* (New York: Oxford University Press, 2006), 149–272.

5. Genesis 30:33–43 (King James Version). Irigoyen-García, *Spanish Arcadia,* 60–66, traces the use of Jacob's breeding efforts in Francisco de Vallés's *De Sacra Philosophia* (1587), Cosme Gómez Tejada's *El filósofo* (1650), Luis Vélez de Guevara's *La hermosura de Raquel* (1615), and Pedro Orrente's painting *Jacob Watering Laban's Sheep before Peeled Branches* (1612–22), among other sources. Thanks to Hannah Marcus for suggesting this connection.

6. Russell, *Like Engend'ring Like,* 12.

7. Russell, *Like Engend'ring Like,* 8–21.

8. On artisanal knowledge, see Edgar Zilsel, *The Social Origins of Modern Science,* Boston Studies in the Philosophy of Science (Dordrecht: Springer Netherlands, 2003); Pamela Long, *Artisan/Practitioners and the Rise of the New Sciences, 1400–1600* (Corvallis: Oregon State University Press, 2011); Antonio Sánchez and Henrique Leitão, eds., "Artisanal Culture in Early Modern Iberian and Atlantic Worlds," special issue, *Centaurus* 60, no. 3 (2018): 135–40.

9. Pamela Smith, *The Body of the Artisan: Art and Experience in the Scientific Revolution* (Chicago: University of Chicago Press, 2006), 17.

10. For a recent overview from which I draw the following summary, see Justin Smith, "Introduction," in *The Problem of Animal Generation in Early Modern Philosophy,* ed. Justin Smith (New York: Cambridge University Press, 2006), esp. 1–21.

11. Russell, *Like Engend'ring Like,* 31–32. Aristotle's role in Renaissance thought was complicated, but he was not displaced. Eugenio Refini, *The Vernacular Aristotle: Translation as Reception in Medieval and Renaissance Italy* (New York: Cambridge University Press, 2020); Charles B. Schmitt, *Aristotle and the Renaissance* (Cambridge, MA: Harvard University Press for Oberlin College, 1983).

12. Paul Studtmann, "Aristotle's Categories," in *Stanford Encyclopedia of Philosophy,* Fall 2018 ed., https://plato.stanford.edu/archives/fall2018/entries/aristotle-categories/.

13. By contrast, pangenesists argued that all parts of the body contain tiny traces of each organ, adopting a vision that Charles Darwin would later explore in his theory of "cell-gemmules" to provide a mechanism of inheritance for evolution.

14. Rebecca Flemming, "Galen's Generations of Seeds," in Hopwood, Flemming, and Kassell, *Reproduction,* 95–108.

15. Nahyan Fancy, "Generation in Medieval Islamic Medicine," in Hopwood, Flemming, and Kassell, *Reproduction,* 129–40.

16. James G. Lennox, "The Comparative Study of Animal Development: William Harvey's Aristotelianism," in Justin Smith, *Problem of Animal Generation,* 21–46.

17. Bertoloni Meli, *Marcello Malpighi and Seventeenth-Century Anatomy* (Balti-

more: Johns Hopkins University Press, 2011); Clara Pinto-Correia, *The Ovary of Eve: Egg and Sperm and Preformation* (Chicago: University of Chicago Press, 1997), 65–103.

18. Lorraine Daston and Katharine Park, *Wonders and the Order of Nature, 1150–1750* (New York: Zone Books, 1998), 173–201.

19. For a survey of the popular adoption of "Aristotle's Masterpiece" for "improved" reproduction, see Rudolph Bell, *How to Do It: Guides to Good Living* (Chicago: University of Chicago Press, 1999), 17–72.

20. Annie Bitbol-Hespériès, "Monsters, Nature, and Generation from the Renaissance to the Early Modern Period," in Justin Smith, *Problem of Animal Generation*, 47–62, esp. 49.

21. Justin Smith, *The Philosopher: A History in Six Types* (Princeton: Princeton University Press, 2016), 1–21.

22. Russell, *Like Engend'ring Like*, 39.

23. Russell, *Like Engend'ring Like*, 38.

24. Smith, *The Body of the Artisan*, 6–8.

25. Anastasija Ropa and Timothy Dawson, eds., *The Horse in Premodern European Culture*, Studies in Medieval and Early Modern Culture (Boston: Walter de Gruyter, 2019), 45–69, 177–203.

26. See, for example, Renaissance saddle, Smithsonian American Art Museum, 1600–1700, 1929.8.435.1; Saddle, ca. 1570–80, Milan, MET 4.3.252.

27. See, for instance, the Neapolitan bit of the 1350s, MET 4.3.478ab. On bits for riding and training, see Filippo Orsi, *Trattato sui freni dei cavalli*, Biblioteca Apostolica Vaticana (hereafter cited as BAV), Urb.lat.270, 1r–85r; "Trattato sui freni e sui capestri dei cavalli con illustrazioni," with illustrations by Silvestro Vanzy, 1620, BAV Urb.lat.271, 2v–47v.

28. Russell, *Like Engend'ring Like*, 216.

29. Personal correspondence with Ryan Kennedy, Susan deFrance, Nora Battermann, and David Orton concerning the zooarchaeology of Renaissance curiosity cabinets, June 2020.

30. Personal conversation with zooarchaeologists Elizabeth Reitz and Barnet Pavao-Zuckerman, June 8, 2020. For projects aiming to tackle this issue, see the forthcoming dissertation on the colonial archaeology of cattle farming by Nicholas Del Sol at the University of Florida.

31. This approach follows Mary Terrall's history of the naturalist Réaumur's efforts to catch nature in the act of reproduction, and subsequent attempts to understand the making of life, from the generation of animals to asexual reproduction. Mary Terrall, *Catching Nature in the Act: Réaumur & the Practice of Natural History in the Eighteenth Century* (Chicago: University of Chicago Press, 2014).

32. Duncan, *Privileged Horses*, 28. This potential for perfection differed from the Spanish emphasis on good *casta* more broadly.

33. Russell, *Like Engend'ring Like*, 24, 87.

34. Duncan, *Privileged Horses*, 28.

35. Anthony Grafton, *Leon Battista Alberti: Master Builder of the Italian Renaissance* (Cambridge, MA: Harvard University Press, 2000), 190. Alberti's influence as a model for Renaissance excellence remained profound. Mantuan legend would even weave him into the court's posterity, suggesting that Andrea Mantegna's court scene in the Camera degli Sposi depicted him as the man leaning down to speak counsel in Ludovico Gonzaga's ear (see fig. 3.3).

36. Grafton, *Leon Battista Alberti*, 190.

37. For later versions of this genre, see Giulio Rutati, "Discorso al Ser.mo principe d'Urbino [Francesco Maria II Della Rovere] mio signore, dove accennandosi quanto sia utile l'imitazione degli antenati si discorse brevemente dell'uso del cavallo," BAV Urb.lat.860, 527r–544r. Many texts from this period translate and recopy the works: for example, Laurentius Rusius (1288–1347), "Della cura dei cavalli," BAV Urb.lat.256, 2r–83r; "Modo et ordine per creare razza di cavalli," BAV Urb.lat.251, 1r–12r; "De' tempi e misure ch'osservar deve il Cavaliere facendo operare i cavalli ne' maneggi per farli giusti e facili," BAV Urb.lat.255, 21r–31r; "Modo che si tiene per imbrigliar cavalli," BAV Urb.lat.269, 1r–36r.

38. Grafton, *Leon Battista Alberti*, 190; Duncan, *Privileged Horses*, 14.

39. Russell, *Like Engend'ring Like*, 23.

40. "Modo et ordine per crear razza de cavalli," BAV Cappon. 172, 224r–232r, esp. 224r. Thanks to Byron Hamann for recommending this source.

41. BAV Cappon. 172, 226v, 229r.

42. BAV Cappon. 172, 230r.

43. BAV Cappon. 172, 232r.

44. Giovanni Bernardino Papa, "Trattato delle Razze dei cavalli, diretto al grand-uca di Toscana nel di 25 marzo 1607," Biblioteca Marciana, 5223, Cod. 24 Cart in 40, sec. S. 7, esp. 6, 8, 11–12, 28.

45. Russell, *Like Engend'ring Like*, 67.

46. Papa, "Trattato delle Razze dei cavalli," 6, 8, 11–12, 28.

47. *De equo animante* survives in two non-autograph fifteenth-century manuscript copies and a Basel edition from 1556. L. B. Alberti, *De equo animante*, ed. C. Grayson, J. Y. U. Boriaud, and F. Furlar, *Albertiana* 2 (1999). For the extensive literature on Alberti, see Elisabetta Di Stefano, "Il *De equo animante* di LB Alberti: una teoría della belezza?," *Albertiana* 1 (2010): 15–26.

48. On the rise of formal veterinary expertise, see Janice Gunther Martin, "Un-burdening the Beasts: Equine Medicine and Expert Healers in Early Modern Castile" (Ph.D. dissertation, University of Notre Dame, 2021). Of the many treatises on equine medicine, see "Alcune espressioni latine relative all'allevamento dei cavalli," BAV Urb.lat.1343; Giordano Ruffo, "Hippiatria" and "Raccolta di dicette per cavalli," BAV Urb.lat.1413, 1r–73v.

49. Papa, "Trattato delle Razze dei cavalli," 31, 32, 36.

50. Russell, *Like Engend'ring Like*, 67.

51. Papa, "Trattato delle Razze dei cavalli."

52. Papa, "Trattato delle Razze dei cavalli," 34.

53. Russell, *Like Engend'ring Like*, 63, 66, 78. On this point, Sarah Duncan cites Marco de Pavari's 1581 treatise. Duncan, *Privileged Horses*, 35.

54. "Libro della razza dei cavalli di Sua Altezza Serenissima [Francesco Maria II della Rovere, duca di Urbino] dal 1614 al 1618," BAV Urb.lat.254, 2r. Thanks to Byron Hamann for recommending this source.

55. "Libro della razza dei cavalli," BAV Urb.lat.254, 1r–6r. For similar records in England, see Russell, *Like Engend'ring Like*, 72–76.

56. Archivio di Stato di Napoli, Dipendenza della Sommaria (hereafter cited as ASN DS), I Serie 38, n. 3, 1542, 17.

57. ASN DS I Serie 38, n. 3, 1543, 288. For example, "Supra anno femine: la figlia de Luchia; La Figlia di Prestera; La figlia di Margante; La figlia di Cratice; La figlia de

Conella; La figlia de Zotunda." And "Yenebi masculi: Lo boi figlio di Spefera; Lo boi figlio di Sansza mla madra; Lo boy figlio di mubrulina; Lo boy figlio di Brance turco; Lo boi figlio di Singnora; Lo boi figlio di Cunena turco."

58. Margócsy, "Horses," 1240.

59. Aristotle, *Historia Animalium*, trans. D. M. Balme and Allan Gotthelf, bk. 4, 18 (Cambridge: Cambridge University Press, 2002). See Conway Zirkle, "Animals Impregnated by the Wind," *Isis* 25, no. 1 (May 1936): 95–130, esp. 97. Varro's writing on breeding emphasized selecting to promote animals' economic value and observable external characteristics. Although widely cited, manuscript editions of Varro were in poor condition with many corruptions. Columella emphasized inheritance patterns from grandparents to sons and noted both breeding value (the animal's ability to generate valuable offspring) and appearance (today described as phenotype) without presuming the two to overlap directly. Russell, *Like Engend'ring Like*, 33, 36.

60. Howard B. Adelmann, ed., *The Embryological Treatises of Hieronymus Fabricius of Aquapendente*, vol. 1 (Ithaca: Cornell University Press, 1942), 92, 171.

61. Virgil, *Georgics*, bk. 3, 270–80. For Porta on Virgil, see Porta, *Della fisionomia dell'uomo*, libri sei, vol. 2, ed. Alfonso Paolella, Edizione nazionale delle opere di Giovan Battista della Porta (Napoli: Edizioni Scientifiche Italiane, 2011), bk. 1, chap. 3. Porta quotes Virgil's *Georgics*, bk. 3, 75–88.

62. Duncan, *Privileged Horses*, 12, 221–37. For the work of the blacksmith/farrier (*marescalco*) and early equine medicine, see "Libro di marescalcia, di Giordano Ruffo," as cited by Duncan. For themes of horse keeping, blacksmithing, and masculinity as they appear in a period comedy set in Renaissance Mantua, see Pietro Aretino, *Il marescalco* (Stampata in Vinegia: Per M. Bernardino de Vitali Veneto, 1533).

63. Papa, "Trattato delle Razze dei cavalli," 2–3.

64. Neil Tarrant, "Giambattista della Porta and the Roman Inquisition: Censorship and the Definition of Nature's Limits in Sixteenth-Century Italy," *British Journal for the History of Science*, 46, no. 4 (December 2013): 601–25, esp. 601.

65. For an overview, see Jan Aertsen, *Nature and Creature: Thomas Aquinas' Way of Thought* (Leiden: E. J. Brill, 1988).

66. On cultures of experimentation and natural secrets, see William Eamon, *Science and the Secrets of Nature: Books of Secrets in Medieval and Early Modern Culture* (Princeton: Princeton University Press, 1994), esp. 91–266.

67. For cases of human love magic in the Venetian archives, see Guido Ruggiero, *Binding Passions: Tales of Magic, Marriage, and Power* (Oxford: Oxford University Press, 1993). Breeding practices did not tend to attract inquisitorial attention, but the practice of "magic arts" was by far the most common accusation in inquisitorial proceedings in Naples. Francisco Bethencourt, *The Inquisition: A Global History, 1478–1834*, trans. Jean Birrell (Cambridge: Cambridge University Press, 2009), table 8.3, "Trials of the Inquisition of Naples," 335. "Magical arts" made up 24 percent of the 735 cases from 1564 to 1590, and 49 percent of the 1,021 cases from 1591 to 1620.

68. Michaela Valente, "Della Porta e l'Inquisizione: Nuovi documenti dell'Archivio del Sant'Uffizio," *Bruniana & Campanelliana* 5, no. 2 (1999): 415–34, esp. 421.

69. Sergius Kodera, "The Laboratory as Stage: Giovan Battista della Porta's Experiments," *Journal of Early Modern Studies* 3 (2014): 15–38. One might also consider the links between Porta and Paracelsus, and the latter's impact on Porta's theories of breeding.

70. Giambattista Della Porta, *Dei miracoli et maravigliasi effetti dalla natura pro-*

dotti, Italian translation of the 1558 edition of *Magia naturalis* (Venice: Ludovico Avanzi, 1560), 1r–v. Quoted in Tarrant, "Giambattista della Porta and the Roman Inquisition," 615.

71. Pliny the Elder, *The Natural History*, trans. John Bostock (London: Taylor and Francis, 1855), chap. 10, "Striking Instances of Resemblance."

72. Giovanni Battista Della Porta, *Magiae naturalis* (Lugd. Batavorum, 1651), bk. 2, chap. 18, cited in Sergius Kodera, "Humans as Animals in Giovan Battista della Porta's *Scienza*," *Zeitsprünge* 17 (2013): 414–32.

73. Porta, *Magia naturalis*, 1561, fol. 33r.

74. Io. Baptista Porta, *Magiae Naturalis, sive de miraculis rerum naturalium*, liber 4 (Antverpiae: Ex officina Christophori Plantini, 1576), 177–78, cited in Kodera, "Humans as Animals," 111. Ruminations on bestiality were not unique to Porta, nor invented by him. Keagan Brewer, "Talking Wolves, Golden Fish, and Lion Sex: The Alterations to Gerald of Wales's *Topographia Hibernica* as Evidence of Audience Disbelief?," *Parergon* 37, no. 1 (2020): 27–53, doi:10.1353/pgn.2020.0057.

75. On mules in Columella, see, for example, Lucius Junius Moderatus Columella, *On Agriculture*, vol. 2, trans. E. S. Forster and Edward H. Heffner (Cambridge, MA: Harvard University Press, 1954), bk. 5, chap. 27, lines 1–4.

76. Aristotle, *De Generatione Animalium*, bk. 2, chap. 8.

77. Aristotle, *De Generatione Animalium*, bk. 2, chap. 8.

78. Porta, *Magiae Naturalis* (1576), 171. Stables were often constructed of brick or stone, were two or three stories tall, and could house up to a hundred horses. Many featured decorations on both their exterior and interior walls. Architects designed stables around good water supply, drainage, and hayrack placement. Duncan, *Privileged Horses*, 18, 89–188. See the summary of stables, including patron, location, architect, dimensions, capacity, water supply, and design features, in Duncan, *Privileged Horses*, 237–53.

79. Porta, *Magiae Naturalis* (1576), 171–73.

80. Porta, *Magiae Naturalis* (1576), 173. See also Paula Findlen, "Humanism, Politics, and Pornography in Renaissance Italy," in *The Invention of Pornography: Obscenity and the Origins of Modernity*, ed. Lynn Hunt (Cambridge, MA: MIT Press, 1996).

81. Giovanni Battista Della Porta, *Della Fisionomia dell'uomo*, libri sei, vol. 2, bk. 1, chap. 3, 16.

82. Federico Grisone, *Gli Ordini di cavalcare* (Venetia: Appresso Vincenzo Valgrisi nella bottega d'Erasmo, 1551), bk. 1. See also Federico Grisone, *Federico Grisone's "The Rules of Riding": An Edited Translation of the First Renaissance Treatise on Classical Horsemanship*, ed. Elizabeth M. Tobey (Tempe: Arizona Center for Medieval and Renaissance Studies, 2014), 76–77.

83. Grisone, *The Rules of Riding*, 82.

84. Frances Gage, "Human and Animal in the Renaissance Eye," in "The Animal in Renaissance Italy," special issue, *Renaissance Studies* 31, no. 2 (2017): 261–72. Gage has shown that eye color in horses posed a problem for practical physiognomy, as clever noble horses were prized for black eyes while oxen, cows, and bulls with black eyes were thought to be slow and stupid.

85. Grisone, *The Rules of Riding*, 84.

86. For astrological prediction and human fates, see Monica Azzolini, *The Duke and the Stars: Astrology and Politics in Renaissance Milan* (Cambridge, MA: Harvard University Press, 2013).

87. Saint Augustine, *The Catholic and Manichaean Ways of Life*, trans. D. A. and I. J. Gallagher, Fathers of the Church, vol. 56 (Washington, DC: Catholic University of America Press, 1966), 102–3, 108–9. For the Renaissance connections, see Meredith Gill, *Augustine in the Italian Renaissance: Art and Philosophy from Petrarch to Michelangelo* (Cambridge: Cambridge University Press, 2005), 199; Erica Fudge, *Brutal Reasoning: Animals, Rationality, and Humanity in Early Modern England* (Ithaca: Cornell University Press, 2006).

88. For efforts to train falcons, see Giancarlo Malacarne, *I signori del cielo: la falconeria a Mantova al tempo dei Gonzaga* (Mantua: Artiglio Editore, 2003); Malacarne, *Le cacce del principe: l'ars venandi nella terra dei Gonzaga* (Modena: Il Bulino, 1998). For contemporary awareness of this training, see the discussion of Castiglione in chapter 3 of this book.

89. Cola Pagano had trained in the courts of King Henry VIII of England and of Philibert of Chalon, prince of Orange and prominent sacker of Rome. Cola's father, another trainer by the name of Monte Pagano, ran the stable of the king of Naples, Frederick of Aragon (1452–1504). Though there is little archival evidence about his life, the genealogies of Grisone's contemporaries Giovan Battista Ferraro (author of *Delle razze disciplina del cavalcare* [1560]) and Pasquale Caracciolo reveal some details. Elizabeth M. Tobey, "Legacy of Grisone," in *The Horse as Cultural Icon: The Real and Symbolic Horse in the Early Modern World*, ed. Peter Edwards, Karl A. E. Enenkel, and Elspeth Graham, Intersections: Interdisciplinary Studies in Early Modern Culture, vol. 18 (Leiden: Brill, 2011), 143–74. Ferrarese riding master Cesare Fiaschi established his own academy in Ferrara in 1534 and then wrote his own book, *Trattato dell'imbrigliare, maneggiare, et ferrare cavalli* (Bologna: Per Anselmo Giaccarelli, 1556). Other equine experts linked to Naples included Giovanni Pontano and Giovaiano Miao. Duncan, *Privileged Horses*, 43nn19–20.

90. On the overlap between humanism and horsemanship, see Hilda Nelson, "Antoine de Pluvinel, Classical Horseman and Humanist," *French Review* 58, no. 4 (March 1985): 514–23; Raber, *Animal Bodies*, 87–92. Thanks to Amanda Miller for sharing her thesis on Pluvinel.

91. Duncan, *Privileged Horses*, 36. Pluvinel's Parisian riding academy also trained the German Frazser and the Spanish Vargas and Paolo d'Aquito.

92. Atti di Governo, Studi parte Antica, 19, Biblioteche spec. inventari di libri 1569, Archivio di Stato di Milano.

93. These books are Francesco Sforzino, *Tre libri de gli uccelli, con un trattato de' cani da caccia* (Vicenza: Il Megietti, 1568); and Giovanni Miranda, *Osservationi della lingua castigliana* (Venezia: Gabriel Giolito de Ferrari, 1566).

94. Grisone's methods of horsemanship did not only circulate in print. The Biblioteca Nacional de España holds three manuscript translations of the text, which had been housed in libraries from Andalusia to Mallorca. There is one rough translation of the Italian text into Spanish, completed in an Iberian humanistic hand in the sixteenth century, and annotated in yet a different hand. The author of *Ordenes de cabalgar* made beautiful copies of several of Grisone's exercises, like in the print book, with many images of bits found at the end of the text. MSS/9999, *Ordenes de cabalgar De Fadrique Grison, cavallero Napolitano; traducidas de la lengua ytaliana en vulgar castellano*, S. 16. Yet another manuscript dedicated to the Conde de Miranda included several of the books, but no images. While the text was not reprinted extensively in the seventeenth century, that did not mean that readers lost interest. A 1553 Venetian

copy found its way to a Mallorcan library by the 1640s. R/11548, Federico Grisone, *Ordini di cavalcare* (Vinegia: Vincenzo Valgrisi alla bottega d'Erasmo, 1553). The most beautiful of such books was found in the library of "the most illustrious Lord Don Alvaro de Luna." MSS/1059, Federico Grisone, *Órdenes de caballear de Federico Grisone, caballero Napolitano, traducido de la lengua italiana en romance castellano.*

95. Barbara Furlotti, *A Renaissance Baron and His Possessions: Paolo Giordano I Orsini, Duke of Bracciano (1541–1585)* (Turnhout, Belgium: Brepols Publishers, 2012), 283–91; see 281–82 for a 1584 inventory of the duke's horses.

96. For example, Grisone, *The Rules of Riding*, 304–5.

97. Sylvia Loch, *The Royal Horse of Europe: The Story of the Andalusian and Lusitano* (London: J. A. Allen, 1986); Loch, *Dressage: The Art of Classical Riding* (London: Trafalgar Square Publishing, 1990). See also Walter A. Liedtke, *The Royal Horse and Rider: Painting, Sculpture and Horsemanship, 1500–1800* (New York: Abaris Books in association with the Metropolitan Museum of Art, 1989).

98. Grisone, *The Rules of Riding*, 357.

99. Grisone, *The Rules of Riding*, 360–62, 366.

100. Grisone, *The Rules of Riding*, 366.

101. Grisone, *The Rules of Riding*, 71.

102. Grisone, *The Rules of Riding*, 356.

103. Clearly, the estate was a beautiful destination in Mantua's Sermide territory, to which many nobles ventured "with great pleasure and contentment" to visit the various races of horses the Gonzaga cultivated in the countryside. ASMN AG b. 2521, c. 346, Gian Giacomo Calandra to da Gonzaga, 3 settembre 1534.

104. ASMN AG b. 258.

105. Columella, *On Agriculture/Res Rustica*, vol. 2, bk. 6, chap. 2, line 12.

106. ASMN AG b. 258.

107. Russell, *Like Engend'ring Like*, 87.

Chapter Two

1. For the line between race and breed, see John Borneman, "Race, Ethnicity, Species, Breed: Totemism and Horse-Breed Classification in America," *Comparative Studies in Society and History* 30, no. 1 (January 1988): 25–51; Harriet Ritvo, "Race, Breed, and the Myths of Origin: Chillingham Cattle as Ancient Britons," *Representations* 39 (Summer 1992): 1–22.

2. Nonhuman animals also react to difference and develop in-group/out-group dynamics: see Jane Goodall, *Through a Window: My Thirty Years with the Chimpanzees of Gombe* (1990; Boston: Mariner Books, 2010), on chimpanzee war (chap. 10) and relationships with baboons (chap. 12); Robert Sapolsky, *Behave: The Biology of Humans at Our Best and Worst* (London: Vintage, 2018), 34, 387–424. Among the many studies on "us"/"them" category making, see A. S. Baron and M. R. Banajii, "The Development of Implicit Attitudes: Evidence of Race Evaluations from Ages 6 and 10 and Adulthood," *Psychological Science* 17, no. 1 (2006): 53–58. For more recent work on animals, see Duncan A. Wilson and Masaki Tomonaga, "Exploring Attentional Bias towards Threatening Faces in Chimpanzees Using the Dot Probe Task," *PLoS One* 13, no. 11 (2018); Marc Kissel and Nam King, "The Emergence of Human Warfare: Current Perspectives," *Yearbook of Physical Anthropology* 168 (2019): 141–63.

3. Spitzer, "Ratio > Race," 129–43; Contini, "I più antichi esempi di razza," 319–27; Contini, "Tombeau de Leo Spitzer," 651–60; Coluccia, "L'etimologia di razza," 325–30.

4. *Oxford English Dictionary Online*, s.v. "race," n. 5. As a mark on an animal, s.v. "race, " n. 4.

5. Charles du Fresne, sieur Du Cange, *Le Glossarium mediae et infimae latinitatis* (Niort: L. Favre, 1883–87) defines *haracium* as "grex equorum, nostris vulgo *haras*; ab *hara* forte—stabulo—vel grege porcorum. Alii vocem ex Italico *la razza*, seu Gallico *la race*, formatam volunt, quod in eiusmodii *Haraciis* equi secundum ortus, generationis, notionis speciem distinguantur. Nemesianus in Cynegetico: . . . delectus equorum / Quos Phrygiae matres Argaeque gramina pasta / Semine Cappidocum sacris praesepibus edunt." Du Cange provides the wrong citation here, which comes not from Nemesianus, a minor third-century poet (though he apparently does talk about horse breeds at some length in the *Cynegeticus*), but rather Claudian's *Laus Serenae Reginae*, a panegyric dedicated to the wife of Stilicho, the fifth-century Roman-Vandalic field marshal. Other medieval citations use *haracium* to refer to those horses belonging to the king or the royal stables. Thanks to John Eldevik for this citation.

6. *Oxford English Dictionary Online*, s.v. "haras, n."

7. Linguists have sought to show the centrality of metaphor to our conceptual system. Lakoff and Johnson suggest that "the essence of metaphor is understanding and experiencing one kind of thing in terms of another" and that "metaphors that are outside a period's conventional conceptual system can offer new understandings of experience." Lakoff and Johnson, *Metaphors We Live By*, 3–5, 139.

8. On ethno-religious-national identity, see Carlos Eduardo Amorim et al., "Understanding 6th-Century Barbarian Social Organization and Migration through Paleogenomics," *Nature Communications* 9, no. 3547 (2018); Patrick Geary, *The Myth of Nations: The Medieval Origins of Europe* (Princeton: Princeton University Press, 2002).

9. For future directions, see RaceB4Race, which is an "ongoing conference series and professional network community by and for scholars of color working on issues of race in premodern literature, history, and culture. RaceB4Race centers the expertise, perspectives, and sociopolitical interests of BIPOC scholars, whose work seeks to expand critical race theory." Arizona Center for Medieval and Renaissance Studies, accessed January 9, 2022, https://acmrs.asu.edu/RaceB4Race. See also the associated book series RaceB4Race: Critical Race Studies of the Premodern, edited by Geraldine Heng and Ayanna Thompson, published by the University of Pennsylvania Press.

10. See Nancy E. van Deusen, *Global Indios: The Indigenous Struggle for Justice in Sixteenth-Century Spain* (Durham: Duke University Press, 2015); van Deusen, "Seeing Indios in Sixteenth-Century Castile," *William and Mary Quarterly* 69, no. 2 (2012): 208–14. In addition, see Jorge Cañizares-Esguerra, "New World, New Stars: Patriotic Astrology and the Invention of Indian and Creole Bodies in Colonial Spanish America, 1600–1650," *American Historical Review* 104 (February 1999): 33–68; Cañizares-Esguerra, "Demons, Stars, and the Imagination: The Early Modern Body in the Tropics," in Eliav-Feldon, Isaac, and Ziegler, *Origins of Racism in the West*, 313–25; Joanne Rappaport, "'Así lo paresçe por su aspeto': Physiognomy and the Construction of Difference in Colonial Bogotá," *Hispanic American Historical Review* 91, no. 4 (November 2011): 601–31, esp. 610–11.

11. For the emergence of an idea of natural slavery centered on a hierarchical vision of human potential in the Americas, see Anthony Pagden, "The Peopling of the New

World: Ethnos, Race, and Empire in the Early-Modern World," in Eliav-Feldon, Isaac, and Ziegler, *Origins of Racism in the West*, 292–312, esp. 299, 300–303, 307, 311.

12. See Justin Smith, *Nature, Human Nature, and Human Difference: Race in Early Modern Philosophy* (Princeton: Princeton University Press, 2015); and Justin Smith, *Embodiment: A History* (New York: Oxford University Press, 2017). Likewise, see Irigoyen-García, *Spanish Arcadia*. See also Michael James and Adam Burgos, "Race," *Stanford Encyclopedia of Philosophy*, Summer 2020 edition, https://plato.stanford.edu /entries/race/; Helmuth Nyborg, "Race as Social Construct," *Psych* 1, no. 1 (2019): 139–65; Ivan Hannaford, *Race: The History of an Idea in the West* (Washington, DC: Woodrow Wilson Center Press, 1996).

13. David Nirenberg, "Was There Race before Modernity? The Example of 'Jewish' Blood in Late Medieval Spain," in Eliav-Feldon, Isaac, and Ziegler, *Origins of Racism in the West*, 71–87.

14. Ana M. Gómez-Bravo, "The Origins of *Raza*: Racializing Difference in Early Spanish," *Interfaces: A Journal of Medieval European Literatures* 7 (2020): 77.

15. Gómez-Bravo, "The Origins of *Raza*," 64.

16. Nirenberg, "Was There Race before Modernity?," 71–87.

17. Renton, "Defining 'Race' in the Spanish Horse," 13–27.

18. Pierre Boulle, "La Construction du concept de race dans la France d'ancien régime," *Outremer: Revue d'histoire*, t. 89, no. 336–37 (2e trim. 2002): 155–75; Guillaume Aubert, "'The Blood of France': Race and Purity of Blood in the French Atlantic World," *William and Mary Quarterly*, 3rd series, 61, no. 3 (July 2004): 439–78; Arlette Jouanna, *L'idée de race en France au XVIème siècle et au début du XVIIème siècle (1498-1614)* (Lille: Université Lille III, 1976). Thank you to Melanie Aimee Marie Lamotte for these suggestions.

19. Charles de Miramon, "Noble Dogs, Noble Blood: The Invention of the Concept of Race in the Late Middle Ages," in Eliav-Feldon, Isaac, and Ziegler, *Origins of Racism in the West*, 200–216, 202–3. For more on the origins of race, see Benjamin Braude, "The Sons of Noah and the Construction of Ethnic and Geographical Identities in the Medieval and Early Modern Periods," *William and Mary Quarterly* 54, no. 1 (January 1997): 103–42; Barbara Fuchs, "The Spanish Race," in *Rereading the Black Legend: The Discourses of Religious and Racial Difference in the Renaissance Empires*, ed. Margaret Greer, Walter Mignolo, and Maureen Quilligan (Chicago: University of Chicago Press, 2007), 88–98.

20. Margócsy, "Culture of Collection," 1239.

21. Florio, *World of Words* (1611), 401. The 1598 version is more expansive about this term: "a beginning, a commencement, an onset, an entrance, a proem, a ground, a fountaine, a well spring, a race, a stocke, or chief originall of a thing." Florio, *A World of Wordes* (London: Arnold Hatfield for Edw. Blount, 1598), 294. See p. 424 of the 1611 edition for *razza* therein.

22. For the debate about the benefits and risks of a capacious definition of modern race applied to the past, see Geraldine Heng, *The Invention of Race in the European Middle Ages* (Cambridge: Cambridge University Press, 2018), esp. 3. Heng turns toward an expansive definition of *race* to explain episodes of articulated difference. For Heng, drawn to the term *race* in order to deploy the toolkit of critical race theory, its definition must be capacious and flexible as "one of the primary names we have—a name we retain for the strategic, epistemological, and political commitments it recognizes—that is attached to a repeating tendency of the gravest import,

to demarcate human beings through differences among humans that are selectively essentialized as absolute and fundamental, in order to distribute positions and powers differentially to human groups." For a critique of this approach, see S. J. Pearce, "The Inquisitor and the Moseret: *The Invention of Race in the European Middle Ages and the New English Colonialism in Jewish Historiography*," *Medieval Encounters* 26, no. 2 (2020): 145–90.

23. Florio, *World of Wordes* (1598), 343.

24. Karl Moore and Susan Reid, "The Birth of Brand: 4000 Years of Branding History," *Business History* 50, no. 4 (2008): 419–32; Ilja Van Damme, "From a 'Knowledgeable' Salesman towards a 'Recognizable Product'?: Questioning Branding Strategies before Industrialization (Antwerp, Seventeenth to Nineteenth Centuries)," in *Concepts of Value in European Material Culture, 1500–1900*, ed. Dries Lyna and Bert De Munck (Farnham, Surrey: Ashgate, 2015), 75–102.

25. For the role of paper in the Spanish monarchy's information overload, see Geoffrey Parker, *The Grand Strategy of Philip II* (New Haven: Yale University Press, 2000), 13–46.

26. On these broader transitions, see Elizabeth Eisenstein, *The Printing Revolution in Early Modern Europe* (New York: Cambridge University Press, 1983); and see Richard Goldthwaite, *The Economy of Renaissance Florence* (Baltimore: Johns Hopkins University Press, 2009), on double-entry accounting, esp. 91–93.

27. Frank S. Fanselow, "The Bazaar Economy or How Bizarre Is the Bazaar Really?," *Man*, new series, 25, no. 2 (1990): 250–65.

28. Andrew Bevan and David Wengrow, *Cultures of Commodity Branding* (New York: Routledge, 2010), 9–33.

29. Covarrubias, *Tesoro* (1611), *marca*, 539.

30. Archivio di Stato di Firenze and Scuola Normale Superiore di Pisa, "Ceramelli Papiani: blasoni delle famiglie toscane descrite nella Raccolta Ceramelli Papiani," accessed December 30, 2021, http://www.archiviodistato.firenze.it/ceramellipapiani /index.php?page=Home.

31. Florio, *World of Words* (1611) gives for "impresa" "an attempt, an enterprise, an undertaking, Also an impresse, a word, a mot or embleme. Also a iewell worne in ones hate, with some devise in it."

32. William Camden, *Remaines of a Greater Work, Concerning Britaine: The Inhabitants Thereof, Their Languages, Names, Surnames, Empreses, Wise Speeches, Poësies, and Epitaphs* (London: Printed by G. E. for Simon Waterson, 1605), 158 et seq.

33. Florio, *World of Words* (1611), "stemma."

34. Michael Baxandall, *Painting and Experience in Fifteenth-Century Italy* (New York: Oxford University Press, 1972), 29–103.

35. Diane Owen Hughes, "Distinguishing Signs: Ear-Rings, Jews and Franciscan Rhetoric in the Italian Renaissance City," *Past & Present* 112, no. 1 (August 1986): 3–59.

36. Compare, for example, the "Britannus" (English Horse) and the "Equus Hispanus" (Spanish Horse) in the Joannes Stradanus (Jan van der Straet) engraving in *Equile Ioannis Austriaci*, also known as *Speculum equorum* (Mirror of Horses) (Antwerp: Phillipe Galle, ca. 1578). To some extent, the *Equile's* horses each displayed their own distinctive personalities, whether as brave warriors or loyal soldiers. Stereotypes from each region trickled into Galle and Stradanus's horses. The Armenian appeared surly and mysterious, the German horse stubborn and unwieldy.

37. Sarah Duncan, "Stable Design and Horse Management at the Italian Renais-

sance Court," in *Animals at Court, Europe, c. 1200–1800*, ed. Mark Hengerer and Nadir Weber (Berlin: De Gruyter, 2019), 129–52.

38. Duncan, *Privileged Horses*, 24.

39. Katrina H. B. Keefer, "Marked by Fire: Brands, Slavery, and Identity," *Slavery and Abolition* 40, no. 4 (2019): 668.

40. Keefer, "Marked by Fire," 659–81, esp. 661, 668–70. Keefer observes that while many livestock branding catalogues remain in various archives, as will be discussed later in this chapter, historians of slavery have not found similar concrete records of slaveholders' marks. For branding of deviants, see Katherine Dauge-Roth, *Signing the Body: Marks on Skin in Early Modern France* (New York: Routledge, 2020), 217–57.

41. Keefer, "Marked by Fire," 659–81.

42. Rappaport, "'Asi lo paresçe por su aspeto.'"

43. Russell, *Like Engend'ring Like*, 63.

44. Renton, "Social and Environmental History of the Horse," 165–67; Renton, "The Knight with No Horse: Defining Nobility in Late Medieval and Early Modern Castile," *Sixteenth Century Journal* 50, no. 1 (2020): 109–28. See also Charles Prior, *The Royal Studs of the Sixteenth and Seventeenth Century* (London: Horse and Hound, 1935).

45. See Geoffrey Parker, *The Army of Flanders and the Spanish Road, 1567–1659*, 2nd ed. (Cambridge: Cambridge University Press, 2004), 70–89.

46. ASN DS I Serie, no. 38, 2: "Exitu deli diti orgi dari al sosto scripsi cavalli dal primo septembre per tutto el mese de aprile." The horses included "Lo liardo ferrandino; Lo bayo principe; Lo bayo [?]rasta Campagnia; Lo liardo gusterra; Lo sauro turco; Lo ubero tripaldo; Lo bayo muntrie; Lo bayo cosmano; Lo sauro sonda[r]o; Lo liardo turco."

47. ASN DS I Serie 38, no. 3: "Io Haniballo Musulino, mast.o de stalla facio fidi como le sup.a ditti ei la verita e affide inesu per scrissi mano propria."

48. ASN DS I Serie 38, no. 3, 1542, 1.

49. ASN DS I Serie 38, no. 3, 1542, 5.

50. ASN DS I Serie 38, no. 3, 1542, 15.

51. Duncan, *Privileged Horses*, 26.

52. Russell, *Like Engend'ring Like*, 64–65.

53. Papa, "Trattato delle Razze dei cavalli," 2–3, 14.

54. "Neapolitanus," in Stradanus, *Equile Ioannis Austriaci*.

55. Elizabeth Tobey, "The Palio Horse in Italy," in *The Culture of the Horse: Status, Discipline, and Identity in the Early Modern World*, ed. K. Raber and T. Tucker (New York: Palgrave Macmillan, 2005), 73–74.

56. Österreichisches Staatsarchiv (hereafter cited as ÖSA), FA Harrah HS 378. Thanks to Thomas and Pia Wallnig for suggesting this manuscript.

57. Società Napoletana di Storia Patria (hereafter cited as SNSP), *Merchi delle razze de Cavalli de Prencipi Duchi Marchesi Conti Baroni de tutte le Provincie del Regno di Napoli*, MS segn. 22.D36, 181, "Merchi stranieri diversi é di molte parti"; 193, "Marchi de cavalli delli ecclesiastici nel Stato Eclesi"; 200, "Alcuní Merchí de Cardinalí"; 209, "Alcuni Merchi della Casa Ottomana Grazie a Signore é suoi ministri."

58. SNSP, *Merchi delle razze de Cavalli*, MS segn. 22.D36, 182v–184r, 209r–v.

59. SNSP, *Merchi delle razze de Cavalli*, MS segn. 22.D36, "Casa del Tufo," fol. 36. The Tufo family had been one of Tommaso Campanella's close allies, and the notori-

ous heretic stayed at their Neapolitan residence in 1589 after he left Calabria; Mario del Tufo was a good friend and protector, and it was to him that Campanella dedicated his first published work. Jean-Paul de Lucca, "Prophetic Representation and Political Allegorisation: The Hospitaller in Campanella's *The City of the Sun*," *Bruniana & Campanelliana* 15, no. 2 (2009): 396. On Tufo horses in Florence, see Medici Archive Project (hereafter cited as MAP), doc. ID 17769, Archivio di Stato di Firenze, Mediceo del Principato, Florence, Italy (hereafter cited as ASF MdP), 4085, 12, Mario del Tufo in Naples to Ferdinando I de' Medici in Florence, May 11, 1595.

60. Papa, "Trattato delle Razze dei cavalli," 16, 19, 20, 21, 23.

61. SNSP, *Merchi delle razze de Cavalli*, MS segn. 22.D36, "Ruoti."

62. Renton, "Social and Environmental History of the Horse," 128–31.

63. Niccolò Nelli was one of the most productive printmakers-cum-publishers in Venice before the plague of 1576. Gert Jan van der Sman, "Print Publishing in Venice in the Second Half of the Sixteenth Century," *Print Quarterly* 17, no. 3 (September 2000): 235–47. Bernardo Giunti, born in Florence in 1540, was the nephew of the founder of the Florentine branch of the famous printers; he had relocated to Venice by 1571. On the frontispiece of *Libro de marchi de cavalli* (1588), one finds his characteristic Medici coat of arms and the Lily of the Giunti to symbolize the press's Florentine ties. Massimo Ceresa, "Giunti (Giunta), Bernardo," in *Dizionario Biografico degli Italiani*, vol. 57 (2001).

64. On jokes about Venetians on horseback, see Baldassare Castiglione, *The Book of the Courtier*, trans. George Bull (New York: Penguin Books, 2003), 159–62.

65. Nelli and Giunti, *Libro*, "razza del S.r Ferante Pignatello, buona razza et sono cavali belli et buoni"; "razza del S.or Cesaro Pignatello, e buona razza il merchio il va alla coscia sinistra"; "razza del S.or Fabritio Pignatello Marchese di Cherchiara la razza è in Calabria."

66. Giunti, *Libro*, "Mercio della razza del zari incenio da chio san, son cavalli boni da fatica, la razza è in Puglia, il merchio va alla coscia destra non son belli ma son utili."

67. Giunti, *Libro*, 20, "Merchio del S.r Antonio di Rugiero, sono picholi cavalli ma sono belli, et buoni."

68. See "zannetti della razza del Principe di Stilliano il merchio va alla coscia destra" in both Nelli and Giunti.

69. Nelli, *Libro*, "razza del Duca di Monte Leone et detta razza en el Regno di Napoli."

70. Nelli, *Libro*, "razza del Conte Ercule di contrarij gentilhuomo Ferarese il M. va alla coscia destra la razza è sul Ferarese è sono buona razza."

71. Nelli, *Libro*, "razza del conte de Altavilla di casa caraffa et sono grandi cavalli di buona razza il merchio va alla coscia destra."

72. Nelli and Giunti, *Libro*, "razza del prencipe di Salerno et è cavalli grandi, belli et buoni. Il Merchio va alla coscia sinistra"; "razza del prencipe di Salerno soti Cavalli gianeti, belli, e boni il merchio va alla cosa sinistra." Only in Giunti does one find mention of the princess's race of horses.

73. Nelli, *Libro*, "razza del S.r Gioanvicenzo da chiosan sono cavallj buoni da fatica la razza e in Puglia il merchio va alla coscia destra non son belli ma son boni cavalli"; "razza del. S. Pierluigi Fernese Duca di Parma"; Giunti, *Libro*, "razza del zari incenio da chio san, son cavalli boni da fatica, la razza è in puglia, il merchio va alla coscia destra non son belli ma son utili"; "razza del Sig.r Pietro Alovisio fernese Duca di Parma."

74. Giunti, *Libro*, "razza de Zuan Iacomo detese la razza è in Basilicada; è buona razza, et son buoni et belli Cavalli"; "razza del Conte Ravasono alcuni grandi, et buoni, et il merchio va alla coscia destra, la razza è in Puglia"; "razza del S.r gi gienari nel Regno de Napoli Cugnato del S.r Vespesian genaro."

75. Nelli, *Libro*, "Regina Bona Regina di Pollonia et Duchessa de Bari la razza è in Puglia il merchio va alla coscia destra et è buona razza"; Giunti, *Libro*, "de bari la razza è in Puglia il merchio va alla coscia destra e buona razza."

76. Nelli and Giunti, *Libro*, "razza del S. Don Franco da Este il merchio va alla coscia destra et son boni et belli cavallj della razza è basalicada nel Regno di Napoli."

77. Nelli and Giunti, *Libro*, "razza del Signor don Ferrante Gonzaga la razza è in Puglia alla Serra Capriola e sono belli e buoni cavalli il Merchio va alla coscia destra."

78. ÖSA, HHStA, Hausarchiv, Sammelbände, 1–2 (Konv.) 1, fol. 67, Maximilian II to Count de Consa, Vienna, April, 17, 1559: "el cavallo que con el nos embiastes que es muy bueno"; fols. 68–68v, Maximilian II to Vicenzio de Ebuli, Vienna, April 17, 1559; fol. 70v, Maximilian II to Cesare Gonzaga, Vienna, May 12, 1559. In the following years, Maximilian continued to receive horses from these houses: see fols. 159v–160, to Principe Stiliano, Vienna, October 17, 1561; fols. 160v–161, to the Count of Consa, Vienna, October 17, 1561; fols. 161–161v, to Vicenzio de Ebuli, Vienna, October 17, 1561.

79. ÖSA, HHStA, Hausarchiv, Sammelbände, 1–2 (Konv.) 1, fols. 66–66v, Maximilian II to the Marchioness (Marquesa) del Gasto, Vienna, April 17, 1559.

80. Liedtke, *Royal Horse and Rider*, 21.

81. MAP doc. ID 19555, ASF MdP 219, 83, Cosimo de' Medici in Pisa to Pedro Afán de (Perafán) Ribera in Naples, March 31, 1563.

82. MAP doc. ID 21664, ASF MdP 515a, 860, Asciano Caracciolo in Naples to Francesco di Cosimo I de' Medici in Florence, May 20, 1565. For Philip II's wider breeding projects, see Renton, "Breeding Techniques and Court Influence," 221–34.

83. John Marino, *Becoming Neapolitan: Citizen Culture in Baroque Naples* (Baltimore: Johns Hopkins University Press, 2010), 103.

84. Duncan, *Privileged Horses*, 47–48.

85. The *chinea* was a white riding horse gifted to the papacy in acknowledgment of its historical claim to suzerainty over the kingdom. As the kingdom was disputed and then passed into the hands of Spanish overlords, the king of Spain inherited responsibility for the ceremonial *chinea*. Charles V negotiated the tribute down to 7,000 ducats, along with the symbolic gift of the white Neapolitan horse. On Saint Peter's Day each year, the Spanish ambassador to the Kingdom of Naples traveled to Rome with the equine gift. In 1567, Juan de Zúniga likely presented the annual tribute to Pope Pius V. That gift was much better received than the annual offering two years later. In 1569, Pius complained that the horse had not been outfitted as it had been in previous years; the ambassador countered that they had spent 60 more *scudi* than normal on its adornments. In 1575, the Spanish ambassador, Juan de Requesens, represented the king of Naples, then Philip II of Spain, as papal vassal, presenting Pius's successor, Pope Gregory XIII, with the ritual gift of the *chinea*. Ambassador Requesens became a regular; in 1579, he fulfilled the same role after a decadent banquet in honor of St. Peter's Day. By 1590, the job had passed to another Spanish ambassador, Enrique de Guzmán y Conchillos, Count of Olivares. See also MAP doc. ID 21868, ASF MdP 3090, 570, Cosimo Bartoli in Venice to Francesco di Cosimo de' Medici in Flor-

ence, July 2 1569; MAP doc. ID 27209, ASF MdP 3082, 272, Aviso [sic] from Venice to Florence relaying news from Rome and Lyon, July 2, 1575; MAP doc. ID 28073, ASF MdP 3082, 712, Aviso [sic] from Venice reporting news from Rome, Venice, Prague, Antwerp, Cologne, July 4, 1579; MAP doc. ID 26050, ASF MdP 4027, 312, Aviso [sic] from Rome, June 27, 1590; MAP person ID 2665; MAP doc. ID 21634, ASF MdP 3080, 104, Avviso from Venice from Cosimo Bartoli to Francesco di Cosimo I de' Medici, July 5 1567; Thomas James Dandelet, *Spanish Rome, 1500–1700* (New Haven: Yale University Press, 2001), 57, 77–78, 104; Tommaso Astarita, *Between Salt Water and Holy Water: A History of Southern Italy* (New York: W. W. Norton, 2005), 71, 207–8. While in 1608, the procession included "around five hundred horses," by 1609 it featured the new ambassador, Francisco de Castro, who led "around six hundred horses." John E. Moore, "Prints, Salami, and Cheese: Savoring the Roman Festival of the Chinea," *Art Bulletin* 77, no. 4 (1995): 584–608. The kingdom maintained the tradition until the end of the eighteenth century, when, amid a rise of anti-Church policies, the Spanish king finally stopped sending the equine tribute to Rome.

86. MAP doc. ID 22314, ASF MdP 518, 178, Alfonso (Importuni) Cambi in Naples to Cosimo I de' Medici in Florence, October 29, 1565.

87. ÖSA, HHStA, Hausarchiv, Sammelbände, 1–2 (Konv.) 1, fols. 117v–118, to the viceroy of Naples, Vienna, September 18, 1560: "los hagaes scoger entre los que huviere algunos dias que estuviere ay en la cavallariza de su Alteza y no fueron muy nuevos porque tengan buenos principios y lleguen aqui havendo de passer tan largo camino de provecho porque estas causas desseo que sean delos que estuvieren ya mas hechas y yo spero que viniendo escogidos de vuestra mano seran tales que declararan la voluntad que confiamos teneis de complacernos."

88. ÖSA, HHStA, Hausarchiv, Sammelbände, 1–2 (Konv.) 1, fols. 136–136v, to the Mardones del Consejo of the king of Spain in Naples, Wiener Neustadt, March 13, 1561: "potros de la raza que su Alteza tiene en esse Reyno."

89. ÖSA, HHStA, Hausarchiv, Sammelbände, 1–2 (Konv.) 1, fols. 135v–136, to the viceroy of Naples, Vienna, March 13, 1561: "porque segun soy informado es aora el major tiempo para traer los cavallos. Embio al portador de esta por los doze de este año y por los viij que faltaron darse el pasado."

90. ÖSA, HHStA, Hausarchiv, Sammelbände, 1–2 (Konv.) 1, fols. 159–159v, to the viceroy of Naples, Vienna, October 14, 1561: "los quales estoy contente no dubdando se me embiaron delos mejores qunque es verdad que yo holgara que entrellos huviere mas ginetes."

91. Duncan, *Privileged Horses*, 131.

92. Francesco Carletti, *My Voyage Around the World: The Chronicles of a 16th Century Florentine Merchant*, trans. Herbert Weinstock (New York: Pantheon Books, 1964), 4–5; "Carletti, Francesco," in *Dizionario Biografico degli Italiani*, vol. 20 (1977).

93. Biblioteca Angelica, MS 1331, 9v–10r; Carletti, *My Voyage Around the World*, 12–13.

94. Biblioteca Angelica, MS 1331, 9v–10r; Carletti, *My Voyage Around the World*, 12–13.

95. Biblioteca Angelica, MS 1331, 9v–10r; Carletti, *My Voyage Around the World*, 12–13.

96. Francesco Carletti, *Viaggio intorno al mondo* (Firenze: Giulio Einaudi, 1958), 52, 68.

Chapter Three

1. Robert W. Hanning and David Rosand, *Castiglione: The Ideal and the Real in Renaissance Culture* (New Haven: Yale University Press, 1983), xxiv.

2. Hanning and Rosand, *Castiglione*, xxi.

3. Peter Burke, *The Fortunes of the* Courtier: *The European Reception of Castiglione's* Cortegiano (1997; repr., Malden, MA: Polity Press, 2007), 22–23.

4. Geoffrey Parker, *Emperor: A New Life of Charles V* (New Haven: Yale University Press, 2019), 206, 208.

5. Burke, *Fortunes of the* Courtier, 57.

6. For the general publishing history, see Burke, *Fortunes of the* Courtier, esp. 139. For a comparative publication history, see University College London Special Collections' Castiglione Collection, https://www.ucl.ac.uk/library/special-collections/a-z/castiglione.

7. Parker, *Emperor*, x, 627n12.

8. Burke, *Fortunes of the* Courtier, 32.

9. Burke, *Fortunes of the* Courtier, 9, 15, 39, 41, 57, 139.

10. Burckhardt, *Civilization*, 43.

11. See Stephen Greenblatt, *Renaissance Self-Fashioning: From More to Shakespeare* (Chicago: University of Chicago Press, 2005).

12. Russell, *Like Engend'ring Like*, 219.

13. Burckhardt, *Civilization*, 157.

14. Burckhardt, *Civilization*, 156–58. See, for instance, Andrea Tonni's use of this passage in "The Renaissance Studs of the Gonzaga of Mantua," in Edwards, Enenkel, and Graham, eds., *The Horse as Cultural Icon*, 263.

15. Giacinto Romano, *Cronaca del soggiorno di Carlo V in Italia (dal 26 luglio 1529 al 25 aprile 1530)* [da Luigi Gonzaga] (Milano: Urico Hoepli, 1892), 238, 253, 255, 263; Parker, *Emperor*, 193–94.

16. Lisa Jardine, *Worldly Goods: A New History of the Renaissance* (London: Macmillan, 1996), 311. The latest analysis of these dwellings can be found in Jérémie Koering, *Le Prince en représentation: Histoire des décors du palais ducal de Mantoue au XVIe siècle* (Paris: Actes Sud, 2013).

17. Romano, *Cronaca*, 244, 262: "con tutti li Principi, S.ri et Gentilhomini et col S.r Marchese a canto, adorno sempre ragionando insieme dilla bellezza di la Terra et di molte altre cose."

18. Giorgio Vasari, *The Lives of the Artists*, trans. Julia Conaway Bondanella and Peter Bondanella (Oxford: Oxford University Press, 1998), 365–66. On the racetrack, see Tonni, "Renaissance Studs," 265.

19. Vasari, *Lives of the Artists*, 265; Giorgio Vasari, "Giulio Romano," in *Le vite de' più eccellenti architetti, pittori, et scultori italiani, da Cimabue insino a' tempi nostri* (Firenze: Giunti, 1568).

20. The Gonzaga developed stables on the Isle of Te, and at Gonzaga, Pietole, Margonara, Marimolo, and Roversella. Duncan, *Privileged Horses*, 241–42.

21. Vasari, *Lives of the Artists*, 367.

22. Tonni, "Renaissance Studs," 270; R. Castagna, "L'Alcanna d'Oriente e i Cavalli di Federico II Gonzaga, Ritratti da Giulio Romano a Palazzo Te," *Civiltà Mantovana*, 2nd series, 27 (1990): 109–10.

23. Castagna, "L'Alcanna d'Oriente," 109–10.

24. Romano, *Cronaca*, 262, 264–65, 253.

25. Dukeship was a great honor. As the structure of European statecraft was re-shaped during the turbulent sixteenth century, the nature of nobility was also chang-ing. Hierarchy was in flux as titles were slowly inflated to make room for the many once-independent cities incorporated into the new imperial system. A duke's earn-ing potential and cultural position was a substantial improvement from that allocated to a marquis. Spanish and Italian dukes were certainly different, but during the early years of Charles V's reign, Castilian dukes earned twice, sometimes three times, the annual income of marquises. Castilian dukes earned on average about 43,000 ducats a year, while marquises earned an average of about 25,000 ducats. J. H. Elliot, *Impe-rial Spain, 1469–1716* (London: Penguin, 1970), 112. Ranking was relative. Thus, when Federico became a duke, he suddenly boasted the same title as the rulers of Milan and Ferrara. However, the Gonzaga's income is not likely to have been affected by their promotion (except for the immense expense of hosting an imperial visit and any nec-essary gifts); if the incomes of the grand dukes of Tuscany (admittedly a much larger state) are anything to go by, the marquis of Mantua probably had much higher reve-nues than a Castilian duke. More importantly, in the Holy Roman Empire, dukes were closer to being regional monarchs (within a hierarchical system in which they were not entirely independent) than were major aristocrats in strong kingdoms like Castile, France, and England. That is, the shared title notwithstanding, the Gonzaga were de facto sovereigns in that they could levy taxes, raise armies, send ambassadors, make war and peace, and so forth, while Castilian dukes could do none of these things. In an era of title inflation and careful attention to hierarchy and precedence, the promotion made a major difference to the rank of the Gonzaga on the international stage. This mattered tremendously in and of itself, but was also practically important in things like marriage negotiations.

26. In translating these sources and navigating this history, I have followed sug-gestions from Little People of America, the world's oldest and largest support group organization for people with dwarfism and their families. On their Frequently Asked Questions page, this organization notes that "such terms as dwarf, little person, LP, and person of short stature are all acceptable, but most people would rather be re-ferred to by their name than by a label." While no one organization or individual can speak on behalf of all, I have avoided other translations of these terms, which, while used in earlier periods, are still perceived to have a derogatory valence. Little People of America, "FAQ," accessed January 8, 2022, https://www.lpaonline.org/faq -#Definition. Recent critical scholarship has offered new insights on disability studies both in the present and in a historic vein. For an introduction, see Rosemarie Gar-land Thomson, "The Beauty and the Freak," *Disability, Art, and Culture (Part II)* 37, no. 3 (Summer 1998): 459–74. See also Stephanie Jenkins, Kelly Struthers Montford, and Chloë Taylor, eds., *Disability and Animality: Crip Perspectives in Critical Animal Studies* (New York: Routledge, 2021).

27. ASMN AG 3000, libro 51, cc. 64r–v, Isabella d'Este to Diana d'Este, Septem-ber 11, 1532: "Io promesi già quarto anni sono a M.ma Illma Renea di voler dare a sua Ex. el primo fruto che ucisse della raza delli mei nanini, dicò de femina: et come V.S. sa, hormai sono dui anni che nacque una putina, la quale anchora non dà speranza di dover restare in tutto cossì piccolo come è la mia Dellia, nondimeno senza alcun dub-bio rimanerà nana, et perchè è hora in termine che senza guida è atta da sé sola andar per tutto sicuramente." On trafficking in dwarves and other humans in the Gonzaga

court, see Alessandro Luzio and Rodolfo Renier, *Buffoni, nani e schiavi dei Gonzaga ai tempi d'Isabella d'Este* (Roma: Tipografia della camera dei deputati, 1891), which includes a transcription of this letter (58). Thanks to Deanna Shemek, who pointed me toward this correspondence in the IDEA project. This letter has been mentioned in Dario A. Franchini, *La scienza a corte: Collezionismo eclettico, natura e immagine a Mantova fra Rinascimento e Manierismo* (Roma: Bulzoni, 1979), 140. On Diana d'Este (a.k.a. Countess Contrari), see Luciano Chiappini, *Gli Estensi: Mille anni di storia* (Ferrara: Corbo Editore, 2001), 351. See also Sarah Cockram, *Isabella d'Este and Francesco Gonzaga: Power Sharing at the Italian Renaissance Court* (New York: Routledge, 2013). See also Isabella d'Este, *Selected Letters*, ed. and trans. Deanna Shemek, The Other Voice in Early Modern Europe: The Toronto Series 54; Medieval and Renaissance Texts and Studies 516 (Toronto: Iter Press; Tempe: Arizona Center for Medieval and Renaissance Studies, 2017), 562.

28. Isabella's instructions to Perugino for *The Battle of Love and Chastity* were so clear that they left little to the artist's imagination; see Stephen J. Campbell, *The Cabinet of Eros: Renaissance Mythological Painting and the Studiolo of Isabella d'Este* (New Haven: Yale University Press, 2004), 173n6. For Isabella's relations with Leonardo da Vinci through Fra Pietro da Novellara, see Martin Kemp, *Leonardo da Vinci: The Marvellous Works of Nature and Man* (Oxford: Oxford University Press, 2006), 207. On the natural history collections in Mantua, see Raffaella Morselli, "Collezionismo, mercato e allestimento: I 'naturalia' di Casa Gonzaga tra modernità e contemporaneità," in *Nuove Alleanze: Diritto ed Economia della Cultura e dell'Arte* (supplement in "Arte e Critica" 80/81) (2015).

29. Isabella d'Este, *Selected Letters*, 558–63.

30. For examples of such portraits, see Agnolo di Cosimo (a.k.a. Bronzino), *Morgante* (1553); Antonio Moro's *Cardinal Granvelle's Dwarf* (1549–60); [follower of] Diego Velázquez, *Dwarf with a Dog* (ca. 1645); and Diego Velázquez, *Las Meninas* (1656). On the bestialization of dwarves, see Touba Ghadessi, *Portraits of Human Monsters in the Renaissance: Dwarves, Hirsutes, and Castrati as Idealized Anatomical Anomalies* (Kalamazoo: Medieval Institute Publications, Western Michigan University, 2018), 70–80.

31. Merry E. Wiesner-Hanks, *The Marvelous Hairy Girls: The Gonzales Sisters and Their Worlds* (New Haven: Yale University Press, 2009). For images of the family, see "Girls and Father" in Joris Hoefnagel, *Animalia Rationalia et Insecta (Ignis)*, plate 2, ca. 1575–80. Likewise, see Lavinia Fontana, *Portrait of the Daughter of Pedro Gonzales* (1583). On the wild man trope, see Surekha Davies, *Renaissance Ethnography and the Invention of the Human: New Worlds, Maps and Monsters* (Cambridge: Cambridge University Press, 2016), 41; and Paula Findlen, *Possessing Nature: Museums, Collecting, and Scientific Culture in Early Modern Italy* (Berkeley: University of California Press, 1994), 309, 345.

32. Lovejoy, *Great Chain of Being*, 58–60. See also Michel Foucault, *The Order of Things: An Archeology of the Human Sciences* (New York: Vintage Books, 1994).

33. Luzio and Renier, *Buffoni, nani e schiavi*, 54.

34. Margócsy, "Horses."

35. Findlen, *Possessing Nature*, 27; Alessandra Russo, "The Curator's Eyes: Sebastiano Biavati, Custodian of a Heterogeneous Artistic World," in *The Significance of Small Things: Essays in Honour of Diana Fane*, ed. Luisa Elena Alcalá and Ken Moser (Madrid: Ediciones El Viso, 2018), 150–58.

36. See Valeria Finucci, *The Prince's Body: Vincenzo Gonzaga and Renaissance Medicine* (Cambridge, MA: Harvard University Press, 2015), 28–61.

37. ASMN AG 3000, libro 51, cc. 155r–v, Isabella d'Este to the Princess of Molfetta, November 24, 1533.

38. ASMN AG b. 2996, L. 31, c. 88r, Isabella d'Este to Rigo Carpesano, April 18, 1515.

39. Luzio and Renier, *Buffoni, nani e schiavi*, 140–45. This correspondence is cited in Paul H. D. Kaplan, "Isabella d'Este and Black African Women," in *Black Africans in Renaissance Europe*, ed. T. F. Earle and K. J. P. Lowe (Cambridge: Cambridge University Press, 2005), 125–54, 134–35.

40. Luzio and Renier, *Buffoni, nani e schiavi*, 140–45.

41. This episode was reconstructed in Rodolfo Signorini, "A Dog Named Rubino," *Journal of the Warburg and Courtauld Institutes* 41 (1978): 317–20.

42. The epitaph, from Ulisse Aldrovandi, *De quadrupedibus digitatis viviparis*, libri tres, et *De quadrupedibus digitatis oviparis*, libri duo (Bologna: apud Nicolaum Tabaldinum, 1637), 525, reads "RUBINUS CATULUS / LONGO ET FIDO AMORE PROBATVS / DOMINO SENIO CONFECTVS SERVATA / STIRPE HIC IACEO / HOC ME HONORE SEPVLCHRIHERVS DIGNATVS EST."

43. Ulisse Aldrovandi, *Aldrovandi on Chickens: The Ornithology of Ulisse Aldrovandi (1600)*, ed. and trans. L. R. Lind (Norman: University of Oklahoma Press, 1963), vol. 2, bk. 14.

44. See Aldrovandi, *De quadrupedibus digitatis*, 482–563 (dogs) and 564–85 (cats); Aldrovandi, *De quadrupedibus solidipedibus volumen integrum* (Bologna: Apud Nicolau Thebaldinum, 1639), 1–291 (horses) and 291–351 (asses).

45. Russell, *Like Engend'ring Like*, 8–21.

46. Giancarlo Malacarne has completed foundational work on the Gonzaga's animal culture. His careful transcriptions provide an excellent guide to key documents about animal breeding in the Mantuan collections. I have relied on the following texts to point me toward the archival materials: Giancarlo Malacarne, *Il mito dei cavalli gonzagheschi: Alle origini del purosangue* (Verona: Promoprint, 1995); Malacarne, *Le cacce del principe*; Malacarne, *Sulla mensa del principe: Alimentazione e banchetti alla Corte dei Gonzaga* (Modena: Il Bulino, 2000); Malacarne, *Le feste del principe: Giochi, divertimenti, spettacoli a corte* (Modena: Il Bulino, 2002); Malacarne, *I signori del cielo: La falconeria a Mantova al tempo dei Gonzaga* (Mantua: Artiglio Editore, 2003).

47. ASMN AG b. 1464, c. 334, Iacopo Malatesta (Venice) to Federico II, January 1530. Similarly, ASMN AG b. 2933, L. 301, c. i., from Federico II Gonzaga, February 4, 1530.

48. ASMN AG b. 795, cc. 81–83, November 5, 1498.

49. ASMN AG b. 2455, c. 153, August 7, 1500. Dogs of different regions were considered exotic gifts. For example, a letter from Urbino emphasized dogs from Friuli, two of which would be worthy of a cardinal: dogs of good stature, biting mouth, and great speed. Archivio di Stato di Udine, Fondo Caimo, b. 86, doc 54, Pompeo Caimo to his brother Eusebio Caimo, June 14, 1603. Thanks to Hannah Marcus for suggesting this source.

50. Sarah Cockram, "Interspecies Understanding: Exotic Animals and Their Handlers at the Italian Renaissance Court," in "The Animal in Renaissance Italy," ed. Stephen Bowd and Sarah Cockram, special issue, *Renaissance Studies* 31, no. 2 (2017): 277–96, esp. 278.

51. Florio, *World of Words* (1611), s.v. "canattiere."

52. Florio, *World of Words* (1611), s.v. "strozziére" and "falconieri."

53. Malacarne, *Le cacce del principe*, 50–51.

54. ASMN AG b. 2472, c. 728, Zo. Francesco, *canatero* da Mantova, August 11, 1508.

55. ASMN AG b. 2991, L. 3, l. 14, Isabella d'Este to Leonello da Baesio, February 9, 1493.

56. Malacarne, *Le cacce del principe*, 35, transcribes a letter in ASMN AG b. 2992, L. 5 and L. 170, June 1495. This account is confirmed in letters from Tortuco da Pavia in Venice in 1499. It is also mentioned by C. Cottafavi, "Cani e gatti alla Corte dei Gonzaga," *Il Ceppo, Quaderno di vita fascista e di cultura edito dal comitato di Mantova dell'opera nazionale balilla* (1934): 4. It is worth noting that many publications on the history of animal breeding and human eugenics were published during the Fascist period in Europe, which saw a corresponding interest in eugenics in the United States. A. Bertolotti, "Curiosità storiche mantovane CXXI, I gatti e la gatta della marchesa di Mantova Isabella d'Este," *Il mendico* (16 aprile 1889): 6–7; Franchini, *Scienza a corte*, 100. On Gonzaga pet keeping, see Stephen Bowd and Sarah Cockram, eds., "The Animal in Renaissance Italy," special issue, *Renaissance Studies* 31, no. 2 (April 2017): 175–318.

57. By November 5, 1498, "two large dogs" had arrived from Constantinople. ASMN AG b. 795, cc. 81–83, November 5, 1498. See also ASMN AG b. 866, c. 807, Matteo Giberto to Federico II Gonzaga, October 13, 1522; ASMN AG b. 759, c. 354r, Andrea Doria from Genova to the court, October 27, 1534.

58. ASMN AG b. 2991, L. 1, c. 76r, Isabella d'Este to Protonotario Stanga, December 31, 1492.

59. ASMN AG b. 2891, L. 64, c. 59v, February 15, 1470.

60. ASMN AG b. 2891, L. 64, c. 62.

61. Correspondence between Antonio de Calcho and Marquis Ludovico II of Mantua, April–August 1477, ASMN AG 805, fol. 377r, April 11, 1477; fol. 435r, May 20, 1478; fol. 437r, August 26, 1478; cited in Tonni, "Renaissance Studs," 267.

62. See Tonni, "Renaissance Studs," 265–66. For a systematic study of Renaissance Italian stables with an emphasis on Mantua, see Duncan, *Privileged Horses*.

63. One finds a Renegato in Silvestro Da Lucca, *Il libro dei palii vinti dai cavalli di Francesco Gonzaga, 1512–1518, mineato da Lauro Padovano e forse altri due miniatori* (The Book of the Palios Won by Francesco Gonzaga's Horses, Illuminated by Lauro from Padua and Possibly Two Other Illuminators). The codex is described in the catalogue of the exhibition that took place at the Victoria and Albert Museum in London from November 1981 to January 1982, titled *Splendours of the Gonzagas*, where it was displayed. It is currently owned by the Falk family and held in their collections at the Palazzo Giustian Recanati alle Zattere a Venezia in Venice. Thanks to the archivist Maurizio Romanò for guiding me around the collection. The horse Renato is depicted on page 43 as a dapple gray who won a great number of races. A descendant of Renegato, Renegato Giovane, appears in the palio book as having won a number of races in the first decade of the 1500s. Yet another Renegato is represented as a gray who won races in Ferrara and the San Petronio in Bologna in an unspecified year. Finally, the youngest Renegato won races in 1517 and 1518. ASMN AG b. 2439, c. 89, March 28, 1490.

64. ASMN AG b. 2439, c. 90, June 9, 1490.

65. ASMN AG b. 2921, L. 231, fol. 4v.

66. Mario Equicola, *Chronica di Mantua* (Mantova, 1521), viii (verso)–vvi (recto), cited in Malacarne, *Il mito dei cavalli gonzagheschi*, xvii.

67. Scholars often rely on Carlo Cavriani, *Le razze gonzaghesche di cavalli nel Mantovano e la loro influenze sul puro sangue inglese* (Rome, 1909). My summary of the Mantuan breeds relies on Tonni, "Renaissance Studs," 263, 265.

68. Florio, *World of Words* (1611), "corsiere," 127.

69. Tonni, "Renaissance Studs," 263–64, 265.

70. Russell, *Like Engendr'ing Like*, 18–20.

71. Russell, *Like Engend'ring Like*, 18.

72. Russell, *Like Engend'ring Like*, 217.

73. Elizabeth Tobey, "The Palio in Italian Renaissance Art, Thought, and Culture" (Ph.D. dissertation, University of Maryland, 2005), 71–72, 76–77, 182, 275, 277; Christian Jaser, *Palio und Scharlach. Städtische Sportkulturen des 15. und frühen 16. Jahrhunderts am Beispiel italienischer und oberdeutscher Pferderennen*, Monographien zur Geschichte des Mittelalters (Stuttgart: Hiersemann, 2021); Tobey, "The Palio Horse in Italy," 65. See Giovanni Butteri's *The Return from the Palio*, in Duncan, *Privileged Horses*, 207. For a focused study of Gonzaga palio racing, see Galeazzo Nosari and Franco Canova, *Il Palio nel Rinascimento: I Cavalli di Razza dei Gonzaga nell'età di Francesco II Gonzaga, 1484–1519* (Reggiolo [Reggio Emilia]: E. Lui, 2003).

74. Da Lucca, *Il libro dei palii vinti*. The codex has been analyzed in Malacarne, *Il mito dei cavalli gonzagheschi*; Tobey, "The Palio in Italian Renaissance Art"; Tobey, "The Palio Horse in Italy," 63–90; and Tonni, "Renaissance Studs," 264.

75. Da Lucca, 14v.

76. Tonni, "Renaissance Studs," 268; Tobey, "The Palio in Italian Renaissance Art," 251. More research remains to be done on the relationship between ideas of lineage and horse breeding in the Islamic world and their impact on European thought. One might approach this question through the Ottoman and Arabic discourses on breeding and race. A compelling venue would be to trace how classical Islam pioneered some of the practices described here, such as the detailed maintenance of horse genealogies and the application of those practices to peoples. On genealogy, Arab "lineage," and the Nasrid Kingdom of Granada, see the work of Mohamad Ballan.

77. Tonni, "Renaissance Studs," 264, 269, citing a letter in ASMN AG b. 795, fol. 40r. September 20, 1492.

78. Tobey, "The Palio Horse in Italy," 7; Malacarne, *Il mito dei cavalli gonzagheschi*, 51–52.

79. For a later history of Arabian horses, see Donna Landry, *Noble Brutes: How Eastern Horses Transformed English Culture* (Baltimore: Johns Hopkins University Press, 2008).

80. *Othello*, I.1.109–12.

81. Pasquale Caracciolo, *La Gloria del Cavallo* (Venice: Gabriel Giulito dei Ferrari, 1566), 323.

82. Malacarne, *Il mito dei cavalli gonzagheschi*, 105–34.

83. Ulisse Aldrovandi, *De quadrupedibus solidipedibus volumen integrum* (Bologna: Giovanni Cornelio Uterwer, 1616), 55–56. Aldrovandi's three-volume *Ornithology* (1599–1603) relied on relationships he cultivated with empirical specialists. When it came to birds, Aldrovandi depended on information from court physicians and naturalists like Alfonso Cataneo, who in turn relied on his friends the falconers at the Este court. Mantua and Ferrara in particular were the best places to locate expert falconers. The same was probably the case for other animals, such as horses. Findlen, *Possessing Nature*, 177.

84. Aldrovandi, *De quadrupedibus solidipedibus volumen integrum* (Uterwer, 1616), 56; cited in Franchini, *La Scienza a Corte*, 98–99; my translation from the 1616 text.

85. James Hankins, *Virtue Politics: Soulcraft and Statecraft in Renaissance Italy* (Cambridge, MA: Belknap Press of Harvard University Press, 2019), 37.

86. Hankins, *Virtue Politics*, 38.

87. There are many scholarly editions of Baldassare Castiglione's *The Book of the Courtier*. I have relied on Baldassare Castiglione, *The Book of the Courtier*, trans. Leonard Eckstein Opdycke (New York: Charles Scribner's Sons, 1901); Baldassare Castiglione, *Il libro del Cortegiano*, A cura di Walter Berberis (Firenze: Einaudi, 2017); and Baldassare Castiglione, *The Courtier*, trans. George Bull (New York: Penguin Books, 2003). All quotes are based on the translation of the first edition of *Il libro del cortegiano del conte Baldesare Castiglione* (Firenze: Per li heredi di Philippo di Giunta, 1528), held in the Biblioteca Berenson Special Collections at Villa I Tatti. Passages from this edition are indicated by book and chapter.

88. 1.14; in English: *The Book of the Courtier*, Opdycke, 22; *The Book of the Courtier*, Bull, 54; *Il libro del cortegiano* (1528), 19r.

89. 1.15; *The Book of the Courtier*, Bull, 55; *Il libro del cortegiano* (1528), 20r.

90. 4.30; *The Book of the Courtier*, Bull, 55–56; *Il libro del cortegiano* (1528), 20r–v.

91. 4.25; *The Book of the Courtier*, Bull, 306; *The Book of the Courtier*, Opdycke, 269.

92. 4.12; *The Book of the Courtier*, Opdycke, 252–53.

93. 4.45; *The Book of the Courtier*, Opdycke, 282.

94. Silvia De Renzi, "Family Resemblance in the Old Regime," in Hopwood, Flemming, and Kassell, *Reproduction*, 241–52. Ludovico Zuccolo (1568–1630) contended that nobility resides neither in possession of wealth nor in noble descent, but in innate disposition to perform honorable deeds. To that effect, nobility is like beauty, mostly born to those of noble/beautiful parents, but sometimes emergent in the children of non-noble/ugly parents. He argued that breaks can occur and nobility can be lost in bad habits. By contrast, Alessandro Tassoni (1565–1635) argued that social disorder would come from such an open definition of nobility and its inheritance, suggesting that children of slaves could then claim nobility. Following Aristotle, he argued that nobility is a quality in family; if not, doctors would come to determine nobility, not historical and legal evidence. He concluded that lineage can be interrupted but then restored.

95. Burke, *Fortunes of the* Courtier, 57, 105.

96. Pedro Guibovich Pérez, "Los libros del inquisidor," *Cuadernos para la historia de la evangelización en América Latina* 4 (1989): 47–64, esp. 41–42. Cited in Burke, *Fortunes of the* Courtier, 105.

97. Huntington Library, HM 1316, "Instruttione per negotij nella Corte di Spagna al S.or Ludouico Orsino mandato Sua M.ta Catt:la Dal S.or Duca di Bracciano," likely dating from 1568–79. See also Furlotti, *A Renaissance Baron*, xvi. Ludovico Orsini was Paolo Giordano Orsini's cousin and courtier, and notoriously hired assassins to kill Paolo Giordano's widow in 1585. Elisabetta Mori, "Orsini, Paolo Giordano," in *Dizionario Biografico degli Italiani*, vol. 79 (2013). Paolo Giordano became duke of Bracciano in 1560, and Antonio Peréz, who is mentioned in the document, was a powerful secretary to King Philip II between 1568 and his arrest in 1579. On Antonio Peréz's case and diplomatic corruption, see Geoffrey Parker, *Imprudent King: A New Life of Philip II* (New Haven: Yale University Press, 2014), 247–63, 273–75.

98. On the analogy between the Spaniards and Romans, and Italians and Greeks,

see Giuseppe Galasso, *Il Mezzogiorno nella storia d'Italia* (Firenze: F. Le Monnier, 1977); Giuseppe Galasso, "Aspetti dei Rapporti tra Italia e Spagni nei Secoli XVI e XVII," in *Il tesoro messicano: Libri e saperi tra Europa e Nuovo mondo*, ed. Maria Eugenia Cadeddu and Marco Guardo (Florence: Leo S. Olschki Editore, 2013).

99. On colonial American courts, see Alejandra B. Osorio, *Inventing Lima: Baroque Modernity in Peru's South Sea Metropolis* (New York: Palgrave Macmillan, 2008), esp. "Lima es Corte: The Viceroy as Cultural Capital," 57–80.

Chapter Four

1. This chapter benefits from recent scholarship on the Florentine Codex. I am particularly grateful to Alanna S. Radlo-Dzur for her guidance on Nahuatl-language sources in both chapter 4 and chapter 5. I rely on Fray Bernardino de Sahagún, *General History of the Things of New Spain: The Florentine Codex*, MS Med. Palat. 220, Biblioteca Medicea Laurenziana, reproduced by the World Digital Library, https://www.wdl.org/en/item/10096/ (hereafter cited as FC); and Bernardino de Sahagún, *General History of the Things of New Spain: The Florentine Codex*, 12 books, trans. and ed. C. E. Dibble and A. J. O. Anderson (Santa Fe, NM: School of American Research/University of Utah, 2012) (hereafter cited as A&D). In many key passages, I return to the original manuscript to compare the Nahuatl text, pictographic text, and Spanish text; these passages are cited using folio numbers. I have relied on dictionaries including Frances Karttunen, *An Analytical Dictionary of Nahuatl* (Norman: University of Oklahoma Press, 1992); J. Richard Andrews, *Introduction to Classical Nahuatl*, rev. ed. (Norman: University of Oklahoma Press, 2003); Joe Campbell, *A Morphological Dictionary of Classical Nahuatl* (Madison, WI: Hispanic Seminary of Medieval Studies, 1985); Alexis Wimmer, Dictionnaire de la Langue Nahuatl Classique, http://sites .estvideo.net/malinal/nahuatl.page.html; Online Nahuatl Dictionary, ed. Stephanie Wood, Wired Humanities Projects, https://nahuatl.uoregon.edu/; Gran Diccionario Náhuatl, https://gdn.iib.unam.mx/. For historical definitions, see Alonso de Molina, *Vocabulario en lengua castellana y mexicana* (Mexico: Antonio de Spinosa, 1571); Marc Thouvenot, *Diccionario náhuatl-español* (Ciudad de México: Universidad Nacional Autónoma de México, 2015). Thanks to Joe Campbell for sharing his database of colonial Nahuatl to trace mentions of *xinach/ximach/xinachtli*.

On Tlaloc, see Bernardino de Sahagún, *Primeros Memoriales*, ca. 1558, 250r, 250v, 261v, cited in Sahagún, *Primeros Memoriales: Paleography of the Nahuatl Text and English Translation*, trans. Thelma Sullivan (Norman: University of Oklahoma Press, 1997), 55–56, 59, 83. The phrase referring to his temple in the main ceremonial precinct is *Tlalocan iteopan epcoatl*, "Tlalocan, Epcoatl's temple" (FC bk. 2), the first site in the temple precinct list (A&D bk. 2, 179). Epcoatl translates as "(mother-of-) pearl-serpent." Representations are discussed in the literature as *ixiptla*. Bernardo Ortiz de Montellano, "Las hierbas de Tláloc," *Estudios de cultura náhuatl* 14 (1980): 287–314.

2. A&D bk. 2, 42–44. The festival *Etzalcualiztli* ("The Eating of *Etzalli* [maize and bean porridge]") included twenty days of activities in different parts of the city and other specific temples. During the final day, human embodiments performed a vigil at Tlaloc's temple in the central precinct before their sacrifice. Sullivan, *Primeros Memoriales*, 55–56, 59, 83n17. On linking Tlaloc and John the Baptist, see Antonio Rubial García, "Icons of Devotion," in *Local Religion in Colonial Mexico*, ed. Martin Nesvig (Albuquerque: University of New Mexico, 2006), 37–61.

3. Nixtamalization using limewater and lye enhanced the flavor and nutritional value of maize and transformed it into the dough from which tortillas were made. See Charles Gibson, *The Aztecs under Spanish Rule: A History of the Indians of the Valley of Mexico* (Stanford: Stanford University Press, 1964), 308, 311, 324; Luis M. Gamboa Cabezas and Robert H. Cobean, "Comments on Cultural Continuities between Tula and the Mexica," in *The Oxford Handbook of the Aztecs*, ed. Deborah L. Nichols and Enrique Rodríguez-Alegría (Oxford: Oxford University Press, 2016).

4. Christopher Morehart, "Aztec Agricultural Strategies: Intensification, Landesque Capital, and the Sociopolitics of Production," in Nichols and Rodríguez-Alegría, *Oxford Handbook of the Aztecs*, 263–79. On cochineal insect domestication, see Deirdre Moore, "The Heart of Red: Cochineal in Colonial Mexico and India" (Ph.D. dissertation, Harvard University Graduate School of Arts and Sciences, 2021).

5. For an overview of these debates, see Nerissa Russell, *Social Zooarchaeology: Humans and Animals in Prehistory* (Cambridge: Cambridge University Press, 2012), 207–58; Alejandro Casas, Adriana Otero-Arnaiz, Edgar Pérez-Negrón, and Alfonso Valiente-Banuet, "In Situ Management and Domestication of Plants in Mesoamerica," *Annals of Botany* 100, no. 5 (2007): 1101–15.

6. Gibson, *The Aztecs Under Spanish Rule*, 194–219.

7. Matthew Restall, *When Moctezuma Met Cortés: The True Story of the Meeting that Changed History* (New York: HarperCollins, 2018), 119–27.

8. On using sixteenth-century chronicles for earlier histories, see Townsend, *Fifth Sun*, 213–15.

9. Building on pre-Columbian traditions of pictographic writing, in which scribes had recorded histories on *amatl* paper codices, *lienzos*, and *tiras*, the Nahuas' pictographic writing tradition remained prominent throughout the sixteenth century. By the 1540s, Nahuatl-speaking officials adopted European characters to write to one another, and by the 1570s, many scribes integrated these new styles of writing and the European legal apparatus into the record making of the emerging Spanish empire. See Gordon Whittaker, *Deciphering Aztec Hieroglyphs: A Guide to Nahuatl Writing* (London: Thames & Hudson, 2021). In this chapter, I use Nahuatl-Spanish alphabetic writing. For the use of pre-Columbian Mesoamerican codices in writing the history of Nahua science, see the work of Helen Ellis.

10. Lockhart, *Nahuas after the Conquest*, table 10.1.

11. *Vocabulario trilingüe* (ca. 1540), Ayer MS 1478, Newberry Library; Alonso de Molina, *Aqui comiença un vocabulario en la lengua castellana y mexicana* (Mexico: Juan Pablos, 1555); Molina, *Vocabulario*.

12. Hermann W. Haller, *John Florio: A Worlde of Wordes*, The Da Ponte Library Series (Toronto: University of Toronto Press, 2013), ix–xli; Frances A. Yates, *John Florio: The Life of an Italian in Shakespeare's England* (Cambridge: Cambridge University Press, 1934); Sebastián de Covarrubias Horozco, *Tesoro de la lengua castellana o española*, edición integral e ilustrada de Ignacio Arellano y Rafael Zafra (Madrid: Universidad de Navarra-Iberoamericana-Vervuert-Real Academia Española, 2006), xi–lx.

13. Antonio de Nebrija, *Vocabulario español-latino* (Salamanca, 1495), fols. 57r, 25v.

14. Byron Ellsworth Hamann, *The Translations of Nebrija: Language, Culture, and Circulation in the Early Modern World* (Boston: University of Massachusetts Press, 2015), 5.

15. Hamann, *Translations of Nebrija*, 44–47, 96–107.

16. Hamann, *Translations of Nebrija*, 48–50.

17. Hamann, *Translations of Nebrija*, 97–98.

18. Barbara E. Mundy, "Ecology and Leadership: Pantitlan and Other Erratic Phenomena," in Jeanette Favrot Peterson and Kevin Terraciano, eds., *The Florentine Codex: An Encyclopedia of the Nahua World in Sixteenth-Century Mexico* (Austin: University of Texas Press, 2019), 125–38, esp. 126; Molly H. Bassett, "Bundling Natural History: Tlaquimilolli, Folk Biology, and Book 11," in Peterson and Terraciano, *Florentine Codex*, 139–51.

19. Kevin Terraciano, "Introduction: An Encyclopedia of Nahua Culture," in Peterson and Terraciano, eds., *Florentine Codex*, 1–20, esp. 3, table 1.1.

20. Lockhart, *Nahuas after the Conquest*, 261–373.

21. Molly H. Bassett, *The Fate of Earthly Things: Aztec Gods and God-Bodies* (Austin: University of Texas Press, 2015), 50.

22. On the Italian reception, see Ida Giovanna Rao, "On the Reception of the Florentine Codex: The First Italian Translation," in Peterson and Terraciano, *Florentine Codex*, 37–44.

23. For a recent overview, see Victoriano de la Cruz Cruz, "Chicomexochitl y El Maíz Entre Los Nahuas de Chicontepec: La Continuidad Del Ritual." *Politeja* 12, no. 38 (2015): 129–47; Helen Ellis, "Maize, Quetzalcoatl, and Grass Imagery: Science in the Central Mexican *Codex Borgia*" (Ph.D. dissertation, University of California, Los Angeles, 2015).

24. Alan Sandstrom, *Corn Is Our Blood: Culture and Ethnic Identity in a Contemporary Aztec Indian Village* (Norman: University of Oklahoma Press, 1991), 376–77, also 128, 216; on selecting corn seed, 120–21; on varieties of major crops grown in Amatilán, 122–24, table 3.2.

25. For modern ceremonial use of maize, see Sandstrom, *Corn Is Our Blood*, 244, 256, table 6.1, 292.

26. FC vol. 3, bk. 11, fols. 246r–v, 247r.

27. FC vol. 3, bk. 11, fol. 249r.

28. Marcy Norton, "Subaltern Technologies and Early Modernity in the Atlantic World," *Colonial Latin American Review* 26, no. 1 (2017): 38. See also Roberto J. González, *Zapotec Science: Farming and Food in the Northern Sierra of Oaxaca* (Austin: University of Texas Press, 2001).

29. "itech pa tlatoa, inçaçoquenami ic motla xinacthli." Ellis, "Maize," 80–81, 287.

30. For corn colors in agricultural festivities, see Dominique Raby, "The Cave-Dwellers' Treasure: Folktales, Morality, and Gender in a Nahua Community in Mexico," *Journal of American Folklore* (Boston, MA) 120 (Fall 2007): 401–44, esp. 412, 414. Diego Durán's account highlights black, purple, white, and yellow corn kernels used to celebrate *xochiquetzal*.

31. FC vol. 3, bk. 11, fol. 247r; A&D bk. 11, 280.

32. FC vol. 3, bk. 11, fol. 249r; A&D bk. 11, 283.

33. FC vol. 3, bk. 11, fol. 246v; A&D bk. 11, 279.

34. FC vol. 3, bk. 10, fol. 91; A&D bk. 10, 130.

35. Alfredo López Austin, *Cuerpo Humano e Ideología: Las Concepciones de los Antiguos Nahuas*, 2 vols. (1980; repr., Ciudad Universitaria, Mexico: Universidad Nacional Autónoma de México, 2004), vol. 1, 190; vol. 2, 16, 59, 176. On sexual relations, see López Austin, *Cuerpo Humano e Ideología*, vol. 1, 328–55. Campbell gives *tlacaxinachtli* as "semen of a man or woman" because *tlaca[tl]* does not render gender. Campbell, *Morphological Dictionary*, 410. López Austin gives "la simiente humana." López Aus-

tin, *Cuerpo Humano e Ideologia*, vol. 2., 187. In classical Nahuatl, maleness is rendered with *oquichtli*, so that one can have *noquicho*, from *no-oquich[tli]-yo*, "my sperm" or "my semen," literally "my inherently possessed male-ness." Likewise, animals are given *oquich-[tli]* as an adverbial prefix to indicate their sex as distinct from female. There are many references to "women's semen" (*tixpampa quiza; tixpampa huetzi; tocihuayo*) and "men's semen" (*tepulayotl*).

36. See these descriptions in A&D bk. 11, 130.

37. Andrew Shryock, Thomas R. Trautmann, and Clive Gamble, "Imagining the Human in Deep Time," in *Deep History: The Architecture of Past and Present*, by Andrew Shryock and Daniel Lord Small (Berkeley: University of California Press, 2011), 21–52.

38. See Don Alonso's biography and Castañeda heritage in Dana Leibsohn, *Script and Glyph: Pre-Hispanic History, Colonial Bookmaking and the Historica Tolteca-Chichimeca* (Washington, DC: Dumbarton Oaks Research Library and Collection, 2009), 58; see also 51, 52, 42. Other royal histories of interest have been studied by Susan Gillespie and Lori Diel.

39. Molina, *Vocabulario*, part 2, fol. 115, col. 1.

40. Julia Madajczak, "Nahuatl Kinship Terminology as Reflected in Colonial Written Sources from Central Mexico: A System of Classification" (Ph.D. thesis, Warsaw, 2014), 304–6; also 320–25.

41. Lisa Sousa, *The Woman Who Turned into a Jaguar, and Other Narratives of Native Women in Archives of Colonial Mexico* (Stanford: Stanford University Press, 2017), 51–52.

42. Gibson, *Aztecs Under Spanish Rule*, 344.

43. Literally "forest-turkey," from *cuauhtitlan*, forest (lit. "among the trees") + *totolin*, "turkey." One could read this simply as "wild bird." Karttunen derives crested guan (*Penelope purpurascens*). Karttunen, *Analytical Dictionary of Nahuatl*, 65.

44. A&D bk. 11, 29, 53–54; FC vol. 3, bk. 11, fols. 56v–57v.

45. These keepers fed the birds *tonacaiutl*. *To-naca[tl]-yo-tl* literally means "our inherently possessed flesh." This term refers to maize because the humans living today were believed to have been originally formed from maize dough (masa), after previous generations of people made from other substances. By this thinking, "our flesh" is maize because the people (Nahuas) are made of masa (and sustained by maize as a primary food source). Wimmer, Dictionnaire, ".*tōnacāyōtl:*."

46. A&D bk. 11, 29, 53–54, 57–58, figs. 745–48; FC vol. 3, bk. 11, fols. 56v–57v.

47. For the swelling effects of the *teopochotl* tree, see FC vol. 3, bk. 11, fols. 57rv, 58v.

48. For the appearance of this New World bird in the Medici court, see Lia Markey, "A Turkey in a Medici Tapestry," in *Imagining the Americas in Medici Florence* (University Park: Pennsylvania State University Press, 2016), 17–28; Sabine Eiche, *Presenting the Turkey: The Fabulous Story of a Flamboyant and Flavourful Bird* (Florence: Centro Di, 2004); Shephard Krech III, "On the Turkey in Rua Nova Dos Mercadores," in *The Global City: On The Streets of Renaissance Lisbon*, ed. Annemarie Jordan Gschwend and K. J. P. Lowe (London: Paul Holberton, 2015), 181; Larry Silver, "World of Wonders: Exotic Animals in European Imagery, 1515–1650," in Pia F. Cuneo, *Animals and Early Modern Identity* (London: Routledge, 2017), 304–5.

49. While scholars of European history might urge us to remember the history of breast-feeding behind the naming of mammals, experimentation with nature living anywhere was intrinsically gendered. Thus, what made European science exceptionally

productive must not have been those shared gender norms. See Londa Schiebinger, "Why Mammals Are Called Mammals: Gender Politics in Eighteenth-Century Natural History," *American Historical Review* 98, no. 2 (April 1993): 382–411.

50. The immense scholarship—including key contributions by Alfred Crosby and Antonello Gerbi—on the impact of the Spanish empire on American nature is beyond the purview of this monograph. For an introduction to the conquest, see Matthew Restall, *Seven Myths of Spanish Conquest* (Oxford: Oxford University Press, 2004).

51. Restall, *When Montezuma Met Cortés*, 127.

52. Restall, *When Montezuma Met Cortés*, 124.

53. Hernán Cortés, second letter to Charles V, 1520, Newberry Library. See also *The Library of Original Sources*, ed. Oliver J. Thatcher, vol. 5, *9th to 16th Centuries* (Milwaukee: University Research Extension, 1907), 317–26; Elizabeth Hill Boone, "This New World Now Revealed: Hernán Cortés and the Presentation of Mexico to Europe," *Word & Image: A Journal of Verbal/Visual Enquiry* 27, no. 1 (2011): 31–46.

54. Bernal Diáz del Castillo, *The True History of the Conquest of New Spain,* trans. Janet Burke and Ted Humphrey (Indianapolis: Hackett Publishing Company, 2012), 200, 204.

55. The accompanying Castilian text translates to "Also they had pages that accompanied them and served them and that they also used: dwarves, hunchbacks, and other monstrous men. Also, they raised [*criavan*] wild animals, eagles, tigers, bears, and wild cats, and all types of birds." FC bk. 8, chap. 10, 19v–20v; A&D bk. 8, 30.

56. A&D bk. 8, 30.

57. A&D bk. 8, 49. See also A&D bk. 8, 29, 37.

58. *Tzapatl* (pl. *tzapame*) is defined as *enano*, or dwarf, in Molina, *Vocabulario*, part 2, fol. 151v, col. 1.

59. A&D bk. 3, 35; Juan de Torquemada, *Segunda parte de los veinte i un libros rituales i monarchia indiana* (Madrid: Nicolas Rodrigo Franco, 1723), 521, cited in A&D's gloss in A&D bk. 4, 43. See also A&D bk. 3, 19–20. They accompanied Quetzalcoatl to climb Popocatepetl and Itztaccihuatl; later Spanish texts described these servants being sacrificed following a master's death, either on the pyre or in some other manner.

60. Restall, *When Montezuma Met Cortés*, 133–36.

61. Restall, *When Montezuma Met Cortés*, 138.

62. Braudel, *The Mediterranean*, 23, 24, 42–43.

63. Liz Herbert McAvoy, Patricia Skinner, and Theresa Tyers, "Strange Fruits: Grafting, Foreigners, and the Garden Imaginary in Northern France and Germany, 1250–1350," *Speculum* 94, no. 2 (April 2019): 467–95, citing the Douay-Rheims Bible. As Leviticus 19:19 demanded, "Thou shalt not make thy cattle to gender with beasts of any other kind. Thou shalt not sow thy field with different seeds."

64. For grafting in the Ottoman tradition, see Aleksandar Shopov, "Grafting in Sixteenth-Century Mamluk and Ottoman Agriculture and Literature," in *History and Society During the Mamluk Period (1250–1517)*, ed. Bethany J. Walker and Abdelkader Al Ghouz (Göttingen: Bonne University Press, 2021), 381–406, esp. 405, citing Ottoman scholar and bureaucrat Gelibolulu Mustafâ ʿAlî's *Kühnhü'l-ahbār*. This tradition includes techniques of "grafting with cleft," relying on beeswax and budding, or "grafting with bud," wherein the bud of the desired plant is inserted beneath the bark of a stock plant. More broadly, on the practical and intellectual ramifications of agriculture, see Aleksandar Shopov, "Between the Pen and the Fields: Books on Farming,

Changing Land Regimes, and Urban Agriculture in the Ottoman and Eastern Mediterranean, ca. 1500–1700" (Ph.D. dissertation, Harvard University, 2016).

65. Marriage policies on the mainland reacted to policies in the Caribbean islands. Consider the marriage policies orchestrated by Nicolás de Ovando (1460–1511), Commander of Lares of the Order of Alcantara, who received the title of Governor of the Indies on September 3, 1501. The policies Ovando put into place, which largely monitored marriage licenses rather than children born of less permanent unions, decided if and how the colonists should form social unions with the indigenous inhabitants, largely the Taíno. Originally, Ovando had declared that Indian women ought not to be retained against their wishes. By 1504, he set a maximum number for mixed marriages. Finally, in 1514, Spaniards were once again free to marry Indians. Karen Anderson-Córdova, *Surviving Spanish Conquest: Indian Fight, Flight, and Cultural Transformation in Hispaniola and Puerto Rico* (Montgomery: University of Alabama Press, 2017), 29–53.

66. On horse breeding in the Americas, see Renton, "Social and Environmental History of the Horse," 89–183. In Mexico, as they had in the Caribbean, Spaniards married or paired off with local women, and together they created a mixed or mestizo population dependent on the new colonial order, while also establishing an agricultural system akin to that of Europe. On *mestizaje* and marriages in Latin America, see Vinson, *Before Mestizaje*, 18–34; Karen Graubert, *With Our Labor and Sweat: Indigenous Women and the Formation of Colonial Society in Peru, 1550–1700* (Stanford: Stanford University Press, 2007); Rappaport, *Disappearing Mestizo*, 61–94.

67. Horns are *cuacuahuitl* (head [*cuaitl*] + wood/sticks [*cuahuitl*]). This term is also used for feathers and antennae, so anything sticking up from the head like a branch. On Nahua animal domestication, see Christopher Valesey, "Managing the Herd: Nahuas, Animals, and Colonialism in Sixteenth-Century New Spain" (Ph.D. dissertation, Pennsylvania State University, 2019).

68. Alanna Radlo-Dzur et al., "The Tira of Don Martin: A Living Nahua Chronicle," *Latin American and Latinx Visual Culture* 3, no. 3 (2021): 7–37, esp. 16–17.

69. Lockhart, *Nahuas after the Conquest*, 283, table 7.14, 408, 428–30; Molina, *Vocabulario*, 55–74. See Mariano Cuevas, *The Codex Saville: America's Oldest Book*, Historical Records and Studies 19 (New York: United States Catholic Historical Society, 1929), 7–20, esp. 16.

70. Parker, *Emperor*, 358–62.

71. The Oztoticpac Lands Map, ca. 1540, 76 × 84 cm, G4414.T54:2O9 1540.O9, Library of Congress Geography and Map Division, Washington, DC. This section relies on Howard F. Cline, "The Oztoticpac Lands Map of Texcoco, 1540," *Quarterly Journal of the Library of Congress* (April 1966): 76–115. When the Inquisition decided to whom these trees belonged following Don Carlos's execution, Bernardino de Sahagún translated the grafting story from Nahuatl to Spanish and acted as a witness. A Spaniard named Alonso de Contreras had been attempting to uproot the entire orchard, and before the trial he had already been prohibited from touching the trees until an outcome had been decided. Indeed, although Pedro de Veraga demanded that his plants be returned, it was unclear what it would mean to return a grafted tree. In the 1530s, he had given Don Carlos the European plants to graft onto the indigenous trees in Don Carlos's orchard in order to receive half of the fruit borne on those trees in return. If the Spanish government had seized the land following the inquisitorial proceedings, then what did that mean for the fused plants? Don Carlos had confirmed this arrangement,

Pedro argued, and thus his goods ought to no longer be embargoed; with the support of Pedro's witnesses, the inquisitorial officials were convinced that the contract had been affirmed by Don Carlos. The outcome of the proceedings was not recorded.

72. Columella, *De Re Rustica* (Lugduni: Apud Seb. Gryphium, 1537), bk. 5. While Columella's advice was ancient, this passage could be found in an annotated edition published three years before the approximate creation of the Oztoticpac Lands Maps.

73. Columella, *De Re Rustica*, bk. 5.

74. René Acuña, *Relaciones Geográficas novohispanas del siglo XVI* (México: Universidad Nacional Autónoma de México–Instituto de Investigaciones Antropológicas, 1982–88) (hereafter cited as *RG*), vol. 8, 107.

75. *RG*, vol. 6, 53–66.

76. For the full description, see *RG*, vol. 6, 125–55.

77. *RG*, vol. 7, 250.

78. Pedro Gutiérrez de Cuevas, Juan de Écija, *RG*, Michoacán, Cuiseo de la Laguna, 28 agosto 1579. See also Abel Alves, *The Animals of Spain: An Introduction to Imperial Perceptions and Human Interaction with Other Animals, 1492–1826* (Leiden: Brill, 2011), 101.

79. *Vocabulario trilingüe*, 40r.

80. *Vocabulario trilingüe*, 81, 83.

81. *Vocabulario trilingüe*, 65.

82. *Vocabulario trilingüe*, 92r.

83. Thouvenot, *Diccionario*, 358.

84. Thouvenot, *Diccionario*, 452.

85. The full list included chia, uauhçatl (modernized as huauhzacatl: a type of tough, leathery herb used as a cover for amaranth seeds; A&D bk. 11, 194), cocotl, teopochotl, uahtzontli, tzitziquilitl, mexixin, petzicatl, and quanacaqujlitl. These plants include a type of amaranth (cocotl, petzicatl) as well as individual parts of the amaranth plant (huauhtzontli), and others that are primarily eaten as greens (tzitziquilitl, mexixin, petzicatl) as well as two whose seeds are primarily used as animal feed (petzicatl, cuanacaquilitl).

86. In modern Nahuatl, *xinachtli* has become ubiquitous, likely in a combination of indigenous American and European agricultural practices. When it came to the ceremonial growing of corn in the Salt Lake City *milla* (garden), students learned Nahuatl grammar through the act of planting corn, using language such as *ticoyazceh huan ticaquechizceh xinachtli; tictlatlalhuilizceh xinachtli tlaixpan; huan Tictocazceh xinachtli*. In education, the Xinactli Project defines its educational ambitions around the "germinating seed." Likewise, the IDIEZ Project has propagated the word *xinachcalli*—"a neologism that we have made to mean 'library' as the 'house of seeds,'" with a more metaphorical meaning in which the books are the seeds," as Abelardo de la Cruz de la Cruz pointed out (personal correspondence, May 6, 2019). See Virginia Lea and Judy Hefland, eds., *Identifying Race and Transforming Whiteness in the Classroom* (New York: Peter Lang, 2006), 263.

87. Molina, *Aqui comiença un vocabulario*: "Tlacaxinachtli. Simiente de varón o de muger. Tlatoca tlacaxinachotl. Generacion de nobles cavalleros. Xinachoa. Nino. Asementarse. P. oninoxinacho." Thouvenot's normalized dictionary, which is based on Molina's dictionary, gives this as a single word: "tlatocatlacaxinachotl: generación de nobles caballeros," from Molina, *Vocabulario*, 2; he also gives a reduplicated version: "tlatocatlacaxinachchotl: genealogía por linaje noble," from Molina, *Vocabulario*, 1.

88. A&D bk. 6, 96. The verb *xinachoa* ("to multiply"; Karttunen, *Analytical Dictionary of Nahuatl*, 325) + nonactive *-lo* & *-z* future suffixes with *ne-* as an indeterminate human object—the quoted phrase indicates that God said for *ne-* (people in general) to *-xinacholoz* "will multiply" and *ne-* (people in general) *-tlapihuiloz* "will increase" *tlalticpac* "on earth." See also A&D bk. 6, 37.

89. This list is meant to be suggestive, not comprehensive. *nopani, nopani +, notahuan +*: Thouvenot, *Diccionario*, 222; *pilli, pillotl, piltoca, nino*: Thouvenot, *Diccionario*, 279.

90. Thouvenot, *Diccionario*, 358.

91. A tear in cloth is defined as *raza del paño* (*ihitlacauhca intilmatli*). Thouvenot, *Diccionario*, 133n2.

92. "Chipahuacanemi, ni: vivir casta, y limpiamente." Thouvenot, *Diccionario*, 81.

93. I rely on the critical translation: Domingo Francisco de San Antón Muñón Chimalpahin Cuauhtlehuanitzin, *Annals of His Time: Don Domingo de San Antón Muñón Chimalpahin Quauhtlehuanitzin*, trans. James Lockhart, Susan Schroeder, and Doris Namala (Stanford: Stanford University Press, 2006). See also María Elena Martínez, "The Black Blood of New Spain: *Limpieza de Sangre*, Racial Violence, and Gendered Power in Early Colonial Mexico," *William and Mary Quarterly*, 3rd series, 61, no. 3 (July 2004): 479–520. On race and blackness, see Daniel Joseph Nemser, "Toward a Genealogy of *Mestizaje*: Rethinking Race in Colonial Mexico" (Ph.D. dissertation in Hispanic Languages and Literatures, University of California, Berkeley, 2011), 122–23. See also Daniel Nemser, *Infrastructures of Race: Concentration and Biopolitics in Colonial Mexico* (Austin: University of Texas Press, 2017), 60–61. Scholars have shown that utopia—the quintessential early modern genre—and dystopia—the twentieth-century term for an imperfect world—must be understood in a dialectic. Both are created through linked acts of imagination. If utopias are the dream of imagined worlds, then dystopias are the nightmares of real ones, tainted by the imperfection of lived experience. By this reading, the resonances across Campanella's utopian vision and Chimalpahin's dystopian one are by no means random, but rather the result of a shared Spanish imperial world. Michael D. Gordin, Helen Tilley, and Gyan Prakash, *Utopia/Dystopia: Conditions of Historical Possibility* (Princeton: Princeton University Press, 2011), 2.

94. Camilla Townsend, *Annals of Native America: How the Nahuas of Colonial Mexico Kept Their History Alive* (Oxford: Oxford University Press, 2017), 14, 141–74.

95. Chimalpahin records the total as 35 executed, while his totals add to 36.

96. Tensions between Spaniards and the growing black population of New Spain had been mounting, fueled by rumors and fears. By 1609, rumors that the Afro-Mexicans would rebel had reached a fever pitch. Velasco sought to stop the rebellion before it could start, dispatching an armed force to Puebla to quash the populations of escaped slaves and rebels, especially those in the maroon colony of Rio Blanco. Gaspar Yanga (1545–1618?) successfully resisted the attack on his colony and continued promoting raids against Spanish settlements. While Velasco *hijo* had finished his tenure as viceroy and taken up a position in Seville as president of the Council of the Indies, his aggression seems to have remained at the forefront of rebels' memories during the 1612 plot that Chimalpahin recorded. On Gaspar Yanga, see Jane G. Landers, "Cimarrón and Citizen: African Ethnicity, Corporate Identity, and the Evolution of Free Black Towns in the Spanish Circum-Caribbean," in *Slaves, Subjects, and Subversives: Blacks in Colonial Latin America*, ed. Jane Lander and Barry Robinson (Albuquerque: University of New Mexico Press, 2006), 111–45.

97. "Diario de Domingo de San Anton Muñón Chimalpáhin," Chimalpahin Cuauhtlehuanitzin (1579–1660), Bibliothèque nationale de France, Mexicain 220, fol. 186.

98. The Spanish loanword *Morisco* is a jarring addition to Chimalpahin's Nahuatl prose. This term evoked slippage between Africans and Moors, mixing religious categories with fixed heredity consolidated under *limpieza de sangre*. Africa, Islam, race, and blood were conflated by this indigenous author, who envisioned no alliance with these rebels. Like other contemporary Nahuatl writers, Chimalpahin regularly used the term *tliltic*—for black color—to refer to black people; the term *tliltic* referred to "something black from Ethiopia" in Alonso de Molina's 1571 Nahuatl dictionary. Molina, *Vocabulario*, fol. 148r, col. 1.

99. Chimalpahin, *Annals of His Time*, 220–21.

100. Chimalpahin, *Annals of His Time*, 220–21.

101. "Diario de Domingo de San Anton Muñón Chimalpáhin," fol. 187.

102. Chimalpahin, *Annals of His Time*, 220–21.

103. Chimalpahin, *Annals of His Time*, 220–21.

104. Chimalpahin, *Annals of His Time*, 220–21.

105. Chimalpahin, *Annals of His Time*, 220–23.

Chapter Five

1. Rappaport, *Disappearing Mestizo*, 4–6.

2. Rappaport, *Disappearing Mestizo*, 29.

3. On *mestizaje* and hybridity, see Ralph Bauer and Marcy Norton, "Introduction: Entangled Trajectories: Indigenous and European Histories," *Colonial Latin American Review* 26, no. 1 (2017): 1–17, esp. 7–8. On nineteenth-century ideas of hybridity and *mestizaje*, see Vinson, *Before Mestizaje*, 187.

4. John Beusterien, *Canines in Cervantes and Velázquez* (New York: Routledge, 2016), 111–21; Stafford Poole, "Criollos and Criollismo," in *Encyclopedia of Mexico: History, Society, and Culture*, ed. Michael S. Werner (Chicago: Fitzroy Dearborn, 1997); Julian Pitt-Rivers, "On the Word Caste," in *The Translation of Culture: Essays to E. E. Evans Pritchard*, ed. T. O. Beidelman (London: Routledge, 1971), 234–35.

5. Thomas M. Stephens, *Dictionary of Latin American Racial and Ethnic Terminology* (Gainesville: University Press of Florida, 1999), 161–65, *criollo*.

6. On the insult *perro-mulatto*, see Chris Garces, "The Interspecies Logic of Race in Colonial Peru: San Martín de Porres's Animal Brotherhood," in *Sainthood and Race: Marked Flesh, Holy Flesh*, ed. Molly Bassett and Vincent Lloyd (New York: Routledge, 2014), 82–101.

7. Covarrubias, *Tesoro* (1611), *mulato*.

8. *Coyote* continues to be used today in reference to people who facilitate migration; then again, in indigenous communities today, *coyote* can be used to refer to outsiders/Others of various degrees. Stephens, *Dictionary*, *coyote*, 158–59; *lobo*, 297–98; *mechino* (offspring of *coyote* + *lobo*, late nineteenth century), 320; *mestizo*, 323–28.

9. Other insults were also animal terms. *Moro* referred to the former people of the state of Mauretania, an area that covered northern Morocco, Algeria, Ceuta, and Melilla during Roman times, as Covarrubias's dictionary defines it: Covarrubias, *Tesoro* (1611), 115v. However, the word took on an increasingly derogatory meaning by the sixteenth century, when it often it referred to "dog" or "Moorish dog." Andrew Hess, *The Forgotten Frontier: A History of the Sixteenth-Century Ibero-African Frontier* (Chi-

cago: University of Chicago Press, 1978), 181; Stuart Schwartz, *All Can Be Saved: Religious Tolerance and Salvation in the Iberian Atlantic World* (New Haven: Yale University Press, 2008), 48–49.

10. Beusterien, *Canines in Cervantes and Velázquez*, 17–24, 28, 40–44. See also John Grier Varner, *Dogs of the Conquest* (Norman: University of Oklahoma Press, 1983). In the media, see James Gordon, "How Old Is the Maltese, Really?," *New York Times*, October 4, 2021, https://www.nytimes.com/2021/10/04/science/dogs-DNA -breeds-maltese.html.

11. For an introduction to the extensive scholarship about Hernández, see Simon Varey, ed., *The Mexican Treasury: The Writings of Dr. Francisco Hernández*, trans. Rafael Chabrá, Cynthia L. Chamberlin, and Simon Varey (Stanford: Stanford University Press, 2000); and Simon Varey, *Sustaining Literature: Essays on Literature, History, and Culture, 1500–1800: Commemorating the Life and Work of Simon Varey* (Lewisburg: Bucknell University Press, 2007), 41–56. See also Ernesto Capanna, "South American Mammal Diversity and Hernandez's Novae Hispaniae Thesaurus," *Rendiconti Lincei* 20, no. 1 (2009): 39–60; Germán Somolino d'Ardois, *Vida y obra de Francisco Hernández* (México: Universidad Nacional Autónoma de México, 1960); José Maria Lopez Piñero and José Pardo-Tomás, *La influencia de Francisco Hernández (1515–1587) en la constitución de la botánica y la materia médica modernas* (Valencia: Instituto de Estudios Documentales e Históricos sobre la Ciencia, 1996); José Maria Lopez Piñero and José Pardo-Tomás, *Nuevos materiales y noticias sobre la "Historia de las plantas de Nueva España," de Francisco Hernández* (Valencia: Instituto de Estudios Documentales e Históricos sobre la Ciencia, 1994); Sandra I. Ramos Maldonado, "Tradición pliniana en la Andalucía del siglo XVI: A propósito de la labor filológica del Doctor Francisco Hernández," in *Las raíces clásicas de Andalucía*, ed. M. Rodríguez-Pantoja (Córdoba: Actas del IV congreso Andaluz de Estudios Clásicos, 2006); Jacqueline Durand-Forest, *Aperçu de l'histoire naturelle de la Nouvelle-Espagne d'après Hernández, les informateurs indigènes de Sahagun et les auteurs du Codex Badianus, Nouveau monde et renouveau de l'histoire naturelle* (París: Publicaciones de la Sorbonne, 1986); Jesus Bustamente Garcia, "Un Libro, Tres Modelos, y el Atlántico: Los Datos de una historia: los antecedentes y el proyecto," in Cadeddu and Guardo, *Il tesoro messicano*, 27. See Jaime Marroquín Arredondo, "The Method of Francisco Hernández: Early Modern Science and the Translation of Mesoamerica's Natural History," in *Translating Nature: Cross-Cultural Histories of Early Modern Science*, ed. Jaime Marroquin Arredondo and Ralph Bauer (Philadelphia: University of Pennsylvania Press, 2019), 45–69. For the later circulation of Hernández's writings, see Nardo Antonio Recchi, *Rerum medicarum Novae Hispaniae thesaurus seu plantarum animalium mineralium Mexicanorum* (Rome: Typographeio Vitalis Mascardi, 1651); Sabina Brevaglieri, *Storie naturali di Roma* (Rome: Viella, La corte dei Papi, forthcoming); Marco Guardo, "Nell'Officina del Tesoro Messicano: Il Ruolo Misconosciuto di Marco Antonio Petilio nel sodalizio Linceo," in Cadeddu and Guardo, *Il tesoro messicano*; Mar Rey Bueno and María Esther Alegre Pérez, "Renovación de la terapéutica real: Los destiladores de su majestad, maestros simplicistas y médicos herbolarios de Felipe II," *Asclepio* 53, no. 1 (2001).

12. Jose Pardo-Tomás, "Viajes de ida o de vuelta: La circulación de la obra de Francisco Hernández en Mexico, 1576–1672," in Cadeddu and Guardo, *Il tesoro messicano*.

13. Francisco Hernández, Rafael Chabrán, and Simon Varey, "'An Epistle to Arias Montano': An English Translation of a Poem by Francisco Hernández," *Huntington Library Quarterly* 55, no. 4 (Autumn 1992): 620–34, lines 110–11. The Spanish trans-

lation is more true to the original: F. Navarro Antolín and J. Solís de los Santos, "La epístola latina en verso de Francisco Hernández a Benito Arias Montano (Madrid, Biblioteca del Ministerio de Hacienda, MS FA 931)," *Myrtia: Revista de filología clásica,* 29 (2014): 201–45.

14. Alfred Crosby, *The Columbian Exchange: Biological and Cultural Consequences of 1492* (Westport, CT: Greenwood Press, 1972) and *Ecological Imperialism: The Biological Expansion of Europe, 900–1900* (Cambridge: Cambridge University Press, 1986). For a succinct historiographic summary, see Rebecca Earle, "The Columbian Exchange," in *The Oxford Handbook of Food History*, ed. Jeffrey Pilcher (Oxford: Oxford University Press, 2012), 341–57. See also Kirkpatrick Sale, *The Conquest of Paradise: Christopher Columbus and the Columbian Legacy* (New York: Alfred A. Knopf, 1990). Scholars such as John Robert McNeill complicated this narrative by arguing that rather than enabling conquest, introduced diseases such as malaria and yellow fever kept Europeans from fulfilling their dreams of empire in the Americas, particularly in the Greater Caribbean. McNeill, *Mosquito Empires: Ecology and War in the Greater Caribbean, 1620–1914* (New York: Cambridge University Press, 2010). On diseases that spread between humans and animals, see Karl Appuhn, "Ecologies of Beef: Eighteenth-Century Epizootics and the Environmental History of Early Modern Europe," *Environmental History* 15, no. 2 (April 2010): 268–87.

15. The tomato common in Europe (and the US) is the *xitomatl* (*jitomate* in Spanish), although there are many varieties. David Gentilcore, *Pomodoro: A History of the Tomato in Italy* (New York: Columbia University Press, 2010). See Renton, "Social and Environmental History of the Horse," on the *cimarrón*; see also Norton, *Sacred Gifts, Profane Pleasures.*

16. Alves, *Animals of Spain.*

17. A new wave of scholarship is emerging that envisions animals as both agents and objects within the Columbian Exchange. For the leading account, see Marcy Norton, "The Chicken or the *Iegue*: Human-Animal Relationships and the Columbian Exchange," *American Historical Review* 120, no. 1 (February 2015): 28–60. I rely on the term "mixture events" to explain moments when two populations interacted and interbred, as per Reich, *Who We Are.*

18. For one of many examples, see the discussion of potatoes in Charles Mann, *1493: Uncovering the New World Columbus Created* (New York: Knopf, 2011), 197–98, 202–11.

19. See, for instance, Rebecca Earle, *The Body of the Conquistador: Food, Race and the Colonial Experience in Spanish America, 1492–1700* (Cambridge: Cambridge University Press, 2012), 160.

20. Heidi G. Parker et al., "Genomic Analyses Reveal the Influence of Geographic Origin, Migration, and Hybridization on Modern Dog Breed Development," *Cell Reports* 19, no. 4 (April 2017): 698–708. Based on genetic evidence, the Chinese crested is believed to have developed in the Americas. There are reports that the Apaches used dogs as beasts of burden, though less is known about their breeding and selection traditions.

21. Traditional historical analyses have recently been supplemented by genetic evidence that permits conclusions about dogs' origins. Scholars currently believe that the ancestors of American dogs had traveled with humans to the Americas from East Asia more than ten thousand years before. Guo-Dong Wang et al., "Out of Southern East Asia: The Natural History of Domestic Dogs Across the World," *Cell Research*

26 (2016): 21–33. On canine migration and intermixing in the Americas, see Angela Perri et al., "New Evidence of the Earliest Domestic Dogs in the Americas," *American Antiquity* 84, no. 1 (January 2019): 68–87; Máire ní Leathlobhair et al., "The Evolutionary History of Dogs in the Americas," *Science* 361, no. 6397 (July 2018): 81–85.

22. Parker et al., "Genomic Analyses."

23. Parker et al., "Genomic Analyses."

24. Research in population genetics has likewise complicated the history of human demography in the Caribbean. For the human Taíno population that had been thought completely lost, see Miguel Vilar, "Genographic Project DNA Results Reveal Details of Puerto Rican History," *National Geographic: Voices*, posted July 25, 2014; Miguel Vilar et al., "Genetic Diversity in Puerto Rico and Its Implications for the Peopling of the Island and the West Indies," *American Journal of Physical Anthropology* 155 (2014): 352–68. While dogs are the protagonists in this chapter, it bears noting that, contemporaneously, feral swine with an up-regulation of endocrine transformed into razorback pigs, horses into mustangs, and cattle into longhorns. For wild horses in the Americas, see Renton, "Social and Environmental History of the Horse," 123–57.

25. "Seated Female Figure with Dog in Lap Tlatilco, 1200–900 BCE," Ceramics, Los Angeles County Museum of Art, M.2019.45.88. Thank you to Alanna Radlo-Dzur for this recommendation and the subsequent Classic Maya example.

26. "Vessel, Mythological Scene," Attributed to the Metropolitan Painter (active 7th–8th century CE), Guatemala Mexico, Metropolitan Museum of Art, 1978.412.206.

27. By the 1580s, wolves were increasing in northern Mexico and parts of present-day Arizona, New Mexico, and Texas. For the relationship between Xolotl and the wolf, see R. Valadez et al., "Dog-Wolf Hybrid Biotype Reconstruction from the Archaeological City of Teotihuacan in Prehispanic Central Mexico," in *Dogs and People in Social, Working, Economic or Symbolic Interaction,* ed. L. M. Snyder and E. A. Moore (Oxford: Oxbow Books, 2002), 309–29, esp. 321. On wolves interred at Tenochtitlan, see the work of Ximena Chavez Balderas, "Los animals y el recinto sagrado de Tenochtitlan," El Colegio Nacional, November 7, 2018, https://colnal.mx/noticias/sintesis -informativa-los-animales-y-el-recinto-sagrado-de-tenochtitlan-dia-1/. Thank you to Chris Valesey for recommending this presentation.

28. Chavez Balderas, "Los animales y el recinto sagrado de Tenochtitlan."

29. Eduardo Matos Moctezuma and Felipe Solís Olguín, eds., *Aztecs* (London: London Academy of Arts, 2002), exhibition catalogue no. 140, 434. One of the sculptures was found by Batres in 1900 on calle de las Escalerillas, the other was uncovered during excavation of the Metro on calle de Tacuba near calle de la Palma in 1967 (MNA 10-116545). Alanna Radlo-Dzur has pointed out that the primary definition of *xolotl* (with short vowels) is a twin, and it is in this context that the term is used as the name for the deity Xolotl, the twin of Quetzalcoatl, who is either dog-headed or takes the form of a xoloitzcuintli, which is why the dog might be translated as a "twin-dog" and not a "slave-dog."

30. See Elizabeth Hill Boone, *Cycles of Time and Meaning in the Mexican Books of Fate* (Austin: University of Texas Press, 2007), 192, 202, 267, 269; Eloise Quiñones Keber, "Xolotl: Dogs, Death, and Deities in Aztec Myth," *Latin American Indian Literatures Journal* 7, no. 2 (1991): 229–39.

31. Bassett, *Fate of Earthly Things*, 24.

32. Bassett, *Fate of Earthly Things*, 12–14.

33. Alanna Radlo-Dzur, personal correspondence, August 3, 2021, regarding her

forthcoming dissertation "The Invisible in Early Modern Nahua Art" (Ohio State University).

34. Molly Bassett, "The Pre-racial Saint? Ma(r)king Aztec God-Bodies," in Bassett and Lloyd, *Sainthood and Race*, 199-216; on spectrum of animacy, 202-3.

35. Thouvenot, *Diccionario*, 304. Also to be born, to hatch, etc. (coming into life) and it also has a second definition, "to masturbate."

36. Valesey, "Managing the Herd."

37. On "people-eaters" as a category for animals, see Bassett, "Bundling Natural History," 149, citing conversation with Leon García Garagarza. In older Nahuatl, *tlapiyalli* refers to something guarded, preserved, taken care of. Karttunen, *An Analytical Dictionary of Nahuatl*, 292, gives *tlapiyalli* as "domesticated animal," citing Key and Key's vocabulary of twentieth-century Nahuatl spoken in Zacapoaxtla, Puebla, and published in 1953. Thus, it remains to be seen when *tlapiyalli* took on this connection to domesticated animals. See also Justyna Olko, "In Ocelotl, in Tequani: The Mesoamerican Jaguar as Described in the Nahuatl Text of the Florentine Codex," in *Birthday Beasts' Book: Where Human Roads Cross Animal Trails, Cultural Studies in Honour of Jerzy Axer*, ed. Katarzyna Marciniak (Warsaw: Institute for Interdisciplinary Studies "Artes Liberales," 2011), 301-8.

38. Josh Fitzgerald, "Cross-Pollinated Coyote Wisdom: Persistent Memories of Nahua-Animal Reciprocity and Natural Science," forthcoming.

39. A&D bk. 4, 19-20; FC vol. 1, bk. 4, fols. 14v-15r.

40. A&D bk. 4, 19-20; FC vol. 1, bk. 4, fols. 14v-15r.

41. FC vol. 1, bk. 4, fols. 14v-15r.

42. On the interpretation of the *tonalpohualli*, see Boone, *Cycles of Time and Meaning*, 240-43.

43. James Lockhart, *Nahuas after the Conquest*, 120, table 4.2, citing *Aztekischer Zenus*, ed. Hinz et al., Archivo Histórico of the Museo Nacional de Antropología e Historia, Mexico City, Colección Antigua, 549-51.

44. FC vol. 1, bk. 4, fols. 14v-15r.

45. A&D bk. 3, 41. On death ritual, see the description of carrying a little yellow dog with a cotton cord to Chiconahuapan, A&D bk. 3, chap. 1 appendix, 43.

46. FC vol. 3, bk. 11, fols. 16v-17v.

47. A&D bk. 11, 151. *Xochcohcoyotl* (Nahuatl, "flower-coyote") or *xochiocoyotl* (Spanish, "flower-[design] coyote" or "coyote covered in flowers"); *tetlami* or *tetlamin* (both Garibay and Siméon regularize with the final "n"); *tehuitzotl* (s.t. like a dog that is "inherently possessed of stone-thorn[s]"). In modern Nahuatl, *itzcuintli* refers only to a native Mexican dog. Karttunen, *An Analytical Dictionary of Nahuatl*, 108; Andrews, *Introduction to Classical Nahuatl*.

48. FC vol. 3, bk. 11, fols. 16v-17v.

49. Mercedes Montes de Oca Vega, "Los difrasismos: Un rasgo del lenguaje ritual," *Estudios de cultura náhuatl* 39 (2008): 228-29.

50. The modern Nahuatl words for female dog, *zohuachichi* and *cihuaitzincuintli*, are not mentioned in these passages. Karttunen, *An Analytical Dictionary of Nahuatl*, 34, 347.

51. FC vol. 3, bk. 11, fols. 16v-17v.

52. FC vol. 3, bk. 11, fols. 16v-17v.

53. FC vol. 3, bk. 11, fols. 16v-17v.

54. Alanna Radlo-Dzur pointed me toward these passages; the Nahuatl in the main

text in the next paragraph relies on her translation. The verb *itzcuinchichihua* ("to prepare, roast dog [meat]") is used in A&D bk. 4, 123: A&D transcription: *itzcujnmjctia, itzcujnchichinoa, itzcujnchichioa*; normalized text: *itzcuinmictiah, itzcuinchichinoah, itzcuinchihchihuah*; A&D translation: "[they] slew, singed, dressed dogs."

55. A&D bk. 2, 132.

56. A&D bk. 9, 48.

57. A&D bk. 10, 80.

58. See Kevin Terraciano, "Reading Between the Lines of Book Twelve," in Peterson and Terraciano, *The Florentine Codex*, 45–62.

59. Image in T. de Bry, *America*, part 4, fig. 22, in the *Historia* of Girolamo Benzoni, Bodleian Library, Oxford. Cited in Richard C. Trexler, *Sex and Conquest: Gendered Violence, Political Order and the European Conquest of the Americas* (Ithaca: Cornell University Press, 1995). See the work of Fernando Urbina on the murals in Guaviare.

60. Beusterien, *Canines in Cervantes and Velázquez*.

61. A&D bk. 12, 19–20. Nahuas used the term *itzcuintli* for European dogs, although they were very different in size and appearance. Lockhart, *Nahuas after the Conquest*, 265.

62. FC bk. 12, fol. 11v.

63. FC bk. 12, fol. 17r.

64. Diego Durán, *The History of the Indies of New Spain*, trans. and ed. Doris Heyden (Norman: University of Oklahoma Press, 1994), 178. Durán's manuscript is now housed in the Biblioteca Nacional de España Vtr/26/11.

65. Hernán Cortés, second letter, in *Letters from Mexico*, trans. and ed. Anthony Pagden (New Haven: Yale University Press, 2001), 103.

66. Pedro de Villela and Francisco Gorjón Toscano, RG, Chilchotla, Michoacán, vol. 4, 109.

67. *RG*, vol. 4, 109; *RG*, vol. 8, 112.

68. Earle, *Body of the Conquistador*, 120. Future studies might show how food culture persisted or changed outside of the Spanish-language documentary record, perhaps by using other types of evidence, such as colonial-period middens in predominantly or exclusively indigenous communities. At this point, it seems that descriptions of dogs' diets and their association with other "stinking, sour" foods of the bad meat seller in the Florentine Codex were European condemnations of foods they did not understand.

69. *RG*, vol. 6, 194.

70. *RG*, vol. 7, 250.

71. Earle, *Body of the Conquistador*, 120.

72. *Relaciones histórico-geográficas de la gobernación de Yucatán*, Fuentes para el estudio de la cultura maya (México: Universidad Nacional Autónoma de México, 1983), 67.

73. *Relaciones histórico-geográficas de la gobernación de Yucatán*, 78.

74. See Elinor G. K. Melville, *A Plague of Sheep: Environmental Consequences of the Conquest of Mexico* (Cambridge: Cambridge University Press, 1994).

75. Parker et al., "Genomic Analyses," 698–708. See also A. R. Boyko et al., "A Simple Genetic Architecture Underlies Morphological Variation in Dogs," *PLoS Biology* 8, no. 8 (2010); Martin Wallen, *Whose Dog Are You? The Technology of Dog Breeds and the Aesthetics of Modern Human-Canine Relations* (East Lansing: Michigan State University Press, 2017).

76. Today, dog breeders have developed modern variations of the *xoloitzcuintli* and *techichi*. The breeds' required characteristics have been clearly articulated by the American Kennel Club. Some breeders see their work as reviving pre-Columbian dogs and link their projects to older imperial histories. For example, in the AKC's "An Introduction to Mexican Dog Breeds" (accessed August 4, 2021, https://www.akc.org/expert-advice/lifestyle/oloitz-dog-breeds-xoloitzcuintli-chihuahua/), Jan Reisen, who has authored other articles on dog keeping, writes: "Around 1,000 years ago, the Chi's ancestor was the larger Techichi, which was the breed of choice for the Toltecs. The Aztecs, who conquered the Toltecs in the 12th century, are responsible for refining the Techichi into a smaller, lighter dog. The breed we know today gets its name from the Mexican state of Chihuahua."

77. For similar forgetting, see Marcy Norton, "The Quetzal Takes Flight: Microhistory, Mesoamerican Knowledge, and Early Modern Natural History," in Marroquín Arredondo and Bauer, *Translating Nature*, 119–47.

78. The Maltese is an animal metric in descriptions of the *xoloitzcuintli*, the *itzcuintepotzotli*, and the *eluro*. On Hernández, see Francisco Hernández, *Obras completas de Francisco Hernández* (Mexico: Universidad Nacional Autónoma de México, 1960), online edition 2015, http://www.franciscohernandez.unam.mx/home.html; Francisco Hernández, *Francisci Hernandi, medici atque historici Philippi II, hispan et indiar: Regis, et totius novi orbis archiatri: Opera, cum edita, tum medita, ad autobiographi fidem et jusu regio*, ed. Casimiro Gómez Ortega (Madrid: Ex Typographia Ibarrae Heredum, 1790). Wimmer, Dictionnaire, notes that *xoloitzcuintli* also appears in Hernández as an alternate designation for the *cuetlachtli* (Mexican wolf).

79. Jeremy Paden, "The Iguana and the Barrel of Mud: Memory, Natural History, and Hermeneutics in Oviedo's *Sumario de la natural historia de las Indias*," *Colonial Latin American Review* 16, no. 2 (2007): 203–26.

80. Enrique Alvarez López, "El perro mudo Americano (El problema del perro mudo de Fernández de Oviedo)," *Boletín de la Real Sociedad Española de Historia Natural* 11 (1942): 411–17.

81. Columbus letter, October 29, cited in Antonello Gerbi, *Nature in the New World: From Christopher Columbus to Gonzalo Fernández de Oviedo* (Pittsburgh: University of Pittsburgh Press, 2010), 17.

82. Michele de Cuneo, "Michele de Cuneo's Letter on the Second Voyage, 28 October 1495," in *Journals and Other Documents on the Life and Voyages of Christopher Columbus*, ed. and trans. Samuel Eliot Morison (Norwalk, CT: Easton Press, 1993), 217.

83. Nicolò Scillacio's letters, C. Merkel, *L'opuscolo "De insulis nuper inventis" del messinese Nicolò Scillacio* (Milan, 1901), 88, cited in Gerbi, *Nature in the New World*, 29. Before Columbian contact, rabies was relegated to bats in the Americas. See Andres Velasco-Villa, "The History of Rabies in the Western Hemisphere," *Antiviral Research* 146 (October 2017): 221–32.

84. The trope of the silent Indian appears in Moteuczoma's "inaction" in the face of Cortés's attack. Rebecca Dufendach, "'As If His Heart Died': A Reinterpretation of Moteuczoma's Cowardice in the Conquest History of the Florentine Codex," *Ethnohistory* 66, no. 4 (October 2019): 624–45.

85. Oviedo, cited in Gerbi, *Nature and the New World*, 295–96.

86. Gonzalo Fernández de Oviedo, *Historia general de las Indias* (Seville: Emprenta de Iuan Cromberger, 1535), bk. 2, 355a.

87. Oviedo, *Historia general de las Indias*, bk. 12, 390b.

88. Pliny the Elder, *The Natural History*, chap. 83.

89. Oviedo, *Historia general de las Indias*, bk. 12, 390b. Alas, Oviedo's other theories concerning animal transportation were similarly flawed. In addition to his silent dog debacle, he had studied the iguana by tying it to a tree for forty days, but its diet remained difficult to ascertain. Aristotle had taught that lower life-forms like reptiles originated from mud via abiogenesis. Therefore, Oviedo sent the iguana to Giovanni Battista Ramusio, a Venetian humanist, packed in with what he thought was its food: a barrel of dirt. Unfortunately, the creature that arrived was only a shriveled corpse. Paden, "The Iguana and the Barrel of Mud," 203–26.

90. Pliny the Elder, *Historia Natural*, trans. Francisco Hernández, Biblioteca Nacional de España, Madrid, Mss/2868, XXI. These notes rely on pagination from José Pardo Tomás, who generously shared his notes on this key source. My engagement with Hernández is deeply informed by his expertise. See Enrique Alvarez López, "El Dr. Francisco Hernández y sus comentarios a Plinio," *Revista de Indias* 3, no. 8 (1942): 251–90. Prince Don Carlos, King Philip's son and heir apparent, even had a hairless dog. Pliny, *Historia Natural*, bk. 8, 409.

91. Pliny, *Historia Natural*, bk. 8, 410.

92. Pliny, *Historia Natural*, bk. 8, 410.

93. Hernández, *Francisci Hernandi*, 1020; Hernández, *Obras completas*, vol. 3, *Historia Natural de la Nueva España* 2, ch. 20. One might compare this to Fabricius's research on animal speech: Girolamo Fabrici, *De brutorum loquela* (Padua: Lorenzo Pasquati, 1603).

94. Hernández, *Obras completas*, vol. 3, *Historia Natural de la Nueva España* 2, ch. 20.

95. *Historia Natural*, bk. 8, 409. On the imagined wonders of Cíbola and the Coronado expedition's attempts to reach them, see Richard Flint and Shirley Cushing Flint, *A Most Splendid Company: The Coronado Expedition in Global Perspective* (Albuquerque: University of New Mexico Press, 2019), 43, 84, 123, 232, 313.

96. For a parallel history, see the mention of *cholos* in Guaman Poma de Ayala and Garcilaso Inca de la Vega's writing: Beusterien, *Canines in Cervantes and Velázquez*, 120–21; Hernández, *Obras completas*, vol. 3, *Historia Natural de la Nueva España* 2, ch. 20. Thomas M. Stephens shows the wide range of use of the word *cholo* for various mixed identities across Latin America today, but the word has the broadest range of definitions and uses in the Andes. See, for instance, the Aymara (and Quechua) use of the word *chola* referring to peasant women of the Bolivian and Peruvian highlands. Stephens, *Dictionary of Latin American Racial and Ethnic Terminology*, 124–30.

97. Hernández, *Obras completas*, vol. 3, *Historia Natural de la Nueva España* 2, chs. 20 and 31.

98. Hernández, *Obras completas*, vol. 3, *Historia Natural de la Nueva España* 2, ch. 21.

99. The Gran Diccionario Nahuatl has an entry from Clavijero's 1780 *Vocabulario* for *techichi*: "Cierto cuadrúpedo que había antiguamente en México semejante al perro; pero hoy se da este nombre a los verdaderos perros." Hernández, *Obras completas*, vol. 3, *Historia Natural de la Nueva España* 2, ch. 20. The jury is out about whether *tepeitzcuintli* was really a dog. Wimmer, Dictionnaire, defines *tepeitzcuintli* as "a mammalian rodent that grunts like a pig whose flesh is well appreciated." The Gran Diccionario Nahuatl records the use of *tepeitzcunitli* as a wild dog in modern Nahuatl

from 1984. Rémi Siméon and Josefina Oliva De Coll, *Diccionario de la lengua náhuatl o mexicana* (México: Siglo Veintiuno, 1992), 495, define it as "Cuadrúpedo feroz muy parecido al perro (Hern., Clav.) R. *tepetl, itzcuintli.*"

100. Valadez et al., "Dog-Wolf Hybrid Biotype," 321.

101. For Faber's animals, see Sabina Brevaglieri, *Natural desiderio di sapere: Roma barocca fra vecchi e nuovi mondi* (Rome: Viella, La corte dei Papi, 2019), 83-136. The Lincei acquired the manuscripts from Recchi's nephew, Marco Antonio Petilio. See Guardo, "Nell'Officina del Tesoro Messicano," 67.

102. Glover M. Allen, "Dogs of the American Aborigines," *Bulletin of the Museum of Comparative Zoology at Harvard College* 63 (1919-20): 431-517.

103. Several members of the Lincei added commentaries to Hernández: Joannes Terentius (1576-1630), Joannes Faber (1574-1629), Fabio Colonna (1567-1650), and Federico Cesi (1585-1630).

104. Recchi, *Rerum medicarum Novae Hispaniae*, 466.

105. Recchi, *Rerum medicarum Novae Hispaniae*, 466.

106. Recchi, *Rerum medicarum Novae Hispaniae*, 476.

107. Recchi, *Rerum medicarum Novae Hispaniae*, 476.

108. Francesco Saverio Clavigero, *The History of Mexico: Collected from Spanish and Mexican Historians, from Manuscripts and Ancient Paintings of the Indians*, trans. Charles Cullen, vol. 2 (London: G. G. J. and J. Robinson, 1787), 282-83.

109. Georges Louis Leclerc, Comte de Buffon, *Natural History: General and Particular, by the Count de Buffon, translated into English* (Edinburgh: Printed for William Creech, 1780-85), "The Dog," 10, 13-14, 30-31.

110. *HNMI*, bk. 4, chap. 34. In bk. 1, chap. 11, Acosta also mentioned Pliny's account of the large number of dogs (*canes*) on Canaria, the largest of the Canary Islands, in Pliny the Elder, *The Natural History*, trans. John Bostock (London: Taylor and Francis, 1855), 6.23.

111. John Low, *The New and Complete American Encyclopedia* (New York, 1806), 307. As Alexander von Humboldt replaced Hernández as the authority on Mesoamerican and South American nature in the early nineteenth century, one might follow the dogs' natural history through his writings rather than ending with Buffon. Although Humboldt had noted interbreeding in ducks, he did not dwell on it in relation to dogs. Mostly, he was concerned with *techichi*, dogs that the inhabitants of Mexico ate. Neither *xolos* nor other types of canines appear in his account. Rather than emphasizing the breeding of such animals, he focused on their castration to become food items. Castrated animals, after all, tend to be larger than sexually viable ones. He suggested that Peruvians ate their dogs (*runalco*) and that Mexicans sold the flesh of mute *techichi* in their markets. Alexander de Humboldt, *Essai Politique sur le Royaume de la Nouvelle-Espagne* (Paris: Chez F. Schoell, 1811), chap. 10, 223.

112. Clavigero, *History of Mexico*, 282-83. Clavigero urged that "these three species, which Count de Buffon has unjustly taken from America, ought to be restored to it."

Chapter Six

1. Francisco de Jerez, *Conquista del Peru: Verdadera Relación de la conquista del Perú* (Salamanca: por Juan de Junta, 1547), fols. vi (recto) and vii (verso).

2. Jerez's account was originally published in 1534: Jerez, *Conquista del Peru*, fol. vii (verso). See Francisco de Jerez, "Report of Francisco de Xerez, Secretary to Fran-

cisco Pizarro," in *Reports of the Discovery of Peru*, trans. and ed. Clements R. Markham (London: Hakluyt Society, 1872), 30.

3. See Miguel Ángel Borrego Soto, *La revuelta mudéjar y la conquista cristiana de Jerez (1261–1267)* (Madrid: Peripecias Libros, 2016). On the Andean side, see Sabine MacCormack, *Religion in the Andes: Vision and Imagination in Early Colonial Peru* (Princeton: Princeton University Press, 1991), 51. For elephants and ideas of Moriscos, see Abel Alves, "Iberia's Imagined Elephant: Animal Behavior through a Human Prism in the Sixteenth Century," *Romance Notes* 60, no. 3 (2020): 425–35. For the depiction of camels through the lens of horses, see Dániel Margócsy, "The Camel's Head: Representing Unseen Animals in Sixteenth-Century Europe," in *Art and Science in the Early Modern Netherlands*, ed. Eric Jorink and Bart Ramakers (Zwolle, The Netherlands: WBOOKS, 2011), 63–85.

4. Karoline P. Cook, *Forbidden Passages: Muslims and Moriscos in Colonial Spanish America* (Philadelphia: University of Pennsylvania Press, 2016), 9.

5. For an extensive study on historical and archaeological sources, see Duccio Bonavia, *The South American Camelids* (Los Angeles: Cotsen Institute of Archaeology, University of California, 2008).

6. On the material changes to shepherding and its cultural impact, see Irigoyen-García, *Spanish Arcadia*, 151–218.

7. Hess, *Forgotten Frontier*, 25; Hillgarth, *Mirror of Spain*, 53.

8. For an introduction to the vast scholarly corpus that documents these transitions: Brian A. Catlos, *Kingdoms of Faith: A New History of Islamic Spain* (New York: Basic Books, 2018); Mark D. Meyerson, *The Muslims of Valencia in the Age of Fernando and Isabel: Between Coexistence and Crusade* (Berkeley: University of California Press, 1991).

9. Nirenberg, *Communities of Violence*, 8–9, 228, 245.

10. Barbara Fuchs, *Exotic Nation: Maurophilia and the Construction of Early Modern Spain* (Philadelphia: University of Pennsylvania Press, 2008), 2.

11. Fuchs, *Exotic Nation*, 2.

12. Eric Calderwood, *Spain and the Making of Modern Moroccan Culture* (Cambridge, MA: Harvard University Press, 2018), 34–35. Thanks to Erica Feild for pointing me toward this passage.

13. Hess, *Forgotten Frontier*, 3.

14. Hess, *Forgotten Frontier*, 143.

15. Parker, *Emperor*, 180–200, 342–90; Parker, *Imprudent King*, 80–99, 140–55, 201–2, 208.

16. Hess, *Forgotten Frontier*, 143, 146–48, 181–87.

17. Martínez, *Genealogical Fictions*, 33. As Philip II consolidated control over Portugal by moving to Lisbon in 1581, he entertained different solutions to his Moorish problem beyond resettlement. During the Council of Lisbon, he determined to expel the Moors from Spain. He was, however, rather preoccupied with wars on many fronts. The final royal decree expelled the Moriscos only in 1609, as the Twelve Years' Truce was agreed upon with the Dutch; the last Hispano-Muslims were supposed to have been expelled from Iberia by 1614. L. P. Harvey, *Muslims in Spain, 1500–1614* (Chicago: University of Chicago Press, 2004), 295.

18. Jared Diamond, *Guns, Germs, and Steel: The Fates of Human Societies* (New York: W. W. Norton, 1997), 91.

19. Today, South American camelids are classified as order Artiodactyla, suborder

Tylopoda, and family Camelidae, but subdivided into Lamini and Camelini at the tribe level. Two New World genera, *Lama* and *Vicugna*, and one Old World genus, *Camelus*, are recognized. Jane C. Wheeler, "Evolution and Present Situation of the South American Camelidae," *Biological Journal of the Linnaean Society* 54 (1995): 289, fig. 3. Thanks to Matt Velasco, Nerissa Russell, and Susan deFrance for their help in navigating and incorporating the zooarchaeological research.

20. R. Fan et al., "Genomic Analysis of the Domestication and Post-Spanish Conquest Evolution of the Llama and Alpaca," *Genome Biology* 21, no. 159 (2020).

21. Fan et al., "Genomic Analysis." For an introduction to the vast historiography of the llama, see María Cecilia Lozada et al., "Camelid Herders: The Forgotten Specialists in the Coastal Señorío of Chiribaya, Southern Peru" in *Andean Civilization: A Tribute to Michael E. Moseley*, eds. Joyce Marcus and Patrick Ryan Williams (Los Angeles: UCLA Cotsen Institute of Archaeology, 2009), 351–64.

22. On domestication, see Miranda Kadwell et al., "Genetic Analysis Reveals the Wild Ancestors of the Llama and the Alpaca," *Proceedings of the Royal Society B: Biological Sciences* 268, no. 1485 (December 22, 2001): 2575–84. See also Penny Dransart, *Earth, Water, Fleece and Fabric: An Ethnography and Archaeology of Andean Camelid Herding* (London: Routledge, 2002).

23. José M. Capriles Flores and Nicholas Tripcevich, eds., *The Archaeology of Andean Pastoralism* (Albuquerque: University of New Mexico Press, 2016), xiii, 2. In this volume, see Katherine Moore, "Early Domesticated Camelids in the Andes," 17–38; Maria C. Bruno and Christine Hastorf, "Gifts from the Camelid: Archaeobotanical Insights into Camelid Pastoralism through the Study of Dung," 55–66; Claudine Vallières, "Camelid Pastoralism at Ancient Tiwanaku: Urban Provisioning in the Highlands of Bolivia," 67–86.

24. Adam Herring, *Art and Vision in the Inca Empire: Andeans and Europeans at Cajamarca* (New York: Cambridge University Press, 2015), n79.

25. Theresa Lange Topic, Thomas H. McGreevy, and John R. Topic, "A Comment on the Breeding and Herding of Llamas and Alpacas on the North Coast of Peru," *American Antiquity* 52, no. 4 (1987): 832–35.

26. See Elena Philips, "Inka Textile Traditions," in *The Inka Empire: A Multidisciplinary Approach*, ed. Izumi Shimada (Austin: University of Texas Press, 2015), 197–214. Even in textiles, the camelids remain fused together. Proteomic differentiation of species from samples of guanaco, vicuña, alpaca, and llama materials have been attempted without success. The keratins seem identical for all these species. Alpacas and vicuñas are from the same genus (*Vicugna*), and proteomic data is not specific enough beyond the genus level. The same challenge holds for guanaco and llama, as both come from the genus *Lama*. Caroline Solazzo, personal correspondence on identifying llama/alpaca hair, June 18, 2019); and Solazzo, "Characterizing Historical Textiles and Clothing with Proteomics," in "Estudos sobre têxteis históricos/Studies in Historical Textiles," ed. A. Serrano, M. J. Ferreira, and E. C. de Groot, special issue, *Conservar Património* 31 (May 2019): 1–18. See also Elena Phipps and Lucy Commoner, "Investigation of a Colonial Latin American Textile," *Textile Society of America Symposium Proceedings*, no. 358 (2006): 485–93.

27. Jorge A. Flores Ochoa, *Pastores de puna, or Uywamichiq punarunakuna* (Lima: Instituto de estudios peruanos, 1977).

28. The mummies' coats suggest that crossbreeding occurred between llamas and alpacas during the sixteenth century, as the increased coarseness and hairiness is ob-

served in both alpacas and llamas, meaning that the coarser-haired llama breed likely developed from crosses between llamas and alpacas. Jane C. Wheeler et al., "A Measure of Loss: Prehispanic Llama and Alpaca Breeds," *Archivos de Zootecnia* 41, no. 154 (1992): 467–75.

29. Melody Shimada and Izumi Shimada, "Prehistoric Llama Breeding and Herding on the North Coast of Peru," *American Antiquity* 50, no. 1 (1985): 3–26.

30. Pedro de Cieza de León, *Crónica del Perú (1551)* (Lima: Instituto de estudios peruanos, 1967), 36; Gary Urton, "The State of Strings: Khipu Administration in the Inka Empire," in Shimada, *Inka Empire*, 149–64, esp. 151.

31. Daniel W. Gade, "Llamas and Alpacas as 'Sheep' in the Colonial Andes: Zoogeography Meets Eurocentrism," *Journal of Latin American Geography* 12, no. 2 (2013): 225.

32. See, for instance, Silver llama figurine, Inca, ca. 1450–1532 CE, H. 241. cm. American Museum of Natural History B-1618. Courtesy of the Division of Anthropology, American Museum of Natural History. See also Herring, *Art and Vision in the Inca Empire*, n30.

33. Shimada and Shimada, "Llama Breeding and Herding," 3–4.

34. Wheeler, "Evolution and Present Situation of the South American Camelidae," 284.

35. Jane C. Wheeler, "Llamas and Alpacas: Pre-conquest Breeds and Post-conquest Hybrids," *Journal of Archaeological Science* 22, no. 6 (November 1995): 833–40.

36. Kadwell, "Genetic Analysis Reveals the Wild Ancestors."

37. Jane Wheeler's work in the field of Andean zooarchaeology suggested that different llamas and vicuñas in the sample set did possess mtDNA of the vicuña and guanaco, respectively, a result of prevalent hybridization in both the near and distant past. Through the genetic branchings of camelids, two predominant genotypes were found in the two wild camelid species in South America, the vicuñas and guanacos. The domestic camelids, llamas and alpacas, did not exhibit one specific genotype, meaning that neither breed is directly descended from either the vicuña or the guanaco, as previously thought. See Heather Pringle, "Secrets of the Alpaca Mummies: Did the Ancient Inca Make the Finest Woolen Cloth the World Has Ever Known," photographs by Grant Delin, *Discover* 22, no. 4 (April 2001).

38. Introgression in the alpaca genome (36 percent) was higher than that among llamas (5 percent). Fan et al., "Genomic Analysis."

39. Mackenzie Cooley, "Southern Italy and the New World in the Age of Encounters," in *The Discovery of the New World in Early Modern Italy*, ed. Elizabeth Horodowich and Lia Markey (Cambridge: Cambridge University Press, 2017), 169–89.

40. *HNMI*, bk. 4, chaps. 40–41.

41. *HNMI*, bk. 4, chap. 33.

42. *HNMI*, bk. 4, chap. 41. Covarrubias defines *iumento* as a generic name "that includes all beasts that carry cargo on their back." Covarrubias, *Tesoro* (1611), 494.

43. *HNMI*, bk. 4, chap. 33.

44. Wheeler, "Evolution and Present Situation of the South American Camelidae," 275.

45. Girolamo Benzoni, *Historia del Nuovo Mondo* (Venice, 1565), bk. 3.

46. Flora Ochoa, *Pastores de puna*, 1982.

47. *HNMI*, bk. 4, chap. 40.

48. *HNMI*, bk. 4, chap. 9.

49. Susan D. deFrance, "Paleopathology and Health of Native and Introduced Animals on Southern Peruvian and Bolivian Spanish Colonial Sites," *International Journal of Osteoarchaeology* 20, no. 5 (2010): 508–24, esp. 521.

50. For similar phenomena, see MacCormack, *Religion in the Andes*, 51.

51. Gade, "Llamas and Alpacas as 'Sheep,'" 221.

52. Conrad Gesner, *Historia animalium*, liber primus, *De Quadrupedibus viuiparis* (Francofurti: In Bibliopolio Cambieriano, 1602), 149. Note that the 1551 edition published in Zurich skips from *camelopardali* to camels to dromedaries to dogs (*cane*) with no *Allocamelus Scaligeri* between, 160–73.

53. Hannah Marcus, *Forbidden Knowledge: Medicine, Science, and Censorship in Early Modern Italy* (Chicago: University of Chicago Press, 2020), 31, 101.

54. Brian Brege, "The Empire That Wasn't: The Grand Duchy of Tuscany and Empire, 1574–1609" (Ph.D. dissertation, Stanford University, 2014), 260–61. For broader context, see Brege, *Tuscany in the Age of Empire* (Cambridge, MA: Harvard University Press, 2021), 123–70.

55. Brege, "The Empire That Wasn't," 260–61.

56. Carletti, *Viaggi intorno al mondo*, 78–80.

57. *HNMI*, bk. 4, chap. 41.

58. Thanks to Annemarie Jordan Gschwend for suggestions on this research (personal correspondence, November 9, 2016): Sabine Haag, *Echt tierisch!: Die Menagerie des Fürsten: Eine Ausstellung des Kunsthistorischen Museums Wien, Schloss Ambras Innsbruck* (An Exhibition of the Kunsthistorisches Museum, Vienna, and Ambras Castle, Innsbruck), June 18–October 4, 2015 (Wien: KHM-Museumsverband, 2015) discusses a llama in the menagerie of Ferdinand II of Tyrol.

59. Detlef Heikamp, *Mexico and the Medici*, with contributions by Ferdinand Anders (Florence: Edam, 1972), 11.

60. See Helen Cowie, *Llama* (London: Reaktion Books, 2017).

61. Fabienne Pigière and Denis Henrotay, "Camels in the Northern Provinces of the Roman Empire," *Journal of Archeological Science* 39 (2012): 1531–39. Preliminary research suggests that between the Visigoths and their Almoravid successors a half millennium later, the camel took on opposing cultural meanings, delineating a tension between the use of the animal in Christian and Islamic regions. Chroniclers recorded that under the Visigoth kings Sisebut (565–621) and Wamba (643–688), camels were used as tools of mockery, as shamed riders were forced astride them sitting backward. Chronicles of Sisebut (discussing the execution of Brunhilda) and Julian of Toledo's *Historia Wambae Regis*, 30, are cited in Caitlin R. Green, "Camels in Early Medieval Western Europe: Beasts of Burden and Tools of Ritual Humiliation," May 28, 2016, http://www.caitlingreen.org/2016/05/camels-in-early-medieval-western-europe.html.

62. Ibn Khallikān wrote that camels were first imported to Iberia in large numbers by Yūsuf ibn Tāshufīn (1061–1106). Syed M. Imamuddin, *Muslim Spain: 0711–1492 A.D.*, vol. 2 (Leiden, E. J. Brill, 1981), 7.

63. For camel artwork made in Castille-León, see San Baudelio de Berlanga hermitage camel (1129–34), The Cloisters Collection, 1961, 61.219.

64. For the dromedary population of Lanzarote and Gran Canaria in modern times, see Gregorio Mentaberre et al., "A Transversal Study on Antibodies against Selected Pathogens in Dromedary Camels in the Canary Islands, Spain," *Veterinary Microbiology* 167, nos. 3–4 (December 2013): 468–73.

65. Felipe Fernández-Armesto, *The Canary Islands after the Conquest: The Making of a Colonial Society in the Early Sixteenth Century* (Oxford: Clarendon Press, 1982), 171, also 5.

66. *HNMI*, bk. 4, chap. 33. In the nineteenth century, the United States established the US Camel Corp to experiment with camels as pack animals in the Southwest, recognizing their utility before abandoning the experiment during the Civil War. Frank Bishop Lammons, "Operation Camel: An Experiment in Animal Transportation in Texas, 1857–1860," *Southwestern Historical Quarterly* 61 (July 1957): 20–50.

67. Bonavia, *South American Camelids*, 466–67.

68. Antonio de Herrera, *Historia general de los hechos de los castellanos en las Islas y Tierra Firme del mar Océano que llaman Indias Occidentales*, tomo 10 (Buenos Aires, Argentina: Editorial Guarania, 1947), bk. 7, chap. 3, 179, quoted in Bonavia, *South American Camelids*, 467.

69. Bonavia, *South American Camelids*, 466–67.

70. Alexander von Humboldt questioned the Spaniards' failure to use camels. Reflecting back on the conquest, he asserted "how advantageous it would have been had the *conquistadores*, from the beginning of the sixteenth century, peopled America with camels, as they have peopled it with horned cattle, horses, and mules. . . . It seems the more surprising that their introduction was not encouraged by the government at the beginning of the conquest, as long after the taking of Granada, camels, for which the Moors had a great predilection, were still very common in the south of Spain." Humboldt, *Personal Narrative of Travels to the Equinoctial Regions of America during the Years, 1799–1804* (London: Longman, 1814), vol. 2, chap. 1.

71. Luis Millones Figueroa, "Bernabé Cobo's Inquiries in the New World and Native Knowledge," in Marroquín Arredondo and Bauer, *Translating Nature*, 70–93.

72. Cobo, *Historia del Nuevo Mundo*, bk. 10, chap. 18, 420–21, quoted in Bonavia, *South American Camelids*, 468.

73. Garcilaso de la Vega, *Royal Commentaries of the Incas and General History of Peru*, part 1, trans. H. V. Livermore (Austin: University of Texas Press, 1966), bk. 9, chap. 18, 585, quoted in Bonavia, *South American Camelids*, 465–70.

74. Bonavia, *South American Camelids*, 465–70.

75. Reginaldo de Lizárraga, *Descripción breve de toda la tierra del Perú, Tucumán, Río de la Plata y Chile*, tomo 216 (Madrid: Ediciones Atlas, 1968), bk. 2, chap. 11, 118, quoted in Bonavia, *South American Camelids*, 467–70.

76. Juan Bromley, "El Capitán Martín de Estete y Doña María de Escobar 'La Romana,' fundadores de la Villa de Trujillo del Perú," *Revista Histórica* (Lima) 22 (1956): 122–41. Just as *mulatto* came to be a term to designate mixed African-European ancestry, mule trains with black African and Andean muleteers connected the coast of Peru with the north during the Spanish colonial era, linking Andean regional economies around Loja and Quito, Cajamarca and Chachapoyas, Trujillo, and Lima. Rachel Sarah O'Toole, *Bound Lives: Africans, Indians, and the Ma Race in Colonial Peru* (Pittsburgh: University of Pittsburgh Press, 2012), 107–11.

77. Barbara Fuchs, *Exotic Nation*, 88. For the extensive literature on Moorishness, and blackmore and matar identities beyond Spain, see Olivia Remie Constable, *To Live Like a Moor: Christian Perceptions of Muslim Identity in Medieval and Early Modern Spain* (Philadelphia: University of Pennsylvania Press, 2018), esp. 15–62; Ross Brann, "The Moors?," *Medieval Encounters* 15 (2009): 307–18; Israel Burshatin, "The Moor in the Text: Metaphor, Emblem and Silence," in *"Race," Writing and Difference*,

ed. Henry Louis Gates (Chicago: University of Chicago Press, 1986), 98–118; Miguel Ángel de Bunes Ibarra, *La imagen de los musulmanes y del norte de África en la España de los siglos XVI y XVII: Los caracteres de una hostilidad* (Madrid: Consejo Superior de Investigaciones Cientificas, 1989). Iberian horsemanship likewise included a Moorish legacy. For riders, Castilian vocabulary retains Arabic terms like *jinete*. Iberian monarchs displayed power through Moorish riding, just as Ferdinand had entered Granada riding Moorish style, and future king Philip III played the game of canes dressed in red, blue, and white with a Turkish beret wrapped in a Tunisian turban, knees tight to the saddle. When King Duarte I penned his equestrian manual in 1433–38, he had featured the game of canes in which competing teams dressed as Moors, and a wide array of texts reported that *gineta* riding was properly Portuguese and admired by Andalusians, Castilians, and North Africans. Javier Irigoyen-García, *Moors Dressed as Moors: Clothing, Social Distinction and Ethnicity in Early Modern Iberia* (Toronto: University of Toronto Press, 2017), 27–28. Again, this interest in Moorish riding was only a partial embrace. As the horse marked access to privilege and knightly status, under Spanish rule Muslim subjects were prohibited from riding horses. Renton, "The Knight with No Horse," 115.

78. Barbara Fuchs, *Exotic Nation*, 100, 166.

79. Biblioteca Francisco de Zabálburu, Altamira, 141, GD. 1, Gobierno de Felipe II (1569–1594), including especially D.9, D.46, 145–46, and D. 74, 1–4.

80. Carlos Gómez-Centurión Jiménez, "Exóticos pero útiles: Los camellos reales de Aranjuez durante el siglo XVIII," *Cuadernos dieciochistas* 9 (2008): 155–80; Gómez-Centurión Jiménez, *Alhajas para soberanos: Los animales reales en el siglo XVIII: De las leoneras a las mascotas de cámara* (Valladolid: Junta de Castilla-León, 2011). On horse-breeding efforts in Aranjuez, see Renton, "Social and Environmental History of the Horse," 162–64.

81. Archivo General de Simancas (hereafter cited as AGS), Casa y Sitios Reales (hereafter cited as CSR), leg. (legajo) 307, 352, May 22, 1628.

82. One can learn about Tovilla's service from his widow's petition for royal financial assistance following his demise. AGS CSR, leg. 305, 2, 31 febrero (?) de 1608.

83. AGS CSR, leg. 305, 7, 21 de mayo 1608, "Sobre que Francesco de Moreno sobre estante que ha sido de los camellos que VMa tiene en Aranxuez." See also AGS CSR, leg. 305, 207–8, 9 diciembre de 1612; AGS CSR, leg. 305, 249, 14 de junio 1613.

84. AGS CSR, leg. 305, 256, Junta de obras y bosques, September 21, 1613.

85. In 1622, Juana Roman, widow of the head shepherd of the Aranjuez camels, Pedro Perez, petitioned for 40 ducats of compensation. Her husband had spent twenty years of his life caring for Aranjuez's camels, and his sudden death left their family, including two children, poor. The camel herder's death left Juana to continue asking for assistance, as she did when she requested 20 ducats four years later in 1626. Following her precedent, when the next head camel herder died in 1639, his widow also petitioned for support. Juliana Lopez, whose husband had been the *Mayoral* of the Aranjuez camels, petitioned for money, along with a widow of a former guard. Lopez's husband must have been even more expert in camel care than was his predecessor Pedro Perez. He "served more than forty years at that site," tending to the dromedary population of Aranjuez. Juliana Lopez asked that the king send a stipend of one *real* a day to care for her family, especially their four daughters, who had been left "very poor" following her husband's death. Since the husband "served many years to the court's satisfaction," his widow argued that she merited compensation. AGS

CSR, leg. 306, 260, 17 de junio 1622; AGS CSR, leg. 307, 120, 5 de junio 1626; AGS CSR, leg. 310, 63, agosto 1639.

86. Biblioteca Francisco de Zabálburu, MIRO, 18, D.535/1, 1661.

87. AGS CSR, leg. 315, 256, 9 de noviembre 1668; AGS CSR, leg. 315, 310, 3 de febrero 1669.

88. See Marcy Norton, "Going to the Birds: Animals as Things and Beings in Early Modernity," in *Early Modern Things: Objects and Their Histories, 1500–1800*, 2nd ed. (New York: Routledge, 2021), 51–81.

89. Findlen, *Possessing Nature*, 81.

90. Lydia Barnett, *After the Flood: Imagining the Global Environment in Early Modern Europe* (Baltimore: Johns Hopkins University Press, 2019), 3.

91. Saverio Ricci, *Il sommo inquisitore, Giulio Antonio Santori tra autobiografía e storia (1532–1602)* (Roma: Salerno, 2002), 38. If the original ark rose above mountains and hills, the new Catholic ark floated above human arrogance, malice, and greed; unshaken by tempests or waves, it was meant to stay steadfast, unfazed by crime, heresy, sedition, or schism.

92. The cartoons for this image have been attributed to the Flemish painter Michiel Coxcie (1499–1592), as it matches both his style and his known history as a tapestry cartoonist. Coxcie began work with his teacher, Berand Van Orley. The original scenes concerning the "History of Noah" were sent to Spain in 1559, but some pieces were lost during shipping by sea.

93. Alessandra Anselmi, *El diario del viaje a España del Cardenal Francesco Barberini* (Madrid: Fundación Carolina, 2004), 101–2.

94. On the emergence of the Noah myth in the Renaissance, see the Don Cameron Allen classic, *The Legend of Noah: Renaissance Rationalism in Art, Science, and Letters* (Urbana: University of Illinois Press, 1949).

95. Frank Salomon, "Inkas through Texts: The Primary Sources," in Shimada, *Inka Empire*, 23–38, esp. 33.

96. Felipe Guaman Poma de Ayala, *Nueva corónica y buen gobierno* (ca. 1615), Royal Danish Library, GKS 2232 kvart (using pencil-marked pagination) 206, 227, 242, 272, 320, 391, 529, 530, 891, 894, 947, 1160.

97. Guaman Poma, GKS 2232, 22–23.

98. Guaman Poma, GKS 2232, 25.

Chapter Seven

1. Dandelet, *Spanish Rome*, 8–9.

2. Brevaglieri, *Natural desiderio di sapere*, 17. For natural history on the stage of Rome, see 31–83.

3. Claudio M. Burgaleta, S.J., *José de Acosta, S.J. (1540–1600): His Life and Thought* (Chicago: Jesuit Way, Loyola Press, 1999), 83. For an introduction to the vast scholarly corpus on Acosta, see the many works of Fermín del Pino Díaz, including "*Las historias naturales y morales de las Indias* como género: Orden y gestación literaria de la obra de Acosta," *Histórica* (Lima) 24, no. 2 (2000): 295–326; Luigi Guarnieri Caló Caducci, "La consideración de las fuentes indígenas en la *Historia natural y moral de las Indias* de José de Acosta," *Confluenze* 12, no. 2 (2020): 260–77; Alexandre Coello de la Rosa, "Más allá del Incario: Imperialismo e historia en José de Acosta, SJ (1540–1600)," *Colonial Latin American Review* 14, no. 1 (2005): 55–81.

4. Paula Findlen, "How Information Travels: Jesuit Networks, Scientific Knowledge, and the Early Modern Republic of Letters, 1540–1640," in *Empires of Knowledge: Scientific Networks in the Early Modern World*, ed. Paula Findlen (New York: Routledge, 2019), 57–105. For an introduction to Jesuit history during this period, see John O'Malley, *The First Jesuits* (Cambridge, MA: Harvard University Press, 1993); Luke Clossey, *Salvation and Globalization in the Early Jesuit Missions* (New York: Cambridge University Press, 2008). In relation to wider changes in Catholicism, see Robert Bireley, *The Refashioning of Catholicism, 1450–1700* (Washington, DC: Catholic University of America Press, 1999).

5. P. Francisco Mateos, *Obras del P. Jose de Acosta* (Madrid: Ediciones Atlas, 1954), 44; Burgaleta, *José de Acosta*, 57–58. These official visitations brought Acosta to Baeza, Córdoba, Montilla, Granada, Cadíz, Jerez de la Frontera, Trigueros, and Marchena. For an English critical edition of his natural historical writing, see José de Acosta, *Natural and Moral History of the Indies*, ed. Jane E. Mangan, trans. Frances López-Morillas (Durham: Duke University Press, 2002).

6. Burgaleta, *José de Acosta*, 56–70.

7. Burgaleta, *José de Acosta*, 59.

8. "Diario del P. Acosta," a copy of Acosta's original diary from the same period, is held in the Archivo S. I. de la Provincia de Toledo and transcribed in Mateos, *Obras*, 353–68. Acosta subsequently defended himself from the accusation that he was the leader of the *memorialistas* and revisits his actions from this period in *Memorial de apología o descargo dirigido al Papa Clemente VIII*, transcribed in Mateos, *Obras*, 368–86.

9. Mateos, *Obras*, 354.

10. Ricci, *Il sommo inquisitore*, 27. On Santori's path to the papacy, see 338–80; on his correspondence with Philip II, 368; on the dynamics of the papal conclave and the election of Spanish-affiliated candidates, 365, 375. On Aldobrandini, see Agostino Borromeo, "Clemente VIII," in *Dizionario Biografico degli Italiani*, vol. 26 (1982). I am grateful to Bradford Bouley for his insights on papal politics from 1588 to 1592.

11. Mateos, *Obras*, 354. An interesting parallel to "blood" genealogies is botanical representations of faith genealogies; see the Ignatian Tree in "The Horoscopium Catholicum," a fold-out within Athanasius Kircher, *Ars magna lucis et umbrae [. . .]* (Rome: Ludovici Grignani, 1646), reproduced in Marcelo Aranda, "Instruments of Religion and Empire: Spanish Science in the Age of the Jesuits, 1628–1756" (Ph.D. dissertation, Stanford University, 2013), 17–19. See also the Zinacantepec Porteria Mural in Toluca, Mexico.

12. Mateos, *Obras*, 354.

13. Burgaleta, *José de Acosta*, 39–55.

14. José de Acosta and Juan de Atienza, *Doctrina Christiana y catecismo para instrucción de los Indios* (Ciudad de los Reyes: Antonio Ricardo, 1584). See also Burgaleta, *José de Acosta*, 44–49; Gregory J. Shepherd, *José de Acosta's "De procuranda Indorum salute": A Call for Evangelical Reforms in Colonial Peru* (New York: Peter Lang, 2014), 39.

15. *HNMI*, bk. 2, chap. 9. I have principally relied on the John Carter Brown Library copy.

16. *HNMI*, bk. 3, chap. 24; bk. 4, chaps. 4, 6, 15, 21. See Anthony Grafton, *New Worlds, Ancient Texts* (Cambridge, MA: Belknap Press of Harvard University Press, 1992), 1. See also Allison Margaret Bigelow, *Mining Language: Racial Thinking, Indig-*

enous Knowledge, and Colonial Metallurgy in the Early Modern Iberian World, Omohundro Institute of Early American History and Culture (Chapel Hill: University of North Carolina Press, 2020), 247–48; and Molly A. Warsh, *American Baroque: Pearls and the Nature of Empire, 1492–1700*, Omohundro Institute of Early American History and Culture (Chapel Hill: University of North Carolina Press, 2018), 81.

17. *HNMI*, "Prologue to the Reader."

18. *HNMI*, bk. 2, chap. 14, followed by "Note to the Reader," 116. At the end of its second book, Acosta highlights that *Historia natural y moral de las Indias* was written in two places, Peru and Spain, first in Latin, then in the vernacular, with resulting differences in style between the two sections.

19. José de Acosta, *De natura Novi Orbis. Libri duo; et De promulgatione evangelii, apud barbados, sive De procuranda Indorum salute*, libri sex (Salmanticae: Apud Gullelmum Froquel, 1589). *De natura Novi Orbis* was reprinted in Cologne in 1596.

20. José de Acosta, *De Christo reuelato*, libri nouem (Romae: Apud Iacobum Tornerium, 1590); José de Acosta, *De temporibus nouissimis*, libri quatuor (Romae, ex typographia Iacobi Tornerij, 1590). Subsequently, he published *Conciones in Quadragesimam* (Venetiis: apud Io. Bapt. Ciottum Senensem sub signo Aurorae, 1599).

21. *HNMI*, bk. 3, chap. 1.

22. By the early seventeenth century, writings from *De natura Novi Orbis*, *De procuranda Indorum salute*, and *Historia natural y moral de las Indias* had appeared in an Italian translation by Gio. Paolo Galucci Salodiano (Venice: Presso Bernardo Basa, 1596); a French translation (Paris: Chez Marc Orry, 1598 and 1600); a Latin anthology, *Americae nona & postrema pars* (Francofurt: Apud Matth. Beckerum, 1602); two German translations (Gedrückt zu Cölln: Bey Johan Christoffel, 1600, and Gedruckt zu Vrsel, Durch Cornelium Sutorium, 1605), a Dutch translation (Tot Enchuysen: By Jacob Lenaertsz, 1598), and an English translation (London: Printed for Val: Sims, 1604).

23. Like any important thinker, Acosta involved himself in many conversations; the threads of New Thomism run throughout his work, overlapping with his other commitments to the importance of direct observation and experience beyond text alone. More recent scholarship has linked Acosta to the other regions and networks of thought that influenced him, from the Andean highlands to Rome as a center of global Catholicism. Sabine MacCormack, "The Mind of the Missionary: José de Acosta on Accommodation and Extirpation, circa 1590," in *Religion in the Andes: Vision and Imagination in Early Colonial Peru* (Princeton: Princeton University Press, 1991), 249–79.

24. Acosta, *De procuranda Indorum salute*, bk. 1, chaps. 16, 17, 18.

25. Shepherd, *José de Acosta's "De procuranda Indorum salute,"* 87, citing Acosta, *De procuranda Indorum salute*, 1.17. See also José de Acosta, *De procuranda Indorum salute: Educación y Evangelización*, ed. Luciano Pereña et al. (Madrid: Consejo Superior de Investigaciones Científicas, 1984–87).

26. Anthony Pagden, *The Fall of Natural Man: The American Indian and the Origins of Comparative Ethnology* (Cambridge: Cambridge University Press, 1986), 160.

27. Pagden, *Fall of Natural Man*, 150–53.

28. Pagden, *Fall of Natural Man*, 161. On conversion in Europe and the "other Indies" discourse in Jesuit correspondence, see Adriano Prosperi, "'Otras indias': Missionari della Controriforma tra contadini e selvaggi," in *Scienze, credenze occulte, livelli di cultura, Atti del Convegno internazionale di studi* (Florence, 26–30 giugno, 1980),

207–8; Jennifer D. Selwyn, *A Paradise Inhabited by Devils: The Jesuits' Civilizing Mission in Early Modern Naples* (Burlington, VT: Ashgate, 2004), 1–3, 22.

29. Anthony Grafton, "José de Acosta: Renaissance Historiography and New World Humanity," in *The Renaissance World*, ed. John Jeffries Martin (New York: Routledge, 2007), 166–88, esp. 185.

30. Pagden, *Fall of Natural Man*, 163–65. Other scholars reiterate this tripartite argument; see Grafton, "José de Acosta," 178. See also *HNMI*, bk. 6, chap. 19, and bk. 7, chap. 28.

31. The animals are described as *generosos*, meaning illustrious or noble just as a man could be noble, a quality fostered by lineage. Covarrubias highlights the "Cavallo generoso, el castizo y de buena raza." Covarrubias, *Tesoro* (1611), 433v.

32. Pagden, *Fall of Natural Man*, 162.

33. Coello de la Rosa, "Más allá del Incario," 67–69.

34. Lovejoy, *Great Chain of Being*, 58–59. One might compare Acosta's engagement with this topic with that of Roberto Bellarmino, *De ascensione mentis in Deum per scalas creaturam* (Antwerp: Apud Viduam & Filos Io. Moreti, 1615), cited in Lovejoy, 91–92.

35. *HNMI*, bk. 3–4, esp. bk. 4, chap. 1.

36. *HNMI*, bk. 4, chap. 1.

37. *HNMI*, bk. 4, chap. 1.

38. *HNMI*, bk. 4, chaps. 1 and 39.

39. Book 3, chapters 14 and 15, are missing in several editions of *Historia natural y moral*, including Seville 1590 and Barcelona 1590. Mateos, *Obras*, 73.

40. *HNMI*, bk. 1, chap. 1.

41. *HNMI*, bk. 1, chap. 8.

42. *HNMI*, bk. 1, chap. 25.

43. *HNMI*, bk. 1, chap. 20.

44. *HNMI*, bk. 1, chap. 21.

45. *HNMI*, bk. 1, chap. 25.

46. *HNMI*, bk. 1, chap. 20.

47. *HNMI*, bk. 1, chap. 21.

48. *HNMI*, bk. 1, chap. 25.

49. *HNMI*, bk. 1, chap. 20, citing Genesis 7, and bk. 4, chap. 34, citing Genesis 6.

50. *HNMI*, bk. 1, chap. 20.

51. *HNMI*, bk. 1, chap. 21.

52. *HNMI*, bk. 1, chap. 20.

53. *HNMI*, bk. 4, chap. 33.

54. *HNMI*, bk. 4, chap. 33.

55. *HNMI*, bk. 4, chap. 33; Acosta, *Historia naturale morale delle indie* (1596), 88; the Italian version of this chapter deployed similar terminology, writing "et alcune razze di quelli sono cosi bone, come sono i migliori di Castiglia." Renton has argued that while early accounts of American horse breeding emphasized both quality and quality, decline of the *casta* in Latin America became a common complaint. Renton, "Social and Environmental History of the Horse," 155.

56. *HNMI*, bk. 4, chap. 34.

57. See Marcy Norton, *The Tame and the Wild* (Cambridge, MA: Harvard University Press, forthcoming).

58. *HNMI*, bk. 4, chap. 33.

59. *HNMI*, bk. 4, chap. 33.

60. *HNMI*, bk. 4, chap. 35.

61. *HNMI*, bk. 1, chap. 21, and bk. 4, chap. 34.

62. *HNMI*, bk. 4, chap. 34.

63. *HNMI*, bk. 4, chap. 34.

64. *HNMI*, bk. 4, chap. 34.

65. *HNMI*, bk. 4, chap. 36. See Florencia Pierri, "Armadillo: A Creature Called Armadillo," in Cooley, Yıldırım, and Toledano, *Natural Things* (forthcoming, 2023).

66. Acosta, *De natura Novi Orbis*, bk. 1, chap. 22: "Quod non venerit genus Indorum per Atlantidem, ut quidam opinantur."

67. Giovanni Balbi, *Catholicon* (Venice: Petri Liechtenstein, 1506), "genero" and related terms.

68. *HNMI*, bk. 1, chap. 22: "Que no passò el linage de Indios por la Isla Atlantida, como algunos ynmaginan."

69. Acosta, *De natura Novi Orbis*, 1589, 150.

70. An otherwise strong critical edition by Jane Mangan, for example, translates *lineage* as "race" while *species* is translated as "breed." Acosta, *Natural and Moral History of the Indies*, 242.

71. *HNMI*, bk. 4, chap. 36 and bk. 1, chap. 21.

72. *HNMI*, bk. 4, chap. 40: "O si son de aquel genero, serán especies diversas, como el linaje de perros es de diversa especie del mastin, y la del lebrel."

73. *HNMI*, bk. 1, chaps. 16 and 21.

74. Covarrubias, *Tesoro* (1611), *especie*, 376r.

75. Covarrubias, *Tesoro* (1611), *linage*, 525v.

76. Acosta, *Historia naturale morale delle indie* (1596), 93: "Overo se sono di quel genere saranno di specie diversa, como nel genere dei cani, vi è la specie dei mastini, & quella dei leureri."

77. Florio, *World of Wordes* (1598), *génere*, 146.

78. *Lineage* would have been a direct alternative, listed in Florio, *World of Wordes* (1598) as "legnaggio, lineage," 201. Florio expanded this definition in 1611 to "legnággio, a linnage, a pedigree, a kindred, a blood, a stock, a generation": Florio, *World of Words* (1611), 208. To be closer to the Castilian, the translation might have alternatively used *schiatta* or *stirpe*. Florio defines *schiatta* as "a race, a family, a stock, a pedigree, or blood of a house" (1598, 352) and later as "a race, a stock, a blood of a family, a generation, a pedigree" (1611, 475). He defines *stirpe* as "a progenie, a generation, a stock, or kindred, a race, issue, or noblenes of birth, an offspring, a house, a bloud, a kinde, a pedigree. Also the roote, stem, or stalk of a tree, plant, or herb, a yong branch or set" (1598, 398); subsequently, he reverses plant and animal order when he defines it as "the root, the stem, the socke or maine, stalke of any plant, tree or hearbe. Also a race, a stocke, a progenie, an off-spring, a bloud, a kin, a pedigree, or generation of a house" (1611, 535).

79. Acosta, *Histoire naturelle et moralle des Indes* (Paris: Chez Marc Orry, 1598), 201: "Car s'ils sont de ce genre, ce seroit un espece diverse: comme en la race des chiens l'espece du mastin est autre que celle du leurier."

80. Acosta, *Historie naturael ende morael van de Westersche Indien* (Tot Enchuysen: By Jacob Lenaertsz, 1598), 206v; Acosta, *The Naturall and Morall Historie of the East and West Indies* (London: Printed for Val: Sims, 1604), 316.

81. For extinction in early Mesoamerica, see Mackenzie Cooley, "The Giant Remains: Mesoamerican Medicine, Extinction, and Cycles of Empire," *Isis* 112, no. 1 (2021): 45–67.

82. *HNMI*, bk. 4, chap. 36.

83. *HNMI*, bk. 4, chap. 36.

84. For example, one might trace these through the use of "propagation" (69, 310) and "generation" (119, 310) throughout *HNMI*.

85. Mateos, *Obras*, 81, citing *HNMI*, bk. 3, chap. 20.

86. Mateos, *Obras*, 81, citing *HNMI*, bk. 3, chap. 20.

87. Mateos, *Obras*, 82–83, citing *HNMI*, bk. 3, chap. 22.

88. On early modern and ancient population thinking, see Philip Kreager, "The Emergence of Population," in Hopwood, Flemming, and Kassell, *Reproduction*, 253–66; Rebecca Flemming, "States and Populations in the Classical World," in Hopwood, Flemming, and Kassell, *Reproduction*, 67–80; Andrea Rusnock, "Biopolitics and the Invention of Population," in Hopwood, Flemming, and Kassell, *Reproduction*, 333–45.

89. Jeffrey Klaiber, preface to Shepherd, *José de Acosta's* De procuranda Indorum salute, 9.

90. *HNMI*, bk. 1, chap. 23; Mangan, *Natural and Moral History of the Indies*, 69n2.

91. Renzi, "Family Resemblance in the Old Regime," 248.

92. *HNMI*, bk. 1, chap. 23.

93. Burgaleta, *José de Acosta*, 63.

94. Burgaleta, *José de Acosta*, 64–65.

95. Burgaleta, *José de Acosta*, 60; John W. Padberg, Martin D. O'Keefe, and John L. McCarthy, eds., *For Matters of Greater Moment: The First Thirty Jesuit General Congregations: A Brief History and a Translation of the Decrees* (Saint Louis, MO: Institute of Jesuit Sources, 1994), General Congregation 5, Decree 1, 187.

96. Padberg, O'Keefe, and McCarthy, *For Matters of Greater Moment*, 192.

97. Padberg, O'Keefe, and McCarthy, *For Matters of Greater Moment*, 192–93.

98. Padberg, O'Keefe, and McCarthy, *For Matters of Greater Moment*, 194.

99. For an overview of this history, see Robert Aleksander Maryks, *The Jesuit Order as a Synagogue of Jews: Jesuits of Jewish Ancestry and Purity-of-Blood Laws in the Early Society of Jesus* (Leiden: Brill, 2010), 159–214; O'Malley, *The First Jesuits*, 188–92; *Historia de la Compañia de Jesús en la Asistencia de España*, 7 vols. (Madrid: Administración de Razón y Fé, 1912–25), esp. vol. 3.

100. Padberg, O'Keefe, and McCarthy, *For Matters of Greater Moment*, 204.

101. Burgaleta, *José de Acosta*, 65.

102. Thanks to Claudio Burgaleta, S.J., and Tom McCoog, S.J., for their help in finding the original Latin version of the decrees, Instituti Societatis Iesu, Volumen Secundum, *Examen Et Constitutiones Decreta Congregationum Generalium* (Florentiae: Ex Typographia A. SS. Conceptione, 1893), 278–279 on Decree 52, in MS Decr. 68. Padberg et al. translate *genere* as stock.

103. *Examen Et Constitutiones*, 279.

104. Padberg, O'Keefe, and McCarthy, *For Matters of Greater Moment*, 204.

105. On the fraught terminology of *Saracen*, see Shokoofeh Rajabzadeh, "The Depoliticized Saracen and Muslim Erasure." *Literature Compass* 16, no. 9–10 (2019).

106. *Examen Et Constitutiones*, 279.

107. Padberg, O'Keefe, and McCarthy, *For Matters of Greater Moment*, 204–5.

108. Padberg, O'Keefe, and McCarthy, *For Matters of Greater Moment*, 204.

109. *Examen Et Constitutiones*, 302 on Decree 28, in MS Decr. 45.

110. Congregation 6, Decree 28; Padberg, O'Keefe, and McCarthy, *For Matters of Greater Moment*, 231–32.

111. Padberg, O'Keefe, and McCarthy, *For Matters of Greater Moment*, 231–32.

112. Burgaleta, *José de Acosta*, 68, 83.

Chapter Eight

1. Giovanni Aquilecchia, "*In Facie prudentis relucit sapientia*: Appunti sulla letteratura metoposcopica tra Cinque e Seicento," in *Giovan Battista della Porta nell'Europea del suo tempo*, ed. Maurizio Torrini (Napoli: Guida, 1990), 199–229, esp. 222. Thanks to Stefano Gulizia and Manuela Bragagnolo for their suggestions on how to improve this chapter.

2. Ricci, *Il sommo inquisitore*, 19. The path from Inquisition to papal tiara had been more direct for Marcellus II, Paul IV, Pius V, Sixtus V, Urban VII, and Innocent IX.

3. Valente, "Della Porta e l'Inquisizione," 425. See also Folke Gernert, *Fictionalizing Heterodoxy: Various Uses of Knowledge in the Spanish World from the Archpriest of Hita to Mateo Alemán* (Berlin: Walter de Gruyter, 2019), 31–35; Gernert, *Adivinos, medicos y profesores de secretos en la España áurea* (Toulouse: Méridiennes, 2017).

4. Valente, "Della Porta e l'Inquisizione," 423–25.

5. Eamon, *Science and the Secrets of Nature*, 139.

6. G. Romei, "Della Porta, Giovambattista," in *Dizionario biografico degli Italiani*, vol. 37 (1989): 170–82; Sergius Kodera, "Giambattista della Porta," *Stanford Encyclopedia of Philosophy*, Summer 2015 ed., revised Summer 2021, https://plato.stanford.edu/entries/della-porta/; Michaela Valente, "Della Porta, Giovanni Battista," in *Dizionario storico dell'Inquisizione*, ed. Adriano Prosperi, vol. 1.3 (Pisa: Edizione della Normale, 2010).

7. Tarrant, "Giambattista della Porta and the Roman Inquisition," 619–25.

8. Aristotle, *Physiognomics*, trans. T. Loveday and E. S. Forster (Oxford: Oxford University, Clarendon Press, 1913), 1.

9. See Benjamin Arbel, "The Beginnings of Comparative Anatomy and Renaissance Reflections on the Human-Animal Divide," in "The Animal in Renaissance Italy," ed. Stephen Bowd and Sarah Cockram, special issue, *Renaissance Studies* 31, no. 2 (2017): 201–22.

10. Romei, "Della Porta, Giovambattista"; Kodera, "Giambattista della Porta"; Valente, "Della Porta, Giovanni Battista."

11. Covarrubias, *Tesoro* (1611), 406–7.

12. Ruth Hill, "Between Black and White: A Critical Race Theory Approach to Caste Poetry in the Spanish New World," *Comparative Literature* 19, no. 4 (2007): 269–93, esp. 282–86.

13. See Pedro Calderón de la Barca, *El astrologo fingido* (Madrid: Carlos Sanchez a costa de Antonio de Ribero, 1641).

14. [Pseudo-]Aristotle, *Pseudo-Aristotele: Fisiognomica; anonimo Latino: Il trattato di fisiognomica,* trans. Giampiera Raina (Milano: Biblioteca Universale Rizzoli, 1994).

15. Aristotle, *Physiognomics*, chap. 1.

16. Aristotle, *Physiognomics*, 1.15–19.

17. Aristotle, *Physiognomics*, 4.1–15.

18. Porter, *Windows of the Soul*, 53.

19. Aristotle, *Physiognomics*, 5.4.

20. Aristotle, *Physiognomics*, 5.1–10.

21. For an overview, see Porter, *Windows of the Soul*. Also see Sarah Kathryn Matthews, "Matter over Mind: Pietro D'Abano (c. 1316) and the Science of Physiognomy" (Ph.D. dissertation, Graduate College of the University of Iowa, 2015), esp. 121–49. For the wider physiognomic model in relation to human skin, see Maaike Van der Lugt, "La peau noire dans la science médiévale," *Micrologus* 13 (2005): 444–47.

22. Miguele de Medina, *Christianae paraenesis sive de recta in Deum fide*, libri septem (Venice, 1564), 12v, cited in Gernert, *Fictionalizing Heterodoxy*, 62.

23. Genesis 3:19.

24. Juan de Horzco y Covarrubias (1540–1610), *De la verdadera y falsa profecía* (Segovia: Juan de la Cuesta, 1588), 93v–94, cited in Gernert, *Fictionalizing Heterodoxy*, 64–66.

25. See the work of Alessandra Celati on Grataroli among a network of heretical physicians.

26. On the interpretation of the cosmos and links to the body, see Anthony Grafton, *Cardano's Cosmos: The Worlds and Works of a Renaissance Astrologer* (Cambridge, MA: Harvard University Press, 1999), 202. Porta would use Cardano's basic structure in *Celestial Physiognomics*. Giambattista della Porta, *Della celeste fisonomia [. . .]* (Padova: Pietro Paolo Toazzi, 1623).

27. Titian, *Allegory of Prudence*, 1550–65, oil on canvas, National Gallery, London; Titian, *Portrait of Federico II Gonzaga*, ca. 1529, Museo del Prado, Madrid; Titian, *Portrait of Charles V with a dog*, 1533, Museo del Prado, Madrid, after Jakob Seisenegger, portrait of Emperor Charles V (1500–1558) with his English water dog, 1532, Kunsthistorisches Museum Vienna, Kunstkammer, Vienna. See discussions surrounding the exhibition "Titian: Love, Desire, Death," at the National Gallery, 16 March 2020–17 January 2021. Emmelyn Butterfield-Rosen, "Men Are Dogs: On Titian's *Poesie* for Philip II," *Art Forum* 6, no. 8 (April 2022): 172–87. Thanks to conversations with John Lansdowne and Amanda Hilliam at I Tatti for drawing my attention to this connection.

28. Valente, "Della Porta e l'Inquisizione"; Tarrant, "Giambattista della Porta and the Roman Inquisition."

29. Eamon, *Science and the Secrets of Nature*, 139.

30. Tarrant, "Giambattista della Porta and the Roman Inquisition," 614–15.

31. Marcus, *Forbidden Knowledge*, 5.

32. Tarrant, "Giambattista della Porta and the Roman Inquisition," 619–20.

33. Valente, "Della Porta e l'Inquisizione," 423.

34. Giovanni Battista Della Porta, *Libro di fisonomia naturale*, autografo, Biblioteca Nazionale di Napoli, MS.12E6. Porta named his friend Giovanni de Rosa as translator. Porta, *Della fisonomia dell'huomo di Gio. Battista della Porta napolitano*, libri quattro (Napoli: Tarquinino Longo, 1598), n117; Gernert, *Fictionalizing Heterodoxy*, 33–34.

35. Marcus, *Forbidden Knowledge*, 5.

36. Edward Peters, *Inquisition* (New York: Free Press, 1988), 75–121. Marcus, *Forbidden Knowledge*, 78, 271n1. Santori recorded his own motivations in his autobiography. Ricci, *Il sommo inquisitore*, 15.

37. Marcus, *Forbidden Knowledge*, 23, 109.

38. On medical censorship, see Marcus, *Forbidden Knowledge*, 101.

39. Gernert, *Fictionalizing Heterodoxy*, 40–41.

40. For the myriad forms of censorship, including expurgation by fire, ink, blade, and paper, see Marcus, *Forbidden Knowledge*, 167–98.

41. Giambattista Della Porta, *Della fisonomia [. . .]* (Vincenza: Per Pierto Psolo Tozzi, 1615), Rare Books BF844. P83 1615, Cornell University Special Collections, with annotation specifying "della libreria de Capuccini di Lucignano." For similar graffiti, see Porter, *Windows of the Soul*, 255–300.

42. Gernert, *Fictionalizing Heterodoxy*, 40–41.

43. Gernert, *Fictionalizing Heterodoxy*, 40–41.

44. Folke Gernert, "La legitimidad de las ciencias parcialmente ocultas: Fisonomía y quiromancia ante la Inquisición," in *Saberes Humanísticos*, ed. Christoph Strosetski (Veuvert: Universidad de Navarra, Iberoamericana, 2014), 105–28, esp. 119–23.

45. Remy Simonetti, "Michele Savonarola," in *Dizionario Biografico degli Italiani*, vol. 92 (2018).

46. Paolo Portone, "Luigi d'Este," in *Dizionario Biografico degli Italiani*, vol. 43 (1993).

47. G. Campori, "Giovanni Battista Della Porta e il cardinale Luigi d'Este," in *Atti e memorie della Dep. di storia patria per le provv. modenesi e parmensi*, 5–6 (1871): 165–90.

48. Paolo Piccari, *Giovan Battista Della Porta: il filosofo, il retore, lo scienziato* (Milan: Franco Angeli, 2007), 29–34; Louise George Clubb, *Giambattista della Porta, Dramatist* (Princeton: Princeton University Press, 1965), 19–25.

49. Campori, "Giovanni Battista Della Porta e il cardinale Luigi d'Este," 165–69.

50. On the later use of these illustrations and their connections to conventions in portraiture, see A. Paolella, "L'autore delle illustrazioni delle Fisiognomiche di Della Porta e la ritrattistica. Esperienze filologiche," in *La 'mirabile' natura: Magica e scienza in Giovan Battista Della Porta (1615–2015)*, Atti del convegno internazionale Napoli-Vico Equense (Pisa: Fabrizio Serra Editore, 2016), 13–17.

51. Campori, "Giovanni Battista Della Porta e il cardinale Luigi d'Este," 165–69.

52. A. Paolella, "Introduzione," in Giovanni Battista Della Porta, *Della Fisionomia dell'uomo, libri sei*, vol. 2, ed. Alfonso Paolella, Edizione nazionale delle opere di Giovan Battista della Porta (Napoli: Edizioni Scientifiche Italiane, 2011), 12n5.

53. Porta's work circulated in manuscript: see, for example, Porta, *Libro di fisonomia naturale*, autografo, and correspondence with Giovanni Antonio Magini, mathematics professor in Bologna, cited in Gernert, *Fictionalizing Heterodoxy*, 40–41.

54. See Kodera, "Laboratory as Stage."

55. Pamela Long has recently adopted the term "trading zone" to mean "arenas in which the unskilled learned and skilled practitioners exchanged substantive knowledge." Long, *Artisan/Practitioners*, 8.

56. Long, *Artisan/Practitioners*, 7.

57. On the villa and its importance in Renaissance culture and intellectual life, see James S. Ackerman, *The Villa: Form and Ideology of Country Houses* (Princeton: Princeton University Press, 1990).

58. *Lettera di M. Alberto Lollio, nella quale rispondendo ad una di m. Hercole Perinato, egli celebra la villa, et lauda molto l'agricultura. Cosa non meno dotta, che dilettevole* (Venice: Gabriel Giolito, 1544), Biii. S.N.R.I. 103, Biblioteca Ambrosiana.

59. Biblioteca dell'Accademia nazionale dei Lincei e Corsiniana, Archivio Linceo, MS 4, in Giuseppe Gabrieli, *Contributi alla storia della Accademia dei Lincei*, tomo 1

(Roma: Accademia Nazionale dei Lincei, 1989), 678. Cited in Findlen, *Possessing Nature*, 316.

60. Laura Balbiani, "Alte und neue Medien in der Della Porta-Rezeption: Enzyklopädische Nachschlagewerke und World Wide Web," *Morgen-Glantz: Zeitschrift der Christian Knorr von Rosenroth-Gesellschaft* 18 (2008): 191–214; William Eamon and Françoise Paheau, "The Accademia Segreta of Girolamo Ruscelli: A Sixteenth-Century Italian Scientific Society," *Isis* 75 (1984): 327–42. See Clubb, *Giambattista della Porta, Dramatist.*

61. For Porta's theatrical works, see Clubb, *Giambattista della Porta, Dramatist.*

62. Giovanni Battista della Porta, *Villae* (Frankfurt, 1592).

63. Io. Baptistae Portae Neapolitani, *Svae Villae. Pomarivm* (Neapoli: Salvianus & Caesare, 1583).

64. Porta, *Villae*, 88.

65. Porta, *Villae*, "De Pyro," 63.

66. Porta, *Villae*, 142.

67. Porta, *Della fisionomia dell'uomo*, libri sei, vol. 2, bk. 1, chap. 3, 74–85.

68. Porta, *Della fisionomia dell'uomo*, libri sei, vol. 2, bk. 1, chap. 3. Porta quotes Virgil's *Georgics*, bk. 3, lines 75–88. Virgil's three great works helped found three distinct literary traditions that shaped the Renaissance worldview. The *Aeneid* generated an epic poetic tradition. The *Eclogues* yielded pastoral literature, creating a tradition that Cervantes could mock mercilessly in *Don Quixote*. The *Georgics* permeated natural histories and agricultural treatises in the ancient and Renaissance world.

69. Porta, *Della fisionomia dell'uomo*, libri sei, vol. 2, bk. 1, chap. 3, 19, esp. lines 94–98.

70. Porta, *Della fisionomia dell'uomo*, libri sei, vol. 2, bk. 1, chap. 3, 19, lines 104–5.

71. Nicola Baldoni, "I fratelli Della Porta e la cultura magica e astrologica a Napoli nel' 500," *Studi Storici* 1 (1958–60), 677–715.

72. Baldoni, "I fratelli Della Porta," 706. Both Porta's experimental and theoretical work were deeply invested in connecting diverse phenomena that others might classify separately as physics, chemistry, meteorology, and biology through a liberatory openness.

73. Eamon, *Science and the Secrets of Nature*, 197 and 401, 155 and 197.

74. Museums and villas were both venues for contemplation, and their histories have significant overlap. See Gigliola Fragnito, *In museo e in villa: Saggi sul Rinascimento perduto* (Venice: Arsenale, 1988).

75. Porta, *Libro di fisonomia naturale*, autografo: "Qua si mostra il ritratto del capo di can di cerca con il capo di P Catone, il quale ho [sotto] dal Museo di Giovan Vincenzo della Porta fratello dotissimo e diligentissimo investigatore e conservatore delle medagle degli antichi [13]."

76. Porta, *Libro di fisonomia naturale*, autografo, 39v–40r: "Qui habbiamo portato l'imagine di Socrate la qual habbiam ritratto dal Museo di Giovan Vincenzo Della Porta, Fratello, di [una] statua per mostrare la sua testa deformata et capelli rari."

77. Porta, *Libro di fisonomia naturale*, autografo, 49r, also mentioned on 56v. "Reconos[ ] in questa tavoletta l'effigie di C. Caligola Imperatore di gran fronte, come l'habbiamo ritratta dalle medaglie di Giovanni Vincenzo della Porta fratello [19]."

78. Porta, *Libro di fisonomia naturale*, autografo, 30v–31r: "In questa tavoletta vien depinta il capo dell'alocco & il gran capo di Vitellio Imperatore, come l'habbiamo ri-

tratto nella statua di marmo del studio di Hadriano Sparafora mio zio dottossimo, e studiossimo conservatore delle memorie antiche."

79. Porta, *Della fisionomia dell'uomo*, libri sei, vol. 2, bk. 1, chap. 19, 72.

80. Porta, *Della fisionomia dell'uomo*, libri sei, vol. 2, bk. 1, chap. 15, 54.

81. Porta, *Libro di fisonomia naturale*, autografo, 24r.

82. Thierry Buquet, "Hunting with Cheetahs at European Courts: From the Origins to the End of a Fashion," in *Animals and Courts: Europe, c. 1200–1800*, ed. Nadir Weber and Mark Hengerer (Berlin: Water De Gruyter, 2020), 17–42.

83. When Venetian Giulio Cesare Muzio offered Francesco I de' Medici a new pair of leopards, the grand duke of Tuscany declined the offer, writing, "I am already so well supplied with leopards that despite the great desire and thought that you have made to satisfy me, I tell you that you can give license to that merchant to sell them to whomever he wants." MAP Doc ID 19718. ASF MdP 269, 13, Francesco di Cosimo I de' Medici in Florence to Giulio Cesare Muzio in Venice, March 28, 1586. Whoever ended up with the felines had a lot of work on their hands; a predatory cat consumed a great deal of meat. For instance, in 1591, one of King Philip II's cats consumed half a chicken, half a cow, and a half *cuarterón* of oil every day, all of which required ten *maravedíes* of coal to cook, coming to a total cost of 340 *reales* a year. Gómez-Centurión, *Alhajas para soberanos*, 102.

84. Porta, *Libro di fisonomia naturale*, autografo, 24v.

85. See Schiesari, *Beasts and Beauties*, 54–72, esp. 67.

86. Porta, *Della fisionomia dell'uomo*, libri sei, vol. 2, bk. 1, chap. 26, 86–87. On the ample historiography on Conrad Gesner, see Urs Leu and Peter Opitz, *Conrad Gessner (1516–1565): Die Renaissance der Wissenschaften* (Oldenbourg: De Gruyter, 2019); Florike Egmond, *Conrad Gessners "Thierbuch": die Originalzeichnungen* (Darmstadt: WBG Edition, 2018); Fabian Kraemer and Helmut Zedelmaier, "Instruments of Invention in Renaissance Europe: The cases of Conrad Gesner and Ulisse Aldrovandi," *Intellectual History Review* 24, no. 3 (2014): 321–41.

87. Porta, *Libro di fisonomia naturale*, autografo, 23r: "Habbiamo portata in forma sotto la forma del leone qual habbiamo dal vivo delineata da alcuni condotti qui in Napoli accio commodamense possi guardar i membri suoi con quello de l'huomo."

88. Porta, *Libro di fisonomía naturale*, autografo, 23r: "Hor descriveremo la forma del leone, connoscra che le para e tutto 'l corpo del leone dimostrano un modelo fia tutti gli animali delle membra de l'homo."

89. Porta, *Della fisonomia [. . .]* (1615).

90. That academic medicine did not provide the leading voice for Porta is characteristic of medical culture in the Spanish empire, in which, as William Eamon has argued, "science, technology, and medicine were deployed for different purposes than they were in the rest of Europe." Eamon, "The Difference That Made Spain, the Difference That Spain Made," in *Medical Cultures of the Early Modern Spanish Empire*, ed. John Slater, María Luz López-Terrada, and José Pardo-Tomás (Burlington, VT: Ashgate, 2014), 231–43, esp. 239.

91. Porta also portrayed himself as a compass, a fish with eyes on top of its head, a grapevine, and an oak tree. This famous connection is cited in a wide array of scholarship, including Findlen, *Possessing Nature*, 317–18; and Daston and Park, *Wonders and the Order of Nature*, 168–69. Daston and Park read the *impresa* as "an emblem of the preternatural philosopher's exploration of secrets and hidden truth."

92. Gabrieli, *Contributi*, 754.

93. On how the lynx became the emblem of the Accademia dei Lincei, see Paolo Galluzzi, *The Lynx and the Telescope* (Leiden: Brill, 2017), 30, 38–44, 46, 49.

94. This passage is mentioned by David Freedberg, *The Eye of the Lynx: Galileo, His Friends, and the Beginnings of Modern Natural History* (Chicago: University of Chicago Press, 2002), 66, 98–102.

95. Pliny, *Natural History*, 28.32.

96. William B. Ashworth, "Natural History and the Emblematic World View," in *Reappraisals of the Scientific Revolution*, ed. David Charles Lindberg and Robert S. Westman (Cambridge: Cambridge University Press, 1990), 303–32.

97. See Wilma George and William B. Yapp, *The Naming of the Beasts: Natural History in the Medieval Bestiary* (London: Duckworth, 1991).

98. On transformations in natural history, see Florike Egmond, *The World of Carolus Clusius: Natural History in the Making, 1550–1610* (London: Pickering & Chatto, 2010); Peter Mason, *Before Disenchantment: Images of Exotic Animals and Plants in the Early Modern World* (London: Reaktion, 2009); Brian W. Ogilvie, *The Science of Describing Natural History in Renaissance Europe* (Chicago: University of Chicago Press, 2006). On related changes in observation, see Sachiko Kusukawa, *Picturing the Book of Nature: Image, Text, and Argument in Sixteenth-Century Human Anatomy and Medical Botany* (Chicago: University of Chicago Press, 2011). Previously, Alciati's *Book of Emblems* had used the emblem to represent forgetfulness as the "lynx neglects food in favour of other prey" in Emblem 66. Among the many editions, see Andrea Alciati, *Omnia Andreae Alciati emblemata* (Paris, 1602).

99. Porta, *Della fisionomia dell'uomo*, libri sei, vol. 2, bk. 1, chap. 28, 7–13.

100. Lina Bolzoni, "Teatro, pittura e fisiognomica nell'arte della memoria di Giovan Battista della Porta," *Intersezioni* A 8, no. 3 (1988): 477–509.

101. See Porta, *Della celeste fisonomía [. . .]* (Padova: Pietro Paolo Toazzi, 1623). Under the reused frontispiece, the text says, "Blandus, bonus, virtus que; simul de lubra tenebant sed binis templis unica Porta fuit. Tu quoque virtutem coniunctam nactus homor. Ambrom digne Porta vocandus erit."

102. Both Aldrovandi and Gesner wrote extensively on ideas of physiognomy and included them in their zoological writing. From his position as a professor at the University of Bologna, Ulisse Aldrovandi (1522–1605) created a famous museum, with natural specimens from dried alligators to unicorn horns. See Ulisse Aldrovandi, *De quadrupedibus solidipedibus volumen integrum* (Bologna: typis Victorii Benatii, 1616), *Quadrupedum omnium bisulcorum historia* (1621); Aldrovandi, *Monstrorum historia, cum paralipomenis historiae omnium animalium* (Bologna: typis Nicolai Tebaldini, Marco Antonio Bernia, 1642). By contrast, Conrad Gesner (1516–1565) published natural histories throughout the sixteenth century. His books on animals included *Historia animalium* (1551–58). See Sashiko Kusukawa, "The Sources of Gessner's Pictures for the *Historia animalium*," *Annals of Science: The History of Science and Technology*, 67 (2010): 303–28; Alfredo Serrai, Maria Cochetti, and Marco Menato, *Conrad Gesner* (Roma: Bulzoni, 1991).

103. Porta, *Della fisionomia dell'uomo*, libri sei, vol. 2, bk. 1, chap. 28, lines 7–13.

104. Ziegler, "Physiognomy, Science and Proto-racism," 181–86.

105. Porta, *Della fisionomia dell'uomo*, libri sei, vol. 2, bk. 1, chap. 27, 90–91.

106. Recent scientific research has started to describe a correlation between certain features and coloring and animal character. At least with foxes and dogs, certain external characteristics are a strong predictor of behavior. In modern experiments

aimed at breeding tame foxes, selecting for the most docile behaviors yields not only animals who behave in a domesticated way, but whose appearance reflects their domestication. Tame foxes come in different colors than their wild progenitors, with dots and stripes that would rarely be seen if they were left to breed according to their own devices. Lee Alan Dugatkin and Lyudmila Trut, *How to Tame a Fox (and Build a Dog): Visionary Scientists and a Siberian Tale of Jump-Started Evolution* (Chicago: University of Chicago Press, 2017), 23, 75.

107. On degeneration of races and animal breeding, see the work of Claude-Olivier Doron, *L'Homme Altéré: Races et dégénérescence (XVIIᵉ–XIXᵉ Siècles)* (Paris: Editions Champ Vallon, 2016), 174–217.

108. Porta, *Phytognomonica* (Naples: apud Horatium Salvianum, 1588), 5: "nam oculi tam lyncei esse non possunt."

109. Porta, *Phytognomonica*, 38.

110. Porta, *Della fisionomia dell'uomo*, libri sei, vol. 2, bk. 1, chap. 28, lines 7–13.

111. Porta, *Phytognomonica*, 78.

112. G. B. della Porta (Naples) to Teodosio Panizza (Rome), June 27, 1586, in the Archivio Palatino; cited in Campori, "Giovanni Battista Della Porta e il cardinale Luigi d'Este," 189–90.

113. Gernert, *Fictionalizing Heterodoxy*, 34–36.

114. Findlen, *Possessing Nature*, 317–18.

115. On the early Accademia dei Lincei, see Irene Baldriga, *L'occhio della lince: i primi lincei tra arte, scienza e collezionismo (1603–1630)* (Rome: Accademia Nazionale dei Lincei, 2002); Silvia De Renzi, "Storia naturale ed erudizione nella prima età moderna: Johann Faber (1574–1629) medico linceo" (Università degli Studi di Bari, Tesi di Dottorato in Storia Scienza, 1992–93); Gabrieli, *Contributi*.

116. G. B. Porta to Federico Cesi, from Naples to Rome, 28 August 1609. Manoscritto Linceo, 12. Cited in Gabrieli, *Contributi*, 646.

117. Although Galileo turned his empirical approach to sunspots, heavenly bodies, and earthly mechanics, as Cesi doubtless hoped that Porta would continue with his work on distillation and fluids, Porta's own relationship with Galileo was touchy at best. Lynx-eyed as they were, both had come up with designs for the telescope and exchanged lenses with their correspondents, as Porta emphasized in his 1586 letter to Luigi d'Este. The Lincei recognized Porta's invention of the device in 1610, perhaps in order to secure his favor. In a letter to Federico Cesi in 1609, Porta described his telescope: an instrument made of silver, three fingers wide across the diameter, with a lens that allowed one to see something more than a thousand steps away. But the fame fell to Galileo. C. H. Lüthy, "Atomism, Lynceus, and the Fate of Seventeenth-Century Microscopy," *Early Science and Medicine* 1, no. 1 (February 1996): 1–27, specifically 6–11.

118. Porter, *Windows of the Soul*, 120–71.

119. Giuseppe Gabrieli, "Il carteggio Linceo della vecchia accademia di Federico Cesi (1603–1630)," *Memorie della R. Accademia Nazionale dei Lincei, Classe di scenze morali, storiche e filologiche*, ser. 6, vol. 7, fols. 1–4 (1938–42), 112–13.

120. Gabrieli, "Giovan Battista della Porta," in "Il carteggio Linceo," 370. See also Giovanni Battista Della Porta, *De Ea naturalis physiognomoniae parte quae ad manuum lineas spectat*, libri duo, ed. Oreste Trabucco, Edizione nazionale delle opera di Giovan Battista della Porta, vol. 9 (Napoli: Edizioni Scientifiche Italiane, 2003), 11–14.

121. Richard Serjeantson, "Francis Bacon and the 'Interpretation of Nature' in the

Late Renaissance," *Isis: A Journal of the History of Science Society* 104, no. 4 (December 2014): 681–705.

122. Recchi, *Rerum medicarum Novae Hispaniae*, 50.

123. Prince Federico Cesi, Lyncei's *Phytosophicarum tabularum*, in Recchi, *Rerum medicarum Novae Hispaniae*, 905.

Epilogue

1. James Hankins, *Plato in the Italian Renaissance* (Leiden: Brill, 1990), vol. 1, 5–7, 109–10. See also Mario Vegetti and Paolo Pissavino, *I Decembrio e la tradizione della Repubblica di Platone tra medioevo e umanesimo* (Napoli: Bibliopolis, 2005).

2. Cassidy, *Sport of Kings*, 37–38. On the link between modern horse breeding and eugenics, see Rebecca Cassidy, *Horse People: Thoroughbred Culture in Lexington and Newmarket* (Baltimore: Johns Hopkins University Press, 2007), 33–36.

3. See Gabriel N. Rosenberg, *The 4-H Harvest: Sexuality and the State in Rural America* (Philadelphia: University of Pennsylvania Press, 2015); Alexandra Minna Stern, *Eugenic Nation: Faults and Frontiers of Better Breeding in Modern America* (Oakland: University of California Press, 2015); Wendy Kline, *Building a Better Race: Gender, Sexuality, and Eugenics from the Turn of the Century to the Baby Boom* (Berkeley: University of California Press, 2005).

Bibliography

Manuscript Sources

Archivio di Stato di Mantova (ASMN), Mantua, Italy
   Archivio Gonzaga (AG) b. (busta) 258.
   AG b. 759, c. (carta) 354r (recto).
   AG b. 795, c. 81–83.
   AG b. 866, c. 807.
   AG b. 1464, c. 334.
   AG b. 2439, c. 89.
   AG b. 2439, c. 90.
   AG b. 2455, c. 153.
   AG b. 2472, c. 728.
   AG b. 2521, c. 346.
   AG b. 2891, L. (libro) 64, c. 59v (verso).
   AG b. 2891, L. 64, c. 62.
   AG b. 2921, L. 231, fol. 4v.
   AG b. 2933, L. 301, c. i.
   AG b. 2991, L. 1, c. 76r.
   AG b. 2991, L. 3, line 14.
   AG b. 2996, L. 31, c. 88r.
   AG 3000, L. 51, cc. 64r–v.
   AG 3000, L. 51, cc. 155r–v.

Archivio di Stato di Milano, Milan, Italy
   Atti di Governo, Studi parte Antica, 19, Biblioteche spec. inventari di libri 1569.

Archivio di Stato di Napoli (ASN), Naples, Italy
Dipendenza della Sommaria (DS) I Serie 38, no. 2.
   DS I Serie 38, no. 3.
   DS I Serie 38, no. 3, 1542, 1.
   DS I Serie 38, no. 3, 1542, 5.
   DS I Serie 38, no. 3, 1542, 15.
   DS I Serie 38, no. 3, 1542, 17.
   DS I Serie 38, no. 3, 1543, 288.

Archivio di Stato di Udine, Udine, Italy
  Fondo Caimo, b. 86, D. 54, Pompeo Caimo to his brother Eusebio Caimo, June 14,
    1603.

Archivo General de Simancas (AGS), Simancas, Spain
  Casa y Sitios Reales (CSR), leg. 305, 2, 31 febrero (?) de 1608.
  CSR, leg. 305, 7, 21 de mayo 1608.
  CSR, leg. 305, 207–8, 9 diciembre de 1612.
  CSR, leg. 305, 249, 14 de junio 1613.
  CSR, leg. 305 256, Junta de obras y bosques, September 21, 1613.
  CSR, leg. 306, 260, 17 de junio 1622.
  CSR, leg. 307, 120, 5 de junio 1626.
  CSR, leg. 307, 352, 22 de mayo, 1628.
  CSR, leg. 310, 63, agosto 1639.
  CSR, leg. 315, 256, 9 de noviembre 1668.
  CSR, leg. 315, 310, 3 de febrero 1669.

Biblioteca Ambrosiana, Milan, Italy
  Lettera di M. Alberto Lollio, nella quale rispondendo ad una di m. Hercole Perinato, egli
    celebra la villa, et lauda molto l'agricultura. Cosa non meno dotta, che dilettevole.
    Venice: Gabriel Giolito, 1544. Biii. S. N. R. I. 103.

Biblioteca Angelica, Rome, Italy
  MS 1331.

Biblioteca Apostolica Vaticana (BAV), Vatican City
  "Alcune espressioni latine relative all'allevamento dei cavalli." Urb.lat.1343.
  "De' tempi e misure ch'osservar deve il Cavaliere facendo operare i cavalli ne'
    maneggi per farli giusti e facili." Urb.lat.255, 21r–31r.
  "Libro della razza dei cavalli di Sua Altezza Serenissima [Francesco Maria II della
    Rovere, duca di Urbino] dal 1614 al 1618." Urb.lat.254, 1r–6r.
  "Modo che si tiene per imbrigliar cavalli." Urb.lat.269, 1r–36r.
  "Modo et ordine per creare razza di cavalli." Urb.lat.251, 1r–12r.
  "Modo et ordine per crear razza de cavalli." Cappon. 172, 224r–232r.
  Orsi, Filippo. *Trattato sui freni dei cavalli.* Urb.lat.270, 1r–85r.
  Ruffo, Giordano. "Hippiatria" e "Raccolta di dicette per cavalli." Urb.lat.1413, 1r–73v.
  Rusius, Laurentius (1288–1347). "Della cura dei cavalli." Urb.lat.256, 2r–83r.
  Rutati, Giulio. "Discorso al Ser.mo principe d'Urbino [Francesco Maria II della Ro-
    vere] mio signore, dove accennandosi quanto sia utile l'imitazione degli antenati
    si discorse brevemente dell'uso del cavallo." Urb.lat.860, 527r–544r.
  "Trattato sui freni e sui capestri dei cavalli con illustrazioni," with illustrations by Sil-
    vestro Vanzy, 1620. Urb.lat.271, 2v–47v.

Biblioteca Comunale di Trento, Trent, Italy
  Campanella, Tommaso. *La città del sole.* Manoscritto anonimo. 1602.

Biblioteca Francisco de Zabálburu, Madrid, Spain
  MIRO, 18, D. (documento) 535/1, 1661.

Altamira, 141, GD. (grupo de documentos) 1. Gobierno de Felipe II (1569–1594), esp. D. 9, D. 46, 145–46, and D. 74, 1–4.

Biblioteca Marciana, Venice, Italy
Papa, Giovanni Bernardino. "Trattato delle Razze dei cavalli, diretto al granduca di Toscana nel di 25 marzo 1607." 5223, cod. 24, c. 40, sec. S. 7.

Biblioteca Medicea Laurenziana, Florence, Italy
Sahagún, Fray Bernardino de. *General History of the Things of New Spain: The Florentine Codex*. MS Med. Palat. 220. Reproduced by the World Digital Library, https://www.wdl.org/en/item/10096/.

Biblioteca Nacional de España, Madrid, Spain
Campanella, Tomasso. *Compendio della Monarchia di Spagna*. S. 7. MSS/1416.
Ordenes de cabalgar De Fadrique Grison, cavallero Napolitano; traducidas de la lengua ytaliana en vulgar castellano, S. 16. MSS/9999.
Grisone, Federico. *Ordini di cavalcare*. Vinegia: Vincenzo Valgrisi alla bottega d'Erasmo, 1553. R/11548.
Grisone, Federico. *Órdenes de caballear de Federico Grisone, caballero Napolitano, traducido de la lengua italiana en romance castellano*. MSS/1059.

Biblioteca Nazionale di Napoli, Naples, Italy
Giovanni Battista Della Porta, *Libro di fisonomia naturale*, autografo. MS.12E6.

Bibliothèque Nationale de France, Paris, France
Mexicain 220, "Diario de Domingo de San Antón Muñón Chimalpáhin," Chimalpahin Cuauhtlehuanitzin (1579–1660).

Cornell University Library, Special Collections, Ithaca, NY, United States of America
Giambattista della Porta. *Della fisonomia [. . .]*. Vincenza: Per Pierto Psolo Tozzi, 1615. Rare Books BF844. P83 1615.

Falk Family, private collection, Palazzo Giustian Recanati alle Zattere, Venice, Italy
Silvestro Da Lucca, *Il libro dei palii vinti dai cavalli di Francesco Gonzaga, 1512–1518, mineato da Lauro Padovano e forse altri due miniatori*.

Huntington Library, Pasadena, CA, United States of America
HM 1316, "Instruttione per negotij nella Corte di Spagna al S.or Ludouico Orsino mandato Sua M.ta Catt:la Dal S.or Duca di Bracciano," likely dating from 1568–1579.

The Library of Congress Geography and Map Division, Washington, DC, United States of America
The Oztoticpac Lands Map, ca. 1540. 76 × 84 cm. G4414.T54:2O9 1540.O9.

Monasterio de Santa María de la Vid Biblioteca, Burgos, Spain
Bestiario de D. Juan de Austria, original manuscript.

Bestiario de D. Juan de Austria [Texto impreso]: S. XVI: original conservado en la Biblioteca del Monasterio de Sta. Mª de la Vid (Burgos), Monasterio de Santa María de la Vid Biblioteca (Burgos: Gil de Siloé, 1998).

Newberry Library, Chicago, IL, United States of America
Cortés, Hernán. Second letter to Charles V, 1520. Reproduced by the World Digital Library. https://www.wdl.org/en/item/19994/.
Vocabulario trilingüe (ca. 1540). Ayer MS 1478, fols. 81, 83.

Österreichisches Staatsarchiv (ÖSA), Vienna, Austria
FA Harrah HS 378, "Los Hierros de los cavalos que ay hoy dia en Andalusia y Cordova, MDCXCVI" ("The Brands of Horses that there are today in Andalusia and Cordova, 1696").
HHStA (Haus-, Hof- und Staatsarchiv), Hausarchiv, Sammelbände, 1–2 (Konv.) 1, fols. 66–66v. Maximilian II to the Marchioness (Marquesa) del Gasto, Vienna, April 17, 1559.
HHStA, Hausarchiv, Sammelbände, 1–2 (Konv.) 1, fol. 67. Maximilian II to Count de Consa, Vienna, April, 17, 1559.
HHStA, Hausarchiv, Sammelbände, 1–2 (Konv.) 1, fols. 68–68v. Maximilian II to Vicenzio de Ebuli, Vienna, April 17, 1559; fol. 70v, Maximilian II to Cesare Gonzaga, Vienna, May 12, 1559.
HHStA, Hausarchiv, Sammelbände, 1–2 (Konv.) 1, fols. 117v–118. To the Viceroy in Naples, Vienna, September 18, 1560.
HHStA, Hausarchiv, Sammelbände, 1–2 (Konv.) 1, fols. 135v–136. To the Viceroy of Naples, Vienna, March 13, 1561.
HHStA, Hausarchiv, Sammelbände, 1–2 (Konv.) 1, fols. 136–136v. To the Mardones del Consejo of the King of Spain in Naples, Wiener Neustadt, March 13, 1561.
HHStA, Hausarchiv, Sammelbände, 1–2 (Konv.) 1, fols. 159–159v. To the Viceroy of Naples, Vienna, October 14, 1561.
HHStA, Hausarchiv, Sammelbände, 1–2 (Konv.) 1, fols. 159v–160. To Principe Stiliano, Vienna, October 17, 1561.
HHStA, Hausarchiv, Sammelbände, 1–2 (Konv.) 1, fols. 160v–161. To the Count of Consa, Vienna, October 17, 1561.
HHStA, Hausarchiv, Sammelbände, 1–2 (Konv.) 1, fols. 161–161v. To Vicenzio de Ebuli, Vienna, October 17, 1561.

Società Napoletana di Storia Patria (SNSP), Naples, Italy
Merchi delle razze de Cavalli de Prencipi Duchi Marchesi Conti Baroni de tutte le Provincie del Regno di Napoli. MS segn. 22.D36.

Databases

Archivio di Stato di Firenze, Mediceo del Principato (ASF MdP)
Archivio di Stato di Firenze and Scuola Normale Superiore di Pisa, "Ceramelli Papiani: blasoni delle famiglie toscane descrite nella Raccolta Ceramelli Papiani." Accessed December 30, 2021. http://www.archiviodistato.firenze.it/ceramellipapiani/index.php?page=Home.

Medici Archive Project (MAP) doc. ID 17769. ASF MdP 4085, 12. Mario del Tufo in Naples to Ferdinando I de' Medici in Florence, May 11, 1595.

MAP doc. ID 19555. ASF MdP 219, 83. Cosimo de' Medici in Pisa to Pedro Afán de (Perafán) Ribera in Naples, March 31, 1563.

MAP doc. ID 19718. ASF MdP 269, 13. Francesco di Cosimo I de' Medici in Florence to Giulio Cesare Muzio in Venice, March 28, 1586.

MAP doc. ID 21487. ASF MdP 516a, 560. Asciano Caracciolo in Naples to Francesco di Cosimo I de' Medici in Florence, July 1, 1565.

MAP doc. ID 21634, ASF MdP 3080, 104. Avviso from Venice from Cosimo Bartoli to Francesco di Cosimo I de' Medici, July 5, 1567.

MAP doc. ID 21664. ASF MdP 515a, 860. Asciano Caracciolo in Naples to Francesco di Cosimo I de' Medici in Florence, May 20, 1565.

MAP doc. ID 21868. ASF MdP 3090, 570. Cosimo Bartoli in Venice to Francesco di Cosimo de' Medici in Florence, July 2, 1569.

MAP doc. ID 22314. ASF MdP 518, 178. Alfonso (Importuni) Cambi in Naples to Cosimo I de' Medici in Florence, October 29, 1565.

MAP doc. ID 22699. ASF MdP 2964, 67. Diego Hurtado de Mendoza in Venice to Cosimo I de' Medici in Florence, May 8, 1541.

MAP doc. ID 26050. ASF MdP 4027, 312. Aviso [sic] from Rome, June 27, 1590.

MAP doc. ID 27209. ASF MdP 3082, 272. Aviso [sic] from Venice to Florence relaying news from Rome and Lyon, July 2, 1575.

MAP doc. ID 28073. ASF MdP 3082, 712. Aviso [sic] from Venice reporting news from Rome, Venice, Prague, Antwerp, Cologne, July 4, 1579.

MAP person ID 2665.

Dictionaries

Andrews, J. Richard. *Introduction to Classical Nahuatl*. Rev. ed. Norman: University of Oklahoma Press, 2003.

Campbell, Joe. *A Morphological Dictionary of Classical Nahuatl*. Madison, WI: Hispanic Seminary of Medieval Studies, 1985.

Covarrubias Orozco, Don Sebastián de. *Tesoro de la lengua castellana o española*. Madrid: L. Sanchez, 1611.

Covarrubias Horozco, Sebastián de. *Tesoro de la lengua castellana o española*. Edición integral e ilustrada de Ignacio Arellano y Rafael Zafra. Madrid: Universidad de Navarra-Iberoamericana-Vervuert-Real Academia Española, 2006.

Florio, John. *A world of Wordes*. London: Arnold Hatfield for Edw. Blount, 1598.

Florio, John. *A World of Words*. London: Melc. Bradwood for Edw. Blount and William Pavret, 1611.

Gran Diccionario Náhuatl. Universidad Nacional Autónoma de México, Instituto de Investigaciones Bibliográficas/Instituto de Investigaciones Históricas. https://gdn.iib.unam.mx/.

Karttunen, Frances. *An Analytical Dictionary of Nahuatl*. Norman: University of Oklahoma Press, 1992.

Molina, Alonso de. *Aqui comiença un vocabulario en la lengua castellana y mexicana*. Mexico: Juan Pablos, 1555. Medina Mexico 24. Lilly Library call number: PM4066 .M7 A6 1555.

Molina, Alonso de. *Vocabulario en lengua castellana y mexicana*. Mexico: Antonio de Spinosa, 1571.

Online Nahuatl Dictionary. Edited by Stephanie Wood. Wired Humanities Projects. https://nahuatl.uoregon.edu/.

Siméon, Rémi, and Josefina Oliva De Coll. *Diccionario de la lengua náhuatl o mexicana*. México: Siglo Veintiuno, 1992.

Thouvenot, Marc. *Diccionario náhuatl-español*. Ciudad de México: Universidad Nacional Autónoma de México, 2015.

Wimmer, Alexis. Dictionnaire de la Langue Nahuatl Classique. http://sites.estvideo.net/malinal/nahuatl.page.html.

Encyclopedia and Biographical Dictionary Entries

Balbi, Giovanni. *Catholicon*. Venice: Petri Liechtenstein, 1506.

Dizionario Biografico degli Italiani, online, Treccani. https://www.treccani.it/biografico/index.html.

    Borromeo, Agostino. "Clemente VIII." Vol. 26 (1982). https://www.treccani.it/enciclopedia/papa-clemente-viii_%28Dizionario-Biografico%29/.

    "Carletti, Francesco." Vol. 20 (1977). https://www.treccani.it/enciclopedia/francesco-carletti/.

    Ceresa, Massimo. "Giunti (Giunta), Bernardo." Vol. 57 (2001). https://www.treccani.it/enciclopedia/bernardo-giunti_res-1cf840f8-87ee-11dc-8e9d-0016357eee51_%28Dizionario-Biografico%29/.

    Mori, Elisabetta. "Orsini, Paolo Giordano." Vol. 79 (2013). https://www.treccani.it/enciclopedia/paolo-giordano-orsini_res-ac068daf-373d-11e3-97d5-00271042e8d9_(Dizionario-Biografico).

    Portone, Paolo. "Luigi d'Este." Vol. 43 (1993). https://www.treccani.it/enciclopedia/luigi-d-este_(Dizionario-Biografico).

    Romei, G. "Della Porta, Giovambattista," Vol. 37 (1989): 170–82. https://www.treccani.it/enciclopedia/giovambattista-della-porta_(Dizionario-Biografico).

    Simonetti, Remy. "Savonarola, Michele." Vol. 92 (2018). https://www.treccani.it/enciclopedia/michele-savonarola_(Dizionario-Biografico).

Dizionario storico dell'Inquisizione. Edited by Adriano Prosperi. Pisa: Edizione della Normale, 2010.

    Valente, Michaela. "Della Porta, Giovanni Battista." Vol. 1.3.

Stanford Encyclopedia of Philosophy. Edited by Edward N. Zalta.

    Ernst, Germana. "Tommaso Campanella." Fall 2010 edition. https://plato.stanford.edu/entries/campanella/.

    James, Michael, and Adam Burgos. "Race." Summer 2020 edition. https://plato.stanford.edu/entries/race/.

    Klein, Jürgen, and Guido Giglioni. "Francis Bacon." Fall 2020 edition. https://plato.stanford.edu/entries/francis-bacon/.

    Kodera, Sergius. "Giambattista della Porta." Summer 2015 edition, revised Summer 2021. https://plato.stanford.edu/entries/della-porta/.

    Studtmann, Paul. "Aristotle's Categories." Fall 2018 edition, revised Winter 2021. https://plato.stanford.edu/entries/aristotle-categories/.

Printed Sources

Ackerman, James S. *The Villa: Form and Ideology of Country Houses*. Princeton: Princeton University Press, 1990.

Acosta, José de. *Conciones in Quadragesimam*. Venetiis: apud Io. Bapt. Ciottum Senensem sub signo Aurorae, 1599.

Acosta, José de. *De Christo reuelato*, libri nouem. Romae: Apud Iacobum Tornerium, 1590.

Acosta, José de. *De natura Novi Orbis*, libri duo. Coloniae: Agrippinae: In oficina Birckmannica, 1596.

Acosta, José de. *De natura Novi Orbis*, libri duo; et *De promulgatione evangelii, apud barbados, sive De procuranda Indorum salute*, libri sex. Salmanticae: Apud Gullelmum Froquel, 1589.

Acosta, José de. *De procuranda Indorum salute: Educación y Evangelización*. Edited by Luciano Pereña. Madrid: Consejo Superior de Investigaciones Científicas, 1984–87.

Acosta, José de. *De temporibus nouissimis*, libri quatuor. Romae, ex typographia Iacobi Tornerij, 1590.

Acosta, José de. *Histoire naturelle et moralle des Indes*. Paris: Chez Marc Orry, 1598.

Acosta, José de. *Historia naturale morale delle indie*. Translated by Gio. Paolo Galucci Salodiano. Venetia: Presso Bernado Basa, 1596.

Acosta, José de. *Historia natural y moral de las Indias en que se tratan las cosas notables del cielo, y elementos, metales, plantas, y animals dellas: y los ritos, y ceremonias, leyes, y govierno, y guerras de las Indios*. Compuesta por el padre Joseph de Acosta. Sevilla: Iuan de Leon, 1590.

Acosta, José de. *Historie naturael ende morael van de Westersche Indien*. Tot Enchuysen: By Jacob Lenaertsz, 1598.

Acosta, José de. *Natural and Moral History of the Indies*. Edited by Jane E. Mangan. Translated by Frances López-Morillas. Durham: Duke University Press, 2002.

Acosta, José de. *The Naturall and Morall Historie of the East and West Indies*. London: Printed for Val: Sims, 1604.

Acosta, José de, and Juan de Atienza. *Doctrina Christiana y catecismo para instrucción de los Indios*. Ciudad de los Reyes: Antonio Ricardo, 1584.

Acuña, René. *Relaciones Geográficas novohispanas del siglo XVI*. 10 vols. México: Universidad Nacional Autónoma de México–Instituto de Investigaciones Antropológicas, 1982–88.

Adelmann, Howard B. *The Embryological Treatises of Hieronymus Fabricius of Aquapendente*. Vol. 1. Ithaca: Cornell University Press, 1942.

Aertsen, Jan. *Nature and Creature: Thomas Aquinas' Way of Thought*. Leiden: E. J. Brill, 1988.

Alberti, L. B. *De equo animante*. Edited by C. Grayson, J. Y. U. Boriaud, and F. Furlar. *Albertiana* 2 (1999).

Alciati, Andrea. *Omnia Andreae Alciati emblemata*. Paris, 1602.

Aldrovandi, Ulisse. *Aldrovandi on Chickens: The Ornithology of Ulisse Aldrovandi (1600)*, vol. 2, bk. 14. Translated and edited by L. R. Lind. Norman: University of Oklahoma Press, 1963.

Aldrovandi, Ulisse. *De quadrupedibus digitatis viviparis*, libri tres, et *De quadrupedibus digitatis oviparis*, libri duo. Bologna: apud Nicolaum Tabaldinum, 1637.

Aldrovandi, Ulisse. *De quadrupedibus solidipedibus volumen integrum*. Bologna: Giovanni Cornelio Uterwer, 1616.

Aldrovandi, Ulisse. *De quadrupedibus solidipedibus volumen integrum*. Bologna: typis Victorii Benatii, 1616.

Aldrovandi, Ulisse. *De quadrupedibus solidipedibus volumen integrum*. Bologna: Apud Nicolau Thebaldinum, 1639.

Aldrovandi, Ulisse. *Monstrorum historia, cum paralipomenis historiae omnium animalium*. Bologna: typis Nicolai Tebaldini, Marco Antonio Bernia, 1642.

Allen, Don Cameron. *The Legend of Noah: Renaissance Rationalism in Art, Science, and Letters*. Urbana: University of Illinois Press, 1949.

Allen, Glover M. "Dogs of the American Aborigines." *Bulletin of the Museum of Comparative Zoology at Harvard College* 63 (1919–20): 431–517.

Alvarez, G., F. C. Ceballos, and C. Quinteiro. "The Role of Inbreeding in the Extinction of a European Royal Dynasty." *PLoS One* 4, no. 4 (2009): e5174.

Alvarez López, Enrique. "El Dr. Francisco Hernández y sus comentarios a Plinio." *Revista de Indias* 3, no. 8 (1942): 251–90.

Alvarez López, Enrique. "El perro mudo Americano (El problema del perro mudo de Fernández de Oviedo)." *Boletín de la Real Sociedad Española de Historia Natural* 11 (1942): 411–17.

Alves, Abel. *The Animals of Spain: An Introduction to Imperial Perceptions and Human Interaction with Other Animals, 1492–1826*. Leiden: Brill, 2011.

Alves, Abel. "Iberia's Imagined Elephant: Animal Behavior through a Human Prism in the Sixteenth Century." *Romance Notes* 60, no. 3 (2020): 425–35.

Amabile, Luigi. *Fra Tommaso Campanella, la sua congiura, i suoi processi e la sua pazzia*. 3 vols. Naples: Morano, 1882.

Amabile, Luigi. *Fra Tommaso Campanella ne' castelli di Napoli, in Roma ed in Parigi*. 2 vols. Naples: Morano, 1887.

Ambrosoli, Mauro. *The Wild and the Sown: Botany and Agriculture in Western Europe, 1350–1850*. Cambridge: Cambridge University Press, 2009.

Amorim, Carlos Eduardo, Stefania Vai, Cosimo Posth et al. "Understanding 6th-Century Barbarian Social Organization and Migration through Paleogenomics." *Nature Communications* 9, no. 3547 (2018).

Anderson-Córdova, Karen. *Surviving Spanish Conquest: Indian Fight, Flight, and Cultural Transformation in Hispaniola and Puerto Rico*. Montgomery: University of Alabama Press, 2017.

Anselmi, Alessandra. *El diario del viaje a España del Cardenal Francesco Barberini*. Madrid: Fundación Carolina, 2004.

Appuhn, Karl. "Ecologies of Beef: Eighteenth-Century Epizootics and the Environmental History of Early Modern Europe." *Environmental History* 15, no. 2 (April 2010): 268–87.

Aquilecchia, Giovanni. "*In Facie prudentis relucit sapientia*: Appunti sulla letteratura metoposcopica tra Cinque e Seicento." In *Giovan Battista della Porta nell'Europea del suo tempo*, edited by Maurizio Torrini, 199–229. Napoli: Guida, 1990.

Aranda, Marcelo. "Instruments of Religion and Empire: Spanish Science in the Age of the Jesuits, 1628–1756." Ph.D. dissertation, Stanford University, 2013.

Arbel, Benjamin. "The Beginnings of Comparative Anatomy and Renaissance Reflections on the Human-Animal Divide." In "The Animal in Renaissance Italy," edited by Stephen Bowd and Sarah Cockram. Special issue, *Renaissance Studies* 31, no. 2 (2017): 201–22.

Aretino, Pietro. *Il marescalco*. Stampata in Vinegia: Per M. Bernardino de Vitali Veneto, 1533.

Aristotle. *De Generatione Animalium* (On the Generation of Animals). Translated by H. J. Drossart Lulofs. Oxford: Oxford University Press, 1965.

Aristotle. *Historia Animalium*. Translated by D. M. Balme and Allan Gotthelf. Cambridge: Cambridge University Press, 2002.

Aristotle. *Physiognomics*. Translated by T. Loveday and E. S. Forster. Oxford: Oxford University, Clarendon Press, 1913.

[Pseudo-]Aristotle. *Pseudo-Aristotele: Fisiognomica; anonimo Latino: Il trattato di fisiognomica*. Giampiera Raina, translator and commentator. Milano: Biblioteca Universale Rizzoli, 1994.

Ashworth, William B. "Natural History and the Emblematic World View." In *Reappraisals of the Scientific Revolution*, edited by David Charles Lindberg and Robert S. Westman, 303–32. Cambridge: Cambridge University Press, 1990.

Astarita, Tommaso. *Between Salt Water and Holy Water: A History of Southern Italy*. New York: W. W. Norton, 2005.

Aubert, Guillaume. "'The Blood of France': Race and Purity of Blood in the French Atlantic World." *William and Mary Quarterly*, 3rd series, 61, no. 3 (July 2004): 439–78.

Augustine, Saint. *The Catholic and Manichaean Ways of Life*. Translated by D. A. and I. J. Gallagher. Fathers of the Church, vol. 56. Washington, DC: Catholic University of America Press, 1966.

Azzolini, Monica. *The Duke and the Stars: Astrology and Politics in Renaissance Milan*. Cambridge, MA: Harvard University Press, 2013.

Bacon, Francis. *Novum Organum*. Edited by Joseph Devey. New York: P. F. Collier & Son, 1902.

Bacon, Francis, and Tommaso Campanella. *The New Atlantis & The City of the Sun*. Mineola, NY: Dover, 2003.

Bacon, Francis, with William Rawley. "New Atlantis: A Worke Unfinished." In *Sylva sylvarum: or, A natural history*. London: W. Lee, 1627.

Balbiani, Laura. "Alte und neue Medien in der Della Porta-Rezeption: Enzyklopädische Nachschlagewerke und World Wide Web." *Morgen-Glantz: Zeitschrift der Christian Knorr von Rosenroth-Gesellschaft* 18 (2008): 191–214.

Baldoni, Nicola. "I fratelli Della Porta e la cultura magica e astrologica a Napoli nel' 500." *Studi Storici* 1 (1958–60): 677–715.

Baldriga, Irene. *L'occhio della lince: i primi lincei tra arte, scienza e collezionismo (1603–1630)*. Rome: Accademia Nazionale dei Lincei, 2002.

Barker, Graeme. *The Agricultural Revolution in Prehistory: Why Did Foragers Become Farmers?* New York: Oxford University Press, 2006.

Barnett, Lydia. *After the Flood: Imagining the Global Environment in Early Modern Europe*. Baltimore: Johns Hopkins University Press, 2019.

Baron, A. S., and M. R. Banaji. "The Development of Implicit Attitudes: Evidence of Race Evaluations from Ages 6 and 10 and Adulthood." *Psychological Science* 17, no. 1 (2006): 53–58.

Bassett, Molly H. "Bundling Natural History: Tlaquimilolli, Folk Biology, and Book 11." In Peterson and Terraciano, *The Florentine Codex*, 139–51.

Bassett, Molly H. *The Fate of Earthly Things: Aztec Gods and God-Bodies*. Austin: University of Texas Press, 2015.

Bassett, Molly. "The Pre-racial Saint? Ma(r)king Aztec God-Bodies." In Bassett and Lloyd, *Sainthood and Race*, 199–216.

Bassett, Molly, and Vincent Lloyd, eds. *Sainthood and Race: Marked Flesh, Holy Flesh*. New York: Routledge, 2014.

Bauer, Ralph, and Marcy Norton. "Introduction: Entangled Trajectories: Indigenous and European Histories." *Colonial Latin American Review* 26, no. 1 (2017): 1–17.

Baxandall, Michael. *Painting and Experience in Fifteenth-Century Italy*. New York: Oxford University Press, 1972.

Bell, Rudolph. *How To Do It: Guides to Good Living*. Chicago: University of Chicago Press, 1999.

Bellarmino, Roberto. *De ascensione mentis in Deum per scalas creaturam*. Antwerp: Apud Viduam & Filos Io. Moreti, 1615.

Bennett, Joshua. *Being Property Once Myself: Blackness and the End of Man*. Cambridge, MA: Harvard University Press, 2020.

Benzoni, Girolamo. *Historia del Nuovo Mondo*. Venice, 1565.

Bertolotti, A. "Curiosità storiche mantovane CXXI, I gatti e la gatta della marchesa di Mantova Isabella d'Este." *Il mendico* (16 aprile 1889): 6–7.

Bethencourt, Francisco. *The Inquisition: A Global History, 1478–1834*. Translated by Jean Birrell. Cambridge: Cambridge University Press, 2009.

Beusterien, John. *Canines in Cervantes and Velázquez*. New York: Routledge, 2016.

Bevan, Andrew, and David Wengrow. *Cultures of Commodity Branding*. New York: Routledge, 2010.

Bianchi, Lorenzo. "La Magia Naturale a Napoli tra Della Porta e Campanella." In *La magia naturale tra Medioevo e prima età moderna*, edited by Lorenzo Bianchi and A. Sannino, 203–28. Firenze: Sismel, 2018.

Bigelow, Allison Margaret. *Mining Language: Racial Thinking, Indigenous Knowledge, and Colonial Metallurgy in the Early Modern Iberian World*. Omohundro Institute of Early American History and Culture. Chapel Hill: University of North Carolina Press, 2020.

Bireley, Robert. *The Refashioning of Catholicism, 1450–1700*. Washington, DC: Catholic University of America Press, 1999.

Bitbol-Hespériès, Annie. "Monsters, Nature, and Generation from the Renaissance to the Early Modern Period." In Justin Smith, *Problem of Animal Generation*, 47–62.

Bolzoni, Lina. "Teatro, pittura e fisiognomica nell'arte della memoria di Giovan Battista della Porta." *Intersezioni* A 8, no. 3 (1988): 477–509.

Bonansea, Bernardino. *Tommaso Campanella: Renaissance Pioneer of Modern Thought*. Washington, DC: Catholic University of America Press, 1969.

Bonavia, Duccio. *The South American Camelids*. Los Angeles: Cotsen Institute of Archaeology, University of California, 2008.

Boone, Elizabeth Hill. *Cycles of Time and Meaning in the Mexican Books of Fate*. Austin: University of Texas Press, 2007.

Boone, Elizabeth Hill. "This New World Now Revealed: Hernán Cortés and the Presentation of Mexico to Europe." *Word & Image: A Journal of Verbal/Visual Enquiry* 27, no. 1 (2011): 31–46.

Borneman, John. "Race, Ethnicity, Species, Breed: Totemism and Horse-Breed Classification in America." *Comparative Studies in Society and History* 30, no. 1 (January 1988): 25–51.

Borrego Soto, Miguel Ángel. *La revuelta mudéjar y la conquista cristiana de Jerez (1261–1267)*. Madrid: Peripecias Libros, 2016.

Boulle, Pierre. "La Construction du concept de race dans la France d'ancien régime." *Outremer: Revue d'histoire*, t. 89, nos. 336–37 (2e trim. 2002): 155–75.

Bowd, Stephen, and Sarah Cockram, eds. "The Animal in Renaissance Italy." Special issue, *Renaissance Studies* 31, no. 2 (April 2017): 175–318.

Bowler, Peter J. *Evolution: The History of an Idea*. Rev. ed. Berkeley: University of California Press, 2003.

Boyko, A. R., P. Quignon, L. Li, J. J. Schoenebeck, J. D. Degenhardt, K. E. Lohmueller et al. "A Simple Genetic Architecture Underlies Morphological Variation in Dogs." *PLoS Biology* 8, no. 8 (2010).

Brann, Ross. "The Moors?" *Medieval Encounters* 15 (2009): 307–18.

Braude, Benjamin. "The Sons of Noah and the Construction of Ethnic and Geographical Identities in the Medieval and Early Modern Periods," *William and Mary Quarterly* 54, no. 1 (January 1997): 103–42.

Braudel, Fernand. *The Mediterranean and the Mediterranean World in the Age of Philip II*. Translated by Siân Reynolds. 2nd rev. ed. 2 vols. Berkeley: University of California Press, 1995.

Brege, Brian. "The Empire That Wasn't: The Grand Duchy of Tuscany and Empire, 1574–1609." Ph.D. dissertation, Stanford University, 2014.

Brege, Brian. *Tuscany in the Age of Empire*. Cambridge, MA: Harvard University Press, 2021.

Brevaglieri, Sabina. *Natural desiderio di sapere: Roma barocca fra vecchi e nuovi mondi*. Rome: Viella, La corte dei Papi, 2019.

Brewer, Keagan. "Talking Wolves, Golden Fish, and Lion Sex: The Alterations to Gerald of Wales's *Topographia Hibernica* as Evidence of Audience Disbelief?" *Parergon* 37, no. 1 (2020): 27–53, doi:10.1353/pgn.2020.0057.

Bromley, Juan. "El Capitán Martín de Estete y Doña María de Escobar 'La Romana,' fundadores de la Villa de Trujillo del Perú." *Revista Histórica* (Lima) 22 (1956): 122–41.

Bruno, Maria C., and Christine Hastorf. "Gifts from the Camelid: Archaeobotanical Insights into Camelid Pastoralism through the Study of Dung." In Capriles and Tripcevich, *Archaeology of Andean Pastoralism*, 55–66.

Buquet, Thierry. "Hunting with Cheetahs at European Courts: From the Origins to the End of a Fashion." In *Animals and Courts: Europe, c. 1200–1800*, edited by Nadir Weber and Mark Hengerer, 17–42. Berlin: Water De Gruyter, 2020.

Burckhardt, Jacob. *The Civilization of the Renaissance in Italy*. Translated by S. G. C. Middlemore and revised and edited by Irene Gordon. New York: New American Library, 1960.

Burgaleta, Claudio M., S.J. *José de Acosta, S.J. (1540–1600): His Life and Thought*. Chicago: Jesuit Way, Loyola Press, 1999.

Burke, Peter. *The Fortunes of the Courtier: The European Reception of Castiglione's Cortegiano*. 1997. Repr., Malden, MA: Polity Press, 2007.

Burshatin, Israel. "The Moor in the Text: Metaphor, Emblem and Silence." In *"Race," Writing and Difference*, edited by Henry Louis Gates, 98–118. Chicago: University of Chicago Press, 1986.

Bustamente Garcia, Jesus. "Un Libro, Tres Modelos, y el Atlántico: Los Datos de una historia: los antecedentes y el proyecto." In Cadeddu and Guardo, *Il tesoro messicano.*

Butterfield-Rosen, Emmelyn. "Men Are Dogs: On Titian's *Poesie* for Philip II." *Art Forum* 6, no. 8 (April 2022): 172–87.

Cadeddu, Maria Eugenia, and Marco Guardo, eds. *Il tesoro messicano: Libri e saperi tra Europa e Nuovo mondo.* Florence: Leo S. Olschki Editore, 2013.

Calderón de la Barca, Pedro. *El astrologo fingido.* Madrid: Carlos Sanchez a costa de Antonio de Ribero, 1641.

Calderwood, Eric. *Spain and the Making of Modern Moroccan Culture.* Cambridge, MA: Harvard University Press, 2018.

Caló Caducci, Luigi Guarnieri. "La consideración de las fuentes indígenas en la Historia natural y moral de las Indias de José de Acosta." *Confluenze* 12, no. 2 (2020): 260–77.

Camden, William. *Remaines of a Greater Work, Concerning Britaine: The Inhabitants Thereof, Their Languages, Names, Surnames, Empreses, Wise Speeches, Poësies, and Epitaphs.* London: Printed by G. E. for Simon Waterson, 1605.

Campanella, Tommaso. *La città del sole: Dialogo Poetico/The City of the Sun: A Poetical Dialogue.* Translated by Daniel J. Donno. Berkeley: University of California Press, 1981.

Campbell, Stephen J. *The Cabinet of Eros: Renaissance Mythological Painting and the Studiolo of Isabella d'Este.* New Haven: Yale University Press, 2004.

Campori, G. "Giovanni Battista Della Porta e il cardinale Luigi d'Este." In *Atti e memorie della Dep. di storia patria per le provv. modenesi e parmensi,* 5–6 (1871): 165–190.

Cañizares-Esguerra, Jorge. "Demons, Stars, and the Imagination: The Early Modern Body in the Tropics." In Eliav-Feldon, Isaac, and Ziegler, *Origins of Racism in the West,* 313–25.

Cañizares-Esguerra, Jorge. "New Worlds, New Stars: Patriotic Astrology and the Invention of Indian and Creole Bodies in Colonial Spanish America 1600–1650." *American Historical Review* 104 (February 1999): 33–68.

Capanna, Ernesto. "South American Mammal Diversity and Hernandez's Novae Hispaniae Thesaurus." *Rendiconti Lincei* 20, no. 1 (2009): 39–60.

Capriles Flores, José M., and Nicholas Tripcevich, eds. *The Archaeology of Andean Pastoralism.* Albuquerque: University of New Mexico Press, 2016.

Caracciolo, Pasquale. *La Gloria del Cavallo.* Venice: Gabriel Giulito dei Ferrari, 1566.

Carletti, Francesco. *My Voyage Around the World: The Chronicles of a 16th Century Florentine Merchant.* Translated by Herbert Weinstock. New York: Pantheon Books, 1964.

Carletti, Francesco. *Viaggio intorno al mondo.* Firenze: Giulio Einaudi, 1958.

Casas, Alejandro, Adriana Otero-Arnaiz, Edgar Pérez-Negrón, and Alfonso Valiente-Banuet. 2007. "In Situ Management and Domestication of Plants in Mesoamerica." *Annals of Botany* 100 (5): 1101–15.

Cassidy, Rebecca. *Horse People: Thoroughbred Culture in Lexington and Newmarket.* Baltimore: Johns Hopkins University Press, 2007.

Cassidy, Rebecca. *The Sport of Kings: Kinship, Class, and Thoroughbred Breeding in Newmarket.* Cambridge: Cambridge University Press, 2009.

Castagna, R. "L'Alcanna d'Oriente e i Cavalli di Federico II Gonzaga, Ritratti da Giulio Romano a Palazzo Te." *Civiltà Mantovana,* 2nd series, 27 (1990): 109–10.

Castiglione, Baldassare. *The Book of the Courtier.* Translated by George Bull. New York: Penguin Books, 2003.

Castiglione, Baldassare. *The Book of the Courtier*. Translated by Leonard Eckstein Opdycke. New York: Charles Scribner's Sons, 1901.

Castiglione, Baldassare. *Il libro del Cortegiano*. A cura di Walter Berberis. Firenze: Einaudi, 2017.

Castiglione, Baldesare. *Il libro del cortegiano del conte Baldesare Castiglione*. Firenze: Per li heredi di Philippo di Giunta, 1528.

Catlos, Brian A. *Kingdoms of Faith: A New History of Islamic Spain*. New York: Basic Books, 2018.

Cavriani, Carlo. *Le razze gonzaghesche di cavalli nel Mantovano e la loro influenze sul puro sangue inglese*. Rome, 1909.

Chavez Balderas, Ximena. "Los animales y el recinto sagrado de Tenochtitlan." El Colegio Nacional, November 7, 2018. https://colnal.mx/noticias/sintesis-informativa-los-animales-y-el-recinto-sagrado-de-tenochtitlan-dia-1/.

Chiappini, Luciano. *Gli Estensi: Mille anni di storia*. Ferrara: Corbo Editore, 2001.

Chimalpahin Cuauhtlehuanitzin, Domingo Francisco de San Antón Muñón. *Annals of His Time: Don Domingo de San Antón Muñón Chimalpahin Quauhtlehuanitzin*. Translated by James Lockhart, Susan Schroeder, and Doris Namala. Stanford: Stanford University Press, 2006.

Cieza de León, Pedro de. *Crónica del Perú (1551)*. Lima: Instituto de estudios peruanos, 1967.

Clavigero, Francesco Saverio. *The History of Mexico: Collected from Spanish and Mexican Historians, from Manuscripts and Ancient Paintings of the Indians*. Translated by Charles Cullen. 2 vols. London: G. G. J. and J. Robinson, 1787.

Cline, Howard F. "The Oztoticpac Lands Map of Texcoco, 1540." *Quarterly Journal of the Library of Congress* (April 1966): 77–115.

Clossey, Luke. *Salvation and Globalization in the Early Jesuit Missions*. New York: Cambridge University Press, 2008.

Clubb, Louise George. *Giambattista della Porta, Dramatist*. Princeton: Princeton University Press, 1965.

Cockram, Sarah. "Interspecies Understanding: Exotic Animals and Their Handlers at the Italian Renaissance Court." In "The Animal in Renaissance Italy," edited by Stephen Bowd and Sarah Cockram. Special issue, *Renaissance Studies* 31, no. 2 (2017): 277–96.

Cockram, Sarah. *Isabella d'Este and Francesco Gonzaga: Power Sharing at the Italian Renaissance Court*. New York: Routledge, 2013.

Cockram, Sarah, and Andrew Wells. *Interspecies Interactions: Animals and Humans between the Middle Ages and Modernity*. New York: Routledge, 2017.

Coello de la Rosa, Alexandre. "Más allá del Incario: Imperialismo e historia en José de Acosta, SJ (1540–1600)." *Colonial Latin American Review* 14, no. 1 (2005): 55–81.

Coluccia, Rosario. "L'etimologia di razza: Questione aperta o chiusa?" *Studi di Filologia Italiana* 30 (1972): 325–30.

Columella. *De Re Rustica*. Lugduni: Apud Seb. Gryphium, 1537.

Columella, Lucius Junius Moderatus. *On Agriculture*. Vol. 2. Translated by E. S. Forster and Edward H. Heffner. Cambridge, MA: Harvard University Press, 1954.

Constable, Olivia Remie. *To Live Like a Moor: Christian Perceptions of Muslim Identity in Medieval and Early Modern Spain*. Philadelphia: University of Pennsylvania Press, 2018.

Contini, Gianfranco. "I più antichi esempi di razza." *Studi di Filologia Italiana* 17 (1959): 319–27.

Contini, Gianfranco. "Tombeau de Leo Spitzer." In *Varianti e altra linguistica: Una raccolta di saggi (1938–1968),* 651–60. Turin: Einaudi, 1970.

Cook, Karoline P. *Forbidden Passages: Muslims and Moriscos in Colonial Spanish America.* Philadelphia: University of Pennsylvania Press, 2016.

Cooley, Mackenzie. "The Giant Remains: Mesoamerican Medicine, Extinction, and Cycles of Empire." *Isis* 112, no. 1 (2021): 45–67.

Cooley, Mackenzie. "Southern Italy and the New World in the Age of Encounters." In *The Discovery of the New World in Early Modern Italy,* edited by Elizabeth Horodowich and Lia Markey, 169–89. Cambridge: Cambridge University Press, 2017.

Cooley, Mackenzie, Duygu Yıldırım, and Anna Toledano. *Natural Things: Ecologies of Knowledge in the Early Modern World.* Forthcoming.

Correia, Clara Pinto. *The Ovary of Eve: Egg and Sperm and Preformation.* Chicago: University of Chicago Press, 1997.

Cortés, Hernán. *Letters from Mexico.* Translated and edited by Anthony Pagden. New Haven: Yale University Press, 2001.

Cottafavi, C. "Cani e gatti alla Corte dei Gonzaga." *Il Ceppo, Quaderno di vita fascista e di cultura edito dal comitato di Mantova dell'opera nazionale balilla* (1934): 4.

Cowie, Helen. *Llama.* London: Reaktion Books, 2017.

Crosby, Alfred. *The Columbian Exchange: Biological and Cultural Consequences of 1492.* Westport, CT: Greenwood Press, 2003.

Crosby, Alfred. *Ecological Imperialism: The Biological Expansion of Europe, 900–1900.* Cambridge: Cambridge University Press, 1986.

Cuevas, Mariano. *The Codex Saville: America's Oldest Book.* Historical Records and Studies 19. New York: United States Catholic Historical Society, 1929.

Dandelet, Thomas James. *Spanish Rome, 1500–1700.* New Haven: Yale University Press, 2001.

Daston, Lorraine, and Katharine Park. *Wonders and the Order of Nature, 1150–1750.* New York: Zone Books, 1998.

Dauge-Roth, Katherine. *Signing the Body: Marks on Skin in Early Modern France.* New York: Routledge, 2020.

Davies, Surekha. *Renaissance Ethnography and the Invention of the Human: New Worlds, Maps and Monsters.* Cambridge: Cambridge University Press, 2016.

de Bunes Ibarra, Miguel Ángel. *La imagen de los musulmanes y del norte de África en la España de los siglos XVI y XVII: Los caracteres de una hostilidad.* Madrid: Consejo Superior de Investigaciones Científicas, 1989.

de Cuneo, Michele. "Michele de Cuneo's Letter on the Second Voyage, 28 October 1495." In *Journals and Other Documents on the Life and Voyages of Christopher Columbus,* edited and translated by Samuel Eliot Morison, 209–18. Norwalk, CT: Easton Press, 1993.

deFrance, Susan D. "Paleopathology and Health of Native and Introduced Animals on Southern Peruvian and Bolivian Spanish Colonial Sites." *International Journal of Osteoarchaeology* 20, no. 5 (2010): 508–24.

de la Cruz Cruz, Victoriano. "Chicomexochitl y El Maíz Entre Los Nahuas De Chicontepec: La Continuidad Del Ritual." *Politeja* 12, no. 38 (2015): 129–47.

de Lucca, Jean-Paul. "Prophetic Representation and Political Allegorisation: The Hospitaller in Campanella's *The City of the Sun.*" *Bruniana & Campanelliana,* 15, no. 2 (2009): 387–405.

Diamond, Jared. *Guns, Germs, and Steel: The Fates of Human Societies.* New York: W. W. Norton, 1997.

Díaz del Castillo, Bernal. *The True History of the Conquest of New Spain.* Translated by Janet Burke and Ted Humphrey. Indianapolis: Hackett Publishing Company, 2012.

Di Stefano, Elisabetta. Il "*De equo animante* di LB Alberti: Una teoría della belezza?" *Albertiana* 1 (2010): 15–26.

Dolan, Frances. *Digging the Past: How and Why to Imagine Seventeenth-Century Agriculture.* Philadelphia: University of Pennsylvania Press, 2020.

Doron, Claude-Olivier. *L'Homme Altéré: Races et dégénérescence (XVIIᵉ–XIXᵉ Siècles).* Paris: Editions Champ Vallon, 2016.

Dransart, Penny. *Earth, Water, Fleece and Fabric: An Ethnography and Archaeology of Andean Camelid Herding.* London: Routledge, 2002.

Du Cange, Charles du Fresne, sieur. *Le Glossarium mediae et infimae latinitatis.* Niort: L. Favre, 1883–87.

Dufendach, Rebecca. "'As If His Heart Died': A Reinterpretation of Moteuczoma's Cowardice in the Conquest History of the Florentine Codex," *Ethnohistory* 66, no. 4 (October 2019): 624–45.

Dugatkin, Lee Alan, and Lyudmila Trut. *How To Tame a Fox (and Build a Dog): Visionary Scientists and a Siberian Tale of Jump-Started Evolution.* Chicago: University of Chicago Press, 2017.

Duncan, Sarah. *Privileged Horses: The Italian Renaissance Court Stable.* Wimbledon: Stephen Morris, 2020.

Duncan, Sarah. "Stable Design and Horse Management at the Italian Renaissance Court." In *Animals at Court, Europe, c. 1200–1800,* edited by Mark Hengerer and Nadir Weber, 129–52. Berlin: De Gruyter, 2019.

Durán, Diego. *The History of the Indies of New Spain.* Translated and edited by Doris Heyden. Norman: University of Oklahoma Press, 1994.

Durand-Forest, Jacqueline. *Aperçu de l'histoire naturelle de la Nouvelle-Espagne d'après Hernández, les informateurs indigènes de Sahagun et les auteurs du Codex Badianus, Nouveau monde et renouveau de l'histoire naturelle.* Paris: Publicaciones de la Sorbonne, 1986.

Eamon, William. "The Difference That Made Spain, the Difference That Spain Made." In *Medical Cultures of the Early Modern Spanish Empire,* edited by John Slater, María Luz López-Terrada, and José Pardo-Tomás, 231–43. Burlington, VT: Ashgate, 2014.

Eamon, William. "Natural Magic and Utopia in the Cinquecento: Campanella, the della Porta Circle, and the Revolt of Calabria." *Memorie Domenicane* 14 (1995): 369–402.

Eamon, William. *Science and the Secrets of Nature: Books of Secrets in Medieval and Early Modern Culture.* Princeton: Princeton University Press, 1994.

Eamon, William, and Françoise Paheau. "The Accademia Segreta of Girolamo Ruscelli: A Sixteenth-Century Italian Scientific Society." *Isis* 75 (1984): 327–42.

Earle, Rebecca. *The Body of the Conquistador: Food, Race and the Colonial Experience in Spanish America, 1492–1700.* Cambridge: Cambridge University Press, 2012.

Earle, Rebecca. "The Columbian Exchange." In *The Oxford Handbook of Food History,* edited by Jeffrey M. Pilcher, 341–57. Oxford: Oxford University Press, 2011.

Earle, Rebecca. "The Pleasures of Taxonomy: Casta Paintings, Classification, and Colonialism." *William and Mary Quarterly* 73, no. 3 (July 2016): 427–66.

Edwards, Peter, Karl A. E. Enenkel, and Elspeth Graham, eds. *The Horse as Cultural*

Icon: The Real and Symbolic Horse in the Early Modern World. Intersections: Interdisciplinary Studies in Early Modern Culture, vol. 18. Leiden: Brill, 2011.

Egmond, Florike. *Conrad Gessners "Thierbuch": Die Originalzeichnungen*. Darmstadt: WBG Edition, 2018.

Egmond, Florike. *The World of Carolus Clusius: Natural History in the Making, 1550–1610*. London: Pickering & Chatto, 2010.

Eiche, Sabine. *Presenting the Turkey: The Fabulous Story of a Flamboyant and Flavourful Bird*. Florence: Centro Di, 2004.

Eisenstein, Elizabeth. *The Printing Revolution in Early Modern Europe*. New York: Cambridge University Press, 1983.

Eliav-Feldon, Miriam, Benjamin Isaac, and Joseph Ziegler, eds. *The Origins of Racism in the West*. Cambridge: Cambridge University Press, 2009.

Elliot, J. H. *Imperial Spain, 1469–1716*. London: Penguin, 1970.

Ellis, Helen. "Maize, Quetzalcoatl, and Grass Imagery: Science in the Central Mexican *Codex Borgia*." Ph.D. dissertation, University of California, Los Angeles, 2015.

Equicola, Mario. *Chronica di Mantua*. Mantova, 1521.

Ernst, Germana. *Il carcere il politico il profeta: Saggi su Tommaso Campanella*. Pisa: Istituti Editoriali e Poligrafici Internazionali, 2002.

Ernst, Germana. *Religione, ragione e natura: Ricerche su Tommaso Campanella e il tardo Rinascimento*. Milan: Franco Angeli, 1991.

Este, Isabella d'. *Selected Letters*. Edited and translated by Deanna Shemek. The Other Voice in Early Modern Europe: The Toronto Series 54; Medieval and Renaissance Texts and Studies 516. Toronto: Iter Press; Tempe: Arizona Center for Medieval and Renaissance Studies, 2017.

Fabrici, Girolamo. *De brutorum loquela*. Padua: Lorenzo Pasquati, 1603.

Fan, R., Z. Gu, X. Guang et al. "Genomic Analysis of the Domestication and Post-Spanish Conquest Evolution of the Llama and Alpaca." *Genome Biology* 21, no. 159 (2020).

Fancy, Nahyan. "Generation in Medieval Islamic Medicine." In Hopwood, Flemming, and Kassell, *Reproduction*, 129–40.

Fanselow, Frank S. "The Bazaar Economy or How Bizarre Is the Bazaar Really?" *Man*, new series, 25, no. 2 (1990): 250–65.

Fauconnier, Gilles, and Mark Turner. *The Way We Think: Conceptual Blending and the Mind's Hidden Complexities*. New York: Basic Books, 2002.

Fenlon, Iain, and James Haar. *The Italian Madrigal in the Early Sixteenth Century: Sources and Interpretation*. Cambridge: Cambridge University Press, 1988.

Fernández, Diego. *Primera, y segunda parte, de la historia del Peru*. Seuilla: Casa de Hernando Diaz, 1571.

Fernández-Armesto, Felipe. *The Canary Islands after the Conquest: The Making of a Colonial Society in the Early Sixteenth Century*. Oxford: Clarendon Press, 1982.

Feros, Antonio. *Speaking of Spain: The Evolution of Race and Nation in the Hispanic World*. Cambridge, MA: Harvard University Press, 2017.

Ferraro, Giovanni Battista. *Delle razze, disciplina del cavalcare, et altre cose pertinenti ad essercito così fatto*. Napoli: Mattio Cancer, 1560.

Fiaschi, Cesare. *Trattato dell'imbrigliare, maneggiare, et ferrare cavalli: Diviso in tre parti, con alcuni discorsi sopra la natura di cavalli, con disegni di briglie, maneggi, & di cavalieri a cavallo, & de ferri d'esso*. Bologna: Per Anselmo Giaccarelli, 1556.

Figueroa, Luis Millones. "Bernabé Cobo's Inquiries in the New World and Native Knowledge." In Marroquín Arredondo and Bauer, *Translating Nature*, 70–93.

Findlen, Paula. "Anatomy Theaters, Botanical Gardens, and Natural History Collections." In *Early Modern Science*, edited by Katharine Park and Lorraine Daston, 272–88. New York: Cambridge University Press, 2006.

Findlen, Paula. "How Information Travels: Jesuit Networks, Scientific Knowledge, and the Early Modern Republic of Letters, 1540–1640." In *Empires of Knowledge: Scientific Networks in the Early Modern World*, edited by Paula Findlen, 57–105. London: Routledge, 2018.

Findlen, Paula. "Humanism, Politics, and Pornography in Renaissance Italy." In *The Invention of Pornography: Obscenity and the Origins of Modernity*, edited by Lynn Hunt. Cambridge, MA: MIT Press, 1996.

Findlen, Paula. *Possessing Nature: Museums, Collecting, and Scientific Culture in Early Modern Italy*. Berkeley: University of California Press, 1994.

Finucci, Valeria. *The Prince's Body: Vincenzo Gonzaga and Renaissance Medicine*. Cambridge, MA: Harvard University Press, 2015.

Fitzgerald, Josh. "Cross-Pollinated Coyote Wisdom: Persistent Memories of Nahua-Animal Reciprocity and Natural Science." Forthcoming.

Flemming, Rebecca. "Galen's Generations of Seeds." In Hopwood, Flemming, and Kassell, *Reproduction*, 95–108.

Flemming, Rebecca. "States and Populations in the Classical World." In Hopwood, Flemming, and Kassell, *Reproduction*, 67–80.

Flint, Richard, and Shirley Cushing Flint. *A Most Splendid Company: The Coronado Expedition in Global Perspective*. Albuquerque: University of New Mexico Press, 2019.

Flores Ochoa, Jorge A. *Pastores de puna, or Uywamichiq punarunakuna*. Lima: Instituto de estudios peruanos, 1977.

Foucault, Michel. *Les mots et les choses: une archéologie des sciences humaines*. Paris: Gallimard, 1966.

Foucault, Michel. *The Order of Things: An Archeology of the Human Sciences*. New York: Vintage Books, 1994.

Fragnito, Gigliola. *In museo e in villa: Saggi sul Rinascimento perduto*. Venice: Arsenale, 1988.

Franchini, Dario A. *La scienza a corte: Collezionismo eclettico, natura e immagine a Mantova fra Rinascimento e Manierismo*. Roma: Bulzoni, 1979.

Freedberg, David. *The Eye of the Lynx: Galileo, His Friends, and the Beginnings of Modern Natural History*. Chicago: University of Chicago Press, 2002.

Fuchs, Barbara. *Exotic Nation: Maurophilia and the Construction of Early Modern Spain*. Philadelphia: University of Pennsylvania Press, 2008.

Fuchs, Barbara. "The Spanish Race." In *Rereading the Black Legend: The Discourses of Religious and Racial Difference in the Renaissance Empires*, edited by Margaret Greer, Walter Mignolo, and Maureen Quilligan, 88–98. Chicago: University of Chicago Press, 2007.

Fudge, Erica. *Brutal Reasoning: Animals, Rationality, and Humanity in Early Modern England*. Ithaca: Cornell University Press, 2006.

Furlotti, Barbara. *A Renaissance Baron and His Possessions: Paolo Giordano I Orsini, Duke of Bracciano (1541–1585)*. Turnhout, Belgium: Brepols Publishers, 2012.

Gabrieli, Giuseppe. "Il carteggio Linceo della vecchia accademia di Federico Cesi

(1603–1630)." *Memorie della R. Accademia Nazionale dei Lincei, Classe di scienze morali, storiche e filologiche,* ser. 6, vol. 7, fols. 1–4 (1938–42).

Gabrieli, Giuseppe. *Contributi alla storia della Accademia dei Lincei,* tomo 1. Roma: Accademia Nazionale dei Lincei, 1989.

Gade, Daniel W. "Llamas and Alpacas as "Sheep" in the Colonial Andes: Zoogeography Meets Eurocentrism." *Journal of Latin American Geography* 12, no. 2 (2013): 221–43.

Gage, Frances. "Human and Animal in the Renaissance Eye." In "The Animal in Renaissance Italy," edited by Stephen Bowd and Sarah Cockram. Special issue, *Renaissance Studies* 31, no. 2 (2017): 261–72.

Galasso, Giuseppe. "Aspetti dei Rapporti tra Italia e Spagni nei Secoli XVI e XVII." In Cadeddu and Guardo, *Il tesoro messicano.*

Galasso, Giuseppe. *Il Mezzogiorno nella storia d'Italia.* Firenze: F. Le Monnier, 1977.

Galluzzi, Paolo. *The Lynx and the Telescope.* Leiden: Brill, 2017.

Galton, David J. "Greek Theories on Eugenics." *Journal of Medical Ethics* 24, no. 4 (August 1998): 263–67.

Galton, Francis. *Inquiries into Human Faculty and Its Development.* London: J. M. Dent and Sons, 1943.

Gamboa Cabezas, Luis M., and Robert H. Cobean. "Comments on Cultural Continuities between Tula and the Mexica." In Nichols and Rodríguez-Alegría, *Oxford Handbook of the Aztecs,* 53–71.

Garces, Chris. "The Interspecies Logic of Race in Colonial Peru: San Martín de Porres's Animal Brotherhood." In Bassett and Lloyd, *Sainthood and Race,* 82–101.

Geary, Patrick. *The Myth of Nations: The Medieval Origins of Europe.* Princeton: Princeton University Press, 2002.

Gentilcore, David. *Pomodoro: A History of the Tomato in Italy.* New York: Columbia University Press, 2010.

George, Wilma, and William B. Yapp. *The Naming of the Beasts: Natural History in the Medieval Bestiary.* London: Duckworth, 1991.

Gerbi, Antonello. *Nature in the New World: From Christopher Columbus to Gonzalo Fernández de Oviedo.* Pittsburgh: University of Pittsburgh Press, 2010.

Gernert, Folke. *Adivinos, medicos y profesores de secretos en la España áurea.* Toulouse: Méridiennes, 2017.

Gernert, Folke. *Fictionalizing Heterodoxy: Various Uses of Knowledge in the Spanish World from the Archpriest of Hita to Mateo Alemán.* Berlin: Walter de Gruyter, 2019.

Gernert, Folke. "La legitimidad de las ciencias parcialmente ocultas: Fisonomía y quiromancia ante la Inquisición." In *Saberes Humanísticos,* edited by Christoph Strosetski, 105–28. Veuvert: Universidad de Navarra, Iberoamericana, 2014.

Gesner, Conrad. *Historia animalium,* liber primus, *De Quadrupedibus viuiparis.* Francofurti: In Bibliopolio Cambieriano, 1602.

Gesner, Conrad. *Historia animalium,* liber 2, *De quadrupedibus oviparis.* 1554.

Ghadessi, Touba. *Portraits of Human Monsters in the Renaissance: Dwarves, Hirsutes, and Castrati as Idealized Anatomical Anomalies.* Kalamazoo: Medieval Institute Publications, Western Michigan University, 2018.

Gibson, Charles. *The Aztecs under Spanish Rule: A History of the Indians of the Valley of Mexico.* Stanford: Stanford University Press, 1964.

Gill, Meredith. *Augustine in the Italian Renaissance: Art and Philosophy from Petrarch to Michelangelo.* Cambridge: Cambridge University Press, 2005.

Giribet, G., G. Hormiga, and G. D. Edgecombe. "The Meaning of Categorical Ranks in Evolutionary Biology." *Organisms Diversity & Evolution* 16: (2016): 427–30.

Goldthwaite, Richard. *The Economy of Renaissance Florence*. Baltimore: Johns Hopkins University Press, 2009.

Gómez-Bravo, Ana M. "The Origins of *Raza*: Racializing Difference in Early Spanish." *Interfaces: A Journal of Medieval European Literatures*, 7 (2020): 64–114.

Gómez-Centurión Jiménez, Carlos. *Alhajas para soberanos: Los animales reales en el siglo XVIII: De las leoneras a las mascotas de cámara*. Valladolid: Junta de Castilla-León, 2011.

Gómez-Centurión Jiménez, Carlos. "Exóticos pero útiles: Los camellos reales de Aranjuez durante el siglo XVIII." *Cuadernos dieciochistas* 9 (2008): 155–80.

González, Roberto J. *Zapotec Science: Farming and Food in the Northern Sierra of Oaxaca*. Austin: University of Texas Press, 2001.

Goodall, Jane. *Through a Window: My Thirty Years with the Chimpanzees of Gombe*. 1990. Boston: Mariner Books, 2010.

Gordin, Michael D., Helen Tilley, and Gyan Prakash. *Utopia/Dystopia: Conditions of Historical Possibility*. Princeton: Princeton University Press, 2011.

Gordon, James. "How Old Is the Maltese, Really?" *New York Times*, October 4, 2021. https://www.nytimes.com/2021/10/04/science/dogs-DNA-breeds-maltese.html.

Gouwens, Kenneth. "Erasmus, 'Apes of Cicero,' and Conceptual Blending." *Journal of the History of Ideas* 71, no. 4 (2010): 532–45.

Grafton, Anthony. *Cardano's Cosmos: The Worlds and Works of a Renaissance Astrologer*. Cambridge, MA: Harvard University Press, 1999.

Grafton, Anthony. "José de Acosta: Renaissance Historiography and New World Humanity." In *The Renaissance World*, edited by John Jeffries Martin, 166–88. New York: Routledge, 2007.

Grafton, Anthony. *Leon Battista Alberti: Master Builder of the Italian Renaissance*. Cambridge, MA: Harvard University Press, 2000.

Grafton, Anthony. *New Worlds, Ancient Texts*. Cambridge, MA: Belknap Press of Harvard University Press, 1992.

Graubert, Karen. *With Our Labor and Sweat: Indigenous Women and the Formation of Colonial Society in Peru, 1550–1700*. Stanford: Stanford University Press, 2007.

Green, Caitlin R. "Camels in Early Medieval Western Europe: Beasts of Burden and Tools of Ritual Humiliation." May 28, 2016. Dr. Caitlin R. Green, http://www.caitlingreen.org/2016/05/camels-in-early-medieval-western-europe.html.

Greenblatt, Stephen. *Renaissance Self-Fashioning: From More to Shakespeare*. Chicago: University of Chicago Press, 2005.

Greer, Margaret Rich, Walter Mignolo, and Maureen Quilligan, eds. *Rereading the Black Legend: The Discourses of Religious and Racial Difference in the Renaissance Empires*. Chicago: University of Chicago Press, 2008.

Grisone, Federico. *Gli Ordini di cavalcare*. Venetia: Appresso Vincenzo Valgrisi nella bottega d'Erasmo, 1551.

Grisone, Federico. *Federico Grisone's "The Rules of Riding": An Edited Translation of the First Renaissance Treatise on Classical Horsemanship*. Edited by Elizabeth M. Tobey. Tempe: Arizona Center for Medieval and Renaissance Studies, 2014.

Guardo, Marco. "Nell'Officina del Tesoro Messicano: Il Ruolo Misconosciuto di Marco Antonio Petilio nel sodalizio Linceo." In Cadeddu and Guardo, *Il tesoro messicano*.

Guest, Kristen, and Monica Mattfield, eds. *Horse Breeds and Human Society: Purity, Identity and the Making of the Modern Horse*. New York: Routledge, 2020.

Guibovich Pérez, Pedro. "Los libros del inquisidor." *Cuadernos para la historia de la evangelización en América Latina* 4 (1989): 47–64.

Haag, Sabine. *Echt tierisch!: die Menagerie des Fürsten: Eine Ausstellung des Kunsthistorischen Museums Wien, Schloss Ambras Innsbruck* (An exhibition of the Kunsthistorisches Museum Vienna, Ambras Castle Innsbruck), June 18–October 4, 2015. Kunsthistorisches Museum Wien, issuing body; Schlosssammlung Ambras, host institution. Wien: KHM-Museumsverband, 2015.

Haller, Hermann W. *John Florio: A Worlde of Wordes*. Da Ponte Library Series. Toronto: University of Toronto Press, 2013.

Hamann, Byron Ellsworth. *The Translations of Nebrija: Language, Culture, and Circulation in the Early Modern World*. Boston: University of Massachusetts Press, 2015.

Hankins, James. *Plato in the Italian Renaissance*. 2 vols. Leiden: Brill, 1990.

Hankins, James. *Virtue Politics: Soulcraft and Statecraft in Renaissance Italy*. Cambridge, MA: Belknap Press of Harvard University Press, 2019.

Hannaford, Ivan. *Race: The History of an Idea in the West*. Washington, DC: Woodrow Wilson Center Press, 1996.

Hanning, Robert W., and David Rosand. *Castiglione: The Ideal and the Real in Renaissance Culture*. New Haven: Yale University Press, 1983.

Harvey, L. P. *Muslims in Spain, 1500–1614*. Chicago: University of Chicago Press, 2004.

Haugen, Kristine Louise. "Campanella and the Disciplines from Obscurity to Concealment." In *For the Sake of Learning: Essays in Honor of Anthony Grafton*, edited by Ann Blair and Anja-Silvia Goeing, vol. 2, 602–20. Leiden: Brill, 2016.

Heikamp, Detlef. *Mexico and the Medici*. With contributions by Ferdinand Anders. Florence: Edam, 1972.

Hendricks, J. R., E. E. Saupe, C. E. Myers, E. J. Hermsen, and W. D. Allmon. 2014. "The Generification of the Fossil Record." *Paleobiology* 40 (2014): 511–28.

Heng, Geraldine. *The Invention of Race in the European Middle Ages*. Cambridge: Cambridge University Press, 2018.

Hernández, Francisco. *Francisci Hernandi, medici atque historici Philippi II, hispan et indiar: Regis, et totius novi orbis archiatri: Opera, cum edita, tum medita, ad autobiographi fidem et jusu regio*. Edited by Casimiro Gómez Ortega. Madrid: Ex Typographia Ibarrae Heredum, 1790.

Hernández, Francisco. *Obras completas de Francisco Hernández*. Mexico: Universidad Nacional Autónoma de México, 1960.

Hernández, Francisco, Rafael Chabrán, and Simon Varey. "'An Epistle to Arias Montano': An English Translation of a Poem by Francisco Hernández." *Huntington Library Quarterly* 55, no. 4 (Autumn 1992): 620–34.

Herrera, Antonio de. *Historia general de los hechos de los castellanos en las Islas y Tierra Firme del mar Océano que llaman Indias Occidentales*. Tomo 10. Buenos Aires, Argentina: Editorial Guarania, 1947.

Herring, Adam. *Art and Vision in the Inca Empire: Andeans and Europeans at Cajamarca*. New York: Cambridge University Press, 2015.

Herzog, Tamar. *Defining Nations: Immigrants and Citizens in Early Modern Spain and Spanish America*. New Haven: Yale University Press, 2008.

Herzog, Tamar. *Frontiers of Possession: Spain and Portugal in Europe and the Americas*. Cambridge, MA: Harvard University Press, 2015.

Hess, Andrew. *The Forgotten Frontier: A History of the Sixteenth-Century Ibero-African Frontier.* Chicago: University of Chicago Press, 1978.

Hill, Ruth. "Between Black and White: A Critical Race Theory Approach to Caste Poetry in the Spanish New World." *Comparative Literature* 19, no. 4 (2007): 269–93.

Hillgarth, J. N. *The Mirror of Spain, 1500–1700: The Formation of a Myth.* Ann Arbor: University of Michigan Press, 2000.

Historia de la Compañia de Jesús en la Asistencia de España. 7 vols. Madrid: Administración de Razón y Fé, 1912–25.

Hopwood, Nick. "The Keywords 'Generation' and 'Reproduction.'" In Hopwood, Flemming, and Kassell, *Reproduction*, 287–304.

Hopwood, Nick, Rebecca Flemming, and Lauren Kassell, eds. *Reproduction: Antiquity to the Present Day.* Cambridge: Cambridge University Press, 2018.

Hughes, Diane Owen. "Distinguishing Signs: Ear-Rings, Jews and Franciscan Rhetoric in the Italian Renaissance City." *Past & Present* 112, no. 1 (August 1986): 3–59.

Humboldt, Alexander de. *Essai Politique sur le Royaume de la Nouvelle-Espagne.* Paris: Chez F. Schoell, 1811.

Humboldt, Alexander von. *Personal Narrative of Travels to the Equinoctial Regions of America during the Years, 1799–1804.* London: Longman, 1814.

Imamuddin, Syed M. *Muslim Spain: 0711–1492 A.D.* Leiden, E. J. Brill, 1981.

Instituti Societatis Iesu, Volumen Secundum, *Examen Et Constitutiones Decreta Congregationum Generalium.* Florentiae: Ex Typographia A. SS. Conceptione, 1893.

Irigoyen-García, Javier. *Moors Dressed as Moors: Clothing, Social Distinction and Ethnicity in Early Modern Iberia.* Toronto: University of Toronto Press, 2017.

Irigoyen-García, Javier. *The Spanish Arcadia: Sheep Herding, Pastoral Discourse, and Ethnicity in Early Modern Spain.* Toronto: University of Toronto Press, 2014.

Jardine, Lisa. *Worldly Goods: A New History of the Renaissance.* London: Macmillan, 1996.

Jaser, Christian. *Palio und Scharlach. Städtische Sportkulturen des 15. und frühen 16. Jahrhunderts am Beispiel italienischer und oberdeutscher Pferderennen.* Monographien zur Geschichte des Mittelalters. Stuttgart: Hiersemann, 2021.

Jenkins, Stephanie, Kelly Struthers Montford, and Chloë Taylor, eds. *Disability and Animality: Crip Perspectives in Critical Animal Studies.* New York: Routledge, 2021.

Jerez, Francisco de. *Conquista del Perú: Verdadera Relación de la conquista del Perú.* Salamanca: por Juan de Junta, 1547.

Jerez, Francisco de. "Report of Francisco de Xerez, Secretary to Francisco Pizarro." In *Reports of the Discovery of Peru*, edited and translated by Clements R. Markham. London: Hakluyt Society, 1872.

Jouanna, Arlette. *L'idée de race en France au XVIème siècle et au début du XVIIème siècle (1498–1614).* Lille: Université Lille III, 1976.

Kadwell, Miranda, M. Fernandez, H. F. Stanley et al. "Genetic Analysis Reveals the Wild Ancestors of the Llama and the Alpaca." *Proceedings of the Royal Society B: Biological Sciences* 268, no. 1485 (December 22, 2001): 2575–84.

Kaplan, Paul H. D. "Isabella d'Este and Black African Women." In *Black Africans in Renaissance Europe*, edited by T. F. Earle and K. J. P. Lowe, 125–54. Cambridge: Cambridge University Press, 2005.

Katzew, Ilona. *Casta Paintings; Images of Race in Eighteenth-Century Mexico.* New Haven: Yale University Press, 2004.

Katzew, Ilona. *Inventing Race: Casta Painting and Eighteenth-Century Mexico.* Exhibition brochure for eponymous LACMA exhibition. Los Angeles: Los Angeles County Museum of Art, 2004.

Keefer, Katrina H. B. "Marked by Fire: Brands, Slavery, and Identity." *Slavery and Abolition* 40, no. 4 (2019): 659–81.

Kemp, Martin. *Leonardo da Vinci: The Marvellous Works of Nature and Man*. Oxford: Oxford University Press, 2006.

Kim, Claire Jean. *Dangerous Crossings: Race, Species, and Nature in a Multicultural Age*. Cambridge: Cambridge University Press, 2015.

Kissel, Marc, and Nam King. "The Emergence of Human Warfare: Current Perspectives." *Yearbook of Physical Anthropology* 168 (2019): 141–63.

Kline, Wendy. *Building a Better Race: Gender, Sexuality, and Eugenics from the Turn of the Century to the Baby Boom*. Berkeley: University of California Press, 2005.

Kodera, Sergius. "Humans as Animals in Giovan Battista della Porta's *Scienza*." *Zeitsprünge* 17 (2013): 414–32.

Kodera, Sergius. "The Laboratory as Stage: Giovan Battista della Porta's Experiments." *Journal of Early Modern Studies* 3 (2014): 15–38.

Koering, Jérémie. *Le Prince en représentation: Histoire des décors du palais ducal de Mantoue au XVIe siècle*. Paris: Actes Sud, 2013.

Kraemer, Fabian, and Helmut Zedelmaier. "Instruments of Invention in Renaissance Europe: The Cases of Conrad Gesner and Ulisse Aldrovandi." *Intellectual History Review* 24, no. 3 (2014): 321–41.

Kreager, Philip. "The Emergence of Population." In Hopwood, Flemming, and Kassell, *Reproduction*, 253–66.

Krech, Shephard III. "On the Turkey in Rua Nova Dos Mercadores." In *The Global City: On The Streets of Renaissance Lisbon*, edited by Annemarie Jordan Gschwend and K. J. P. Lowe. London: Paul Holberton, 2015.

Kusukawa, Sachiko. *Picturing the Book of Nature: Image, Text, and Argument in Sixteenth-Century Human Anatomy and Medical Botany*. Chicago: University of Chicago Press, 2011.

Kusukawa, Sashiko. "The Sources of Gessner's Pictures for the *Historia animalium*." *Annals of Science: The History of Science and Technology* 67 (2010): 303–28.

LaCapra, Dominick. *History and Its Limits: Human, Animal, Violence*. Ithaca and London: Cornell University Press, 2009.

Lakoff, George, and Mark Johnson. *Metaphors We Live By*. Chicago: University of Chicago Press, 1980.

Lammons, Frank Bishop. "Operation Camel: An Experiment in Animal Transportation in Texas, 1857–1860." *Southwestern Historical Quarterly* 61 (July 1957): 20–50.

Landers, Jane G. "Cimarrón and Citizen: African Ethnicity, Corporate Identity, and the Evolution of Free Black Towns in the Spanish Circum-Caribbean." In *Slaves, Subjects, and Subversives: Blacks in Colonial Latin America*, edited by Jane Lander and Barry Robinson, 111–45. Albuquerque: University of New Mexico Press, 2006.

Landry, Donna. "Habsburg Lipizzaners, English Thoroughbreds and the Paradoxes of Purity," In Guest and Mattfield, *Horse Breeds and Human Society*, 27–49.

Landry, Donna. *Noble Brutes: How Eastern Horses Transformed English Culture*. Baltimore: Johns Hopkins University Press, 2008.

Lea, Virginia, and Judy Hefland, eds. *Identifying Race and Transforming Whiteness in the Classroom*. New York: Peter Lang, 2006.

Leathlobhair, Máire ní, Angela R. Perri, Evan K. Irving-Pease et al. "The Evolutionary History of Dogs in the Americas." *Science* 361, no. 6397 (July 2018): 81–85.

Leclerc, Georges Louis, Comte de Buffon. *Natural History: General and Particular, by the Count de Buffon, translated into English.* Edinburgh: Printed for William Creech, 1780–85.

Leibsohn, Dana. *Script and Glyph: Pre-Hispanic History, Colonial Bookmaking and the Historica Tolteca-Chichimeca.* Washington, DC: Dumbarton Oaks Research Library and Collection, 2009.

Lennox, James G. "The Comparative Study of Animal Development: William Harvey's Aristotelianism." In Justin Smith, *Problem of Animal Generation,* edited by Justin Smith, 21–46.

Leu, Urs, and Peter Opitz. *Conrad Gessner (1516–1565): Die Renaissance der Wissenschaften.* Oldenbourg: De Gruyter, 2019.

Libro de marchi de cavalli. Venetia: Appresso Bernardo Giunti, 1588.

Libro de marchi de cavalli con li nomi de tutti li principi et privati signori che hanno razza di cavalli. Venetia: Nicolò Nelli, 1569.

Liedtke, Walter A. *The Royal Horse and Rider: Painting, Sculpture and Horsemanship 1500–1800.* New York: Abaris Books in association with the Metropolitan Museum of Art, 1989.

Lizárraga, Reginaldo de. *Descripción breve de toda la tierra del Perú, Tucumán, Río de la Plata y Chile.* Tomo 216. Madrid: Ediciones Atlas, 1968.

Loch, Sylvia. *Dressage: The Art of Classical Riding.* London: Trafalgar Square Publishing, 1990.

Loch, Sylvia. *The Royal Horse of Europe: The Story of the Andalusian and Lusitano.* London: J. A. Allen, 1986.

Lockhart, James. *The Nahuas after the Conquest: A Social and Cultural History of the Indians of Central Mexico, Sixteenth Through Eighteenth Centuries.* Stanford: Stanford University Press, 1992.

Long, Pamela. *Artisan/Practitioners and the Rise of the New Sciences, 1400–1600.* Corvallis: Oregon State University Press, 2011.

López Austin, Alfredo. *Cuerpo Humano e Ideología: Las Concepciones de los Antiguos Nahuas.* 2 vols. 1980; repr. Ciudad Universitaria, Mexico: Universidad Nacional Autónoma de México, 2004.

López Piñero, José Maria, and José Pardo-Tomás. *La influencia de Francisco Hernández (1515–1587) en la constitución de la botánica y la materia médica modernas.* Valencia: Instituto de Estudios Documentales e Históricos sobre la Ciencia, 1996.

López Piñero, José Maria, and José Pardo-Tomás. *Nuevos materiales y noticias sobre la "Historia de las plantas de Nueva España," de Francisco Hernández.* Valencia: Instituto de Estudios Documentales e Históricos sobre la Ciencia, 1994.

Lovejoy, Arthur Oncken. *The Great Chain of Being: A Study of the History of an Idea.* Cambridge, MA: Harvard University Press, 1936.

Low, John. *The New and Complete American Encyclopedia.* New York, 1806.

Lozada, María Cecilia, Jane E. Buikstra, Gordon Rakita, and Jane C. Wheeler. "Camelid Herders: The Forgotten Specialists in the Coastal Señorío of Chiribaya, Southern Peru." In *Andean Civilization: A Tribute to Michael E. Moseley,* edited by Joyce Marcus and Patrick Ryan Williams, 351–64. Los Angeles: UCLA Cotsen Institute of Archaeology, 2009.

Lüthy, C. H. "Atomism, Lynceus, and the Fate of Seventeenth-Century Microscopy." *Early Science and Medicine* 1, no. 1 (February 1996): 1–27.

Luzio, Alessandro, and Rodolfo Renier. *Buffoni, nani e schiavi dei Gonzaga ai tempi d'Isabella d'Este*. Roma: Tipografia della camera dei deputati, 1891.

MacCormack, Sabine. "The Mind of the Missionary: José de Acosta on Accommodation and Extirpation, circa 1590." In *Religion in the Andes*, 249–79.

MacCormack, Sabine. *Religion in the Andes: Vision and Imagination in Early Colonial Peru*. Princeton: Princeton University Press, 1991.

Madajczak, Julia. "Nahuatl Kinship Terminology as Reflected in Colonial Written Sources from Central Mexico: A System of Classification." Ph.D. thesis, Warsaw 2014.

Malacarne, Giancarlo. *Le cacce del principe: L'ars venandi nella terra dei Gonzaga*. Modena: Il Bulino, ca. 1998.

Malacarne, Giancarlo. *Le feste del principe: Giochi, divertimenti, spettacoli a corte*. Modena: Il Bulino, 2002.

Malacarne, Giancarlo. *Il mito dei cavalli gonzagheschi: Alle origini del purosangue*. Verona: Promoprint, 1995.

Malacarne, Giancarlo. *I signori del cielo: La falconeria a Mantova al tempo dei Gonzaga*. Mantua: Artiglio Editore, 2003.

Malacarne, Giancarlo. *Sulla mensa del principe: Alimentazione e banchetti alla Corte dei Gonzaga*. Modena: Il Bulino, 2000.

Mann, Charles. *1493: Uncovering the New World Columbus Created*. New York: Knopf, 2011.

Marcus, Hannah. *Forbidden Knowledge: Medicine, Science, and Censorship in Early Modern Italy*. Chicago: University of Chicago Press, 2020.

Margócsy, Dániel. "The Camel's Head: Representing Unseen Animals in Sixteenth-Century Europe." In *Art and Science in the Early Modern Netherlands*, edited by Eric Jorink and Bart Ramakers, 63–85. Zwolle, The Netherlands: WBOOKS, 2011.

Margócsy, Dániel. "Horses, Curiosities, and the Culture of Collection at the Early Modern Germanic Courts." *Renaissance Quarterly* 74, no. 4 (2021): 1210–59.

Marino, John. *Becoming Neapolitan: Citizen Culture in Baroque Naples*. Baltimore: Johns Hopkins University Press, 2010.

Markey, Lia. "A Turkey in a Medici Tapestry." In *Imagining the Americas in Medici Florence*, 17–28. University Park: Pennsylvania State University Press, 2016.

Marroquín Arredondo, Jaime. "The Method of Francisco Hernández: Early Modern Science and the Translation of Mesoamerica's Natural History." In Marroquín Arredondo and Bauer, *Translating Nature*, 45–69.

Marroquín Arredondo, Jaime, and Ralph Bauer, eds. *Translating Nature: Cross-Cultural Histories of Early Modern Science*. Philadelphia: University of Pennsylvania Press, 2019.

Martin, Janice Gunther. "Unburdening the Beasts: Equine Medicine and Expert Healers in Early Modern Castile." Ph.D. dissertation, University of Notre Dame, 2018.

Martínez, María Elena. "The Black Blood of New Spain: *Limpieza de Sangre*, Racial Violence, and Gendered Power in Early Colonial Mexico." *William and Mary Quarterly*, 3rd series, 61, no. 3 (July 2004): 479–520.

Martínez, María Elena. *Genealogical Fictions: Limpieza de Sangre, Religion, and Gender in Colonial Mexico*. Stanford: Stanford University Press, 2008.

Maryks, Robert Aleksander. *The Jesuit Order as a Synagogue of Jews: Jesuits of Jewish Ancestry and Purity-of-Blood Laws in the Early Society of Jesus.* Leiden: Brill, 2010.

Mason, Peter. *Before Disenchantment: Images of Exotic Animals and Plants in the Early Modern World.* London: Reaktion, 2009.

Mateos, P. Francisco. *Obras del P. Jose de Acosta.* Madrid: Ediciones Atlas, 1954.

Matos Moctezuma, Eduardo, and Felipe Solís Olguín, eds. *Aztecs.* London: London Academy of Arts, 2002.

Matthews, Sarah Kathryn. "Matter over Mind: Pietro D'Abano (c. 1316) and the Science of Physiognomy." Ph.D. dissertation, Graduate College of the University of Iowa, May 2015.

Mayer, Thomas F. *The Roman Inquisition: On the Stage of Italy, c. 1590–1640.* Philadelphia: University of Pennsylvania Press, 2014.

McAvoy, Liz Herbert, Patricia Skinner, and Theresa Tyers. "Strange Fruits: Grafting, Foreigners, and the Garden Imaginary in Northern France and Germany, 1250–1350." *Speculum* 94, no. 2 (April 2019): 467–95.

McNeill, John Robert. *Mosquito Empires: Ecology and War in the Greater Caribbean, 1620–1914.* New York: Cambridge University Press, 2010.

Meli, Bertoloni. *Marcello Malpighi and Seventeenth-Century Anatomy.* Baltimore: Johns Hopkins University Press, 2011.

Melville, Elinor G. K. *A Plague of Sheep: Environmental Consequences of the Conquest of Mexico.* Cambridge: Cambridge University Press, 1994.

Mentaberre, Gregorio, Carlos Gutiérrez, Noé F. Rodríguez et al. "A Transversal Study on Antibodies against Selected Pathogens in Dromedary Camels in the Canary Islands, Spain." *Veterinary Microbiology* 167, no. 3–4 (December 2013): 468–73.

Meyerson, Mark D. *The Muslims of Valencia in the Age of Fernando and Isabel: Between Coexistence and Crusade.* Berkeley: University of California Press, 1991.

Miramon, Charles de. "Noble Dogs, Noble Blood: The Invention of the Concept of Race in the Late Middle Ages." In Eliav-Feldon, Isaac, and Ziegler, *Origins of Racism in the West*, 200–216.

Miranda, Giovanni. *Osservationi della lingua castigliana.* Vinezia: Gabriel Giolito de Ferrari, 1566.

Montellano, Bernardo Ortiz de. "Las hierbas de Tláloc," *Estudios de cultura náhuatl* 14 (1980): 287–314.

Montes de Oca Vega, Mercedes. "Los difrasismos: Un rasgo del lenguaje ritual." *Estudios de cultura náhuatl* 39 (2008): 225–38.

Moore, Deirdre. "The Heart of Red: Cochineal in Colonial Mexico and India." Ph.D. dissertation, Harvard University Graduate School of Arts and Sciences, 2021.

Moore, John E. "Prints, Salami, and Cheese: Savoring the Roman Festival of the Chinea." *Art Bulletin* 77, no. 4 (1995): 584–608.

Moore, Karl, and Susan Reid. "The Birth of Brand: 4000 Years of Branding History." *Business History* 50, no. 4 (2008): 419–32.

Moore, Katherine. "Early Domesticated Camelids in the Andes." In Capriles and Tripcevich, *Archaeology of Andean Pastoralism*, 17–38.

Morehart, Christopher. "Aztec Agricultural Strategies: Intensification, Landesque Capital, and the Sociopolitics of Production." In Nichols and Rodríguez-Alegría, *Oxford Handbook of the Aztecs*, 263–79.

Morgado García, Arturo. "The Animal World Vision in 17th Century Spain: The Covarrubias Bestiary." *Cuadernos de historia moderna* 36 (2011): 67–88.

Morison, Samuel Eliot, ed. and trans. *Journals and Other Documents on the Life and Voyages of Christopher Columbus*. Norwalk, CT: Easton Press, 1993.

Morselli, Raffaella. "Collezionismo, mercato e allestimento: I 'naturalia' di Casa Gonzaga tra modernità e contemporaneità." In *Nuove Alleanze: Diritto ed Economia della Cultura e dell'Arte* (supplement in "Arte e Critica" 80/81) (2015).

Mukherjee, Siddhartha. *The Gene: An Intimate History*. London: Vintage, 2017.

Mundy, Barbara E. "Ecology and Leadership: Pantitlan and Other Erratic Phenomena." In Peterson and Terraciano, *The Florentine Codex*, 125–38.

Navarro Antolín, F., and J. Solís de los Santos. "La epístola latina en verso de Francisco Hernández a Benito Arias Montano (Madrid, Biblioteca del Ministerio de Hacienda, MS FA 931)." *Myrtia: Revista de filología clásica* 29 (2014): 201–45.

Nebrija, Antonio de. *Vocabulario español-latino*. Salamanca, 1495.

Nelson, Hilda. "Antoine de Pluvinel, Classical Horseman and Humanist." *French Review* 58, no. 4 (March 1985): 514–23.

Nelson, William Max. "Making Men: Enlightenment Ideas of Racial Engineering." *American Historical Review* 115, no. 5 (December 2010): 1364–94.

Nemser, Daniel. *Infrastructures of Race: Concentration and Biopolitics in Colonial Mexico*. Austin: University of Texas Press, 2017.

Nemser, Daniel Joseph. "Toward a Genealogy of *Mestizaje:* Rethinking Race in Colonial Mexico." Ph.D. dissertation in Hispanic Languages and Literatures, University of California, Berkeley, 2011.

Nichols, Deborah L., and Enrique Rodríguez-Alegría, eds. *The Oxford Handbook of the Aztecs*. Oxford: Oxford University Press, 2016.

Nirenberg, David. *Communities of Violence: Persecution of Minorities in the Middle Ages*. Princeton: Princeton University Press, 1996.

Nirenberg, David. "Was There Race before Modernity? The Example of 'Jewish' Blood in Late Medieval Spain." In Eliav-Feldon, Isaac, and Ziegler, *Origins of Racism in the West*, 71–87.

Norton, Marcy. "The Chicken or the *Iegue*: Human-Animal Relationships and the Columbian Exchange." *American Historical Review* 120, no. 1 (February 2015): 28–60.

Norton, Marcy. "Going to the Birds: Animals as Things and Beings in Early Modernity." In *Early Modern Things: Objects and Their Histories, 1500–1800*, 2nd ed., 51–81. New York: Routledge, 2021.

Norton, Marcy. "The Quetzal Takes Flight: Microhistory, Mesoamerican Knowledge, and Early Modern Natural History." In Marroquín Arredondo and Bauer, *Translating Nature*, 119–47.

Norton, Marcy. *Sacred Gifts, Profane Pleasures: A History of Tobacco and Chocolate in the Atlantic World*. Ithaca: Cornell University Press, 2008.

Norton, Marcy. "Subaltern Technologies and Early Modernity in the Atlantic World." *Colonial Latin American Review* 26, no. 1 (2017): 18–38.

Norton, Marcy. *The Tame and the Wild*. Cambridge, MA: Harvard University Press, forthcoming.

Nosari, Galeazzo, and Franco Canova. *Il Palio nel Rinascimento: I Cavalli di Razza dei Gonzaga nell'età di Francesco II Gonzaga, 1484–1519*. Reggiolo (Reggio Emilia): E. Lui, 2003.

Nyborg, Helmuth. "Race as Social Construct." *Psych* 1, no. 1 (2019): 139–65.

Ogilvie, Brian W. *The Science of Describing Natural History in Renaissance Europe*. Chicago: University of Chicago Press, 2006.

Olko, Justyna. "In Ocelotl, in Tequani: The Mesoamerican Jaguar as Described in the Nahuatl Text of the Florentine Codex." In *Birthday Beasts' Book: Where Human Roads Cross Animal Trails, Cultural Studies in Honour of Jerzy Axer*, edited by Katarzyna Marciniak, 301–8. Warsaw: Institute for Interdisciplinary Studies "Artes Liberales," 2011.

O'Malley, John. *The First Jesuits*. Cambridge, MA: Harvard University Press, 1993.

Osorio, Alejandra B. *Inventing Lima: Baroque Modernity in Peru's South Sea Metropolis*. New York: Palgrave Macmillan, 2008.

O'Toole, Rachel Sarah. *Bound Lives: Africans, Indians, and the Making of Race in Colonial Peru*. Pittsburgh: University of Pittsburgh Press, 2012.

Oviedo, Gonzalo Fernández de. *Historia general de las Indias*. Seville: Emprenta de Iuan Cromberger, 1535.

Padberg, John W., Martin D. O'Keefe, and John L. McCarthy, eds. *For Matters of Greater Moment: The First Thirty Jesuit General Congregations: A Brief History and A Translation of the Decrees*. Saint Louis, MO: Institute of Jesuit Sources, 1994.

Paden, Jeremy. "The Iguana and the Barrel of Mud: Memory, Natural History, and Hermeneutics in Oviedo's *Sumario de la natural historia de las Indias*." *Colonial Latin American Review* 16, no. 2 (2007): 203–26.

Pagden, Anthony. *The Fall of Natural Man: The American Indian and the Origins of Comparative Ethnology*. Cambridge: Cambridge University Press, 1986.

Pagden, Anthony. "The Peopling of the New World: Ethos, Race, and Empire in the Early-Modern World." In Eliav-Feldon, Isaac, and Ziegler, *Origins of Racism in the West*, 292–312.

Palumbo, Margherita. *La città del sole: Bibliografia delle edizioni (1623–2002)*. Pisa: Istituti Editoriali e Poligrafici Internazionali, 2004.

Paolella, A. "L'autore delle illustrazioni delle Fisiognomiche di Della Porta e la ritrattistica: Esperienze filologiche." In *La 'mirabile' natura: Magica e scienza in Giovan Battista Della Porta (1615–2015)*, Atti del convegno internazionale Napoli-Vico Equense, 13–17. Pisa: Fabrizio Serra editore, 2016.

Paolella, A. "Introduzione." In *Della Fisionomia dell'uomo*, by Giovanni Battista Della Porta, libri sei, vol. 2, edited by Alfonso Paolella. Edizione nazionale delle opere di Giovan Battista della Porta. Napoli: Edizioni Scientifiche Italiane, 2011.

Pardo-Tomás, José. "Viajes de ida o de vuelta: La circulación de la obra de Francisco Hernández en México, 1576–1672." In Caddeu and Guardo, *Il tesoro messicano*.

Parker, Geoffrey. *The Army of Flanders and the Spanish Road, 1567–1659*. 1972. 2nd ed. Cambridge: Cambridge University Press, 2004.

Parker, Geoffrey. *Emperor: A New Life of Charles V*. New Haven: Yale University Press, 2019.

Parker, Geoffrey. *The Grand Strategy of Philip II*. New Haven: Yale University Press, 2000.

Parker, Geoffrey. *Imprudent King: A New Life of Philip II*. New Haven: Yale University Press, 2014.

Parker, Heidi G., Dayna L. Dreger, Maud Rimbault et al. "Genomic Analyses Reveal the Influence of Geographic Origin, Migration, and Hybridization on Modern Dog Breed Development." *Cell Reports* 19, no. 4 (April 2017): 698–708.

Pearce, S. J. "The Inquisitor and the Moseret: The Invention of Race in the European Middle Ages and the New English Colonialism in Jewish Historiography." *Medieval Encounters* 26, no. 2 (2020): 145–90.

Perri, Angela, Chris Widga, Dennis Lawler et al. "New Evidence of the Earliest Domestic Dogs in the Americas." *American Antiquity* 84, no. 1 (January 2019): 68–87.

Peters, Edward. *Inquisition.* New York: Free Press, 1988.

Peterson, Jeanette Favrot, and Kevin Terraciano, eds. *The Florentine Codex: An Encyclopedia of the Nahua World in Sixteenth-Century Mexico.* Austin: University of Texas Press, 2019.

Philips, Elena. "Inka Textile Traditions." In Shimada, *Inka Empire,* 197–214.

Phipps, Elena, and Lucy Commoner. "Investigation of a Colonial Latin American Textile." *Textile Society of America Symposium Proceedings* 358 (2006): 485–93.

Piccari, Paolo. *Giovan Battista Della Porta: Il filosofo, il retore, lo scienziato.* Milan: Franco Angeli, 2007.

Pigière, Fabienne, and Denis Henrotay. "Camels in the Northern Provinces of the Roman Empire." *Journal of Archeological Science* 39 (2012): 1531–39.

Pino Díaz, Fermín del. "Las historias naturales y morales de las Indias como género: Orden y gestación literaria de la obra de Acosta," *Histórica* (Lima) 24, no. 2 (2000): 295–326.

Pitt-Rivers, Julian. "On the Word Caste." In *The Translation of Culture: Essays to E. E. Evans Pritchard,* edited by T. O. Beidelman, 231–56. London: Routledge, 1971.

Plato. *Complete Works.* Edited by John M. Cooper and D. S. Hutchinson. Indianapolis: Hackett, 1997.

Pliny the Elder. *Historia Natural.* Translated by Francisco Hernández. Biblioteca Nacional de España, Madrid, Mss/2868, XXI.

Pliny the Elder. *The Natural History.* Translated by John Bostock. London: Taylor and Francis, 1855.

Poma de Ayala, Felipe Guaman. *Nueva corónica y buen gobierno* (ca. 1615), Royal Danish Library, GKS 2232 kvart.

Poole, Stafford. "Criollos and Criollismo." In *Encyclopedia of Mexico: History, Society, and Culture,* edited by Michael S. Werner. Chicago: Fitzroy Dearborn, 1997.

Porta, Giovanni Battista Della. *De Ea naturalis physiognomoniae parte quae ad manuum lineas spectat,* libri duo. Edited by Oreste Trabucco. Edizione nazionale delle opera di Giovan Battista della Porta, vol. 9. Napoli: Edizioni Scientifiche Italiane, 2003.

Porta, Giovanni Battista Della. *De humana physiognomonia.* Neapoli: apud Iosephum Cacchium, 1586.

Porta, Giambattista Della. *Dei miracoli et maravigliasi effetti dalla natura prodotti.* Italian translation of the 1558 edition of *Magia naturalis.* Venice: Ludovico Avanzi, 1560.

Porta, Giambattista della. *Della celeste fisonomia [. . .].* Padova: Pietro Paolo Toazzi, 1623.

Porta, Giovanni Battista Della. *Della fisonomia dell'huomo di Gio. Battista della Porta napolitano,* libri quattro. Tradotti da latino in lingua volgare per Giovanni Di Rosa professore di l'una e l'altra letta. Con l'aggiunta di cento ritratti di rame di più di quelli della prima impressione. Napoli: Tarquinino Longo, 1598.

Porta, Giovanni Battista Della. *Della Fisionomia dell'uomo,* libri sei. Vol. 2. Edited by Alfonso Paolella. Edizione nazionale delle opere di Giovan Battista della Porta. Napoli: Edizioni Scientifiche Italiane, 2011.

Porta, Giovanni Battista Della. *Magiae naturalis.* Ludg: Batavorum, 1651.

Porta, Giovanni Battista Della. *Villae.* Frankfurt, 1592.

Porta, Io. Baptista. *Magiae Naturalis, sive de miraculis rerum naturalium,* liber 4. Antverpiae: Ex officina Christophori Plantini, 1576.

Porta, Io. Baptista. *Phytognomonica*. Naples: apud Horatium Salvianum, 1588.

Portae, Io. Baptistae Neapolitani. *Svae Villae. Pomarivm*. Neapoli: Salvianus & Caesare, 1583.

Porter, Martin. *Windows of the Soul: The Art of Physiognomy in European Culture 1470–1760*. Oxford: Clarendon Press, 2005.

Pringle, Heather. "Secrets of the Alpaca Mummies: Did the Ancient Inca Make the Finest Woolen Cloth the World Has Ever Known." Photographs by Grant Delin. *Discover* 22, no. 4 (April 2001).

Prior, Charles. *The Royal Studs of the Sixteenth and Seventeenth Century*. London: Horse and Hound, 1935.

Prosperi, Adriano. "'Otras indias': Missionari della Controriforma tra contadini e selvaggi." In *Scienze, credenze occulte, livelli di cultura, Atti del Convegno internazionale di studi*, Florence, 26–30 giugno, 1980.

Quiñones Keber, Eloise. "Xolotl: Dogs, Death, and Deities in Aztec Myth." *Latin American Indian Literatures Journal* 7, no. 2 (1991): 229–39.

Raber, Karen. *Animal Bodies, Renaissance Culture*. Philadelphia: University of Pennsylvania Press, 2013.

Raby, Dominique. "The Cave-Dwellers' Treasure: Folktales, Morality, and Gender in a Nahua Community in Mexico." *Journal of American Folklore* (Boston, MA) 120 (Fall 2007): 401–44.

RaceB4Race. Arizona Center for Medieval and Renaissance Studies. Accessed January 9, 2022. https://acmrs.asu.edu/RaceB4Race.

Radlo-Dzur, Alanna, Mackenzie Cooley, Emily Kaplan et al. "The Tira of Don Martín: A Living Nahua Chronicle." *Latin American and Latinx Visual Culture* 3, no. 3 (2021): 7–37.

Rajabzadeh, Shokoofeh. "The Depoliticized Saracen and Muslim Erasure." *Literature Compass* 16, no. 9–10 (2019).

Ramos Maldonado, Sandra I. "Tradición pliniana en la Andalucía del siglo XVI: A propósito de la labor filológica del Doctor Francisco Hernández." In *Las raíces clásicas de Andalucía*, edited by M. Rodríguez-Pantoja, 883–91. Córdoba: Actas del IV congreso Andaluz de Estudios Clásicos, 2006.

Rao, Ida Giovanna. "On the Reception of the Florentine Codex: The First Italian Translation." In Peterson and Terraciano, *The Florentine Codex*, 37–44.

Rappaport, Joanne. "'Así lo paresçe por su aspeto': Physiognomy and the Construction of Difference in Colonial Bogotá." *Hispanic American Historical Review* 91, no. 4 (November 2011): 601–31.

Rappaport, Joanne. *The Disappearing Mestizo: Configuring Difference in the Colonial New Kingdom of Granada*. Durham: Duke University Press, 2014.

Recchi, Nardo Antonio. *Rerum medicarum Novae Hispaniae thesaurus seu plantarum animalium mineralium Mexicanorum*. Rome: Typographeio Vitalis Mascardi, 1651.

Refini, Eugenio. *The Vernacular Aristotle: Translation as Reception in Medieval and Renaissance Italy*. New York: Cambridge University Press, 2020.

Reich, David. *Who We Are and How We Got Here: Ancient DNA and the New Science of the Human Past*. New York: Vintage Books, 2018.

Reiss, Timothy J. "Structure and Mind in Two Seventeenth-Century Utopias: Campanella and Bacon." *Yale French Studies* 49 (1973): 82–95.

Relaciones histórico-geográficas de la gobernación de Yucatán. Fuentes para el estudio de la cultura maya. México: Universidad Nacional Autónoma de México, 1983.

Renton, Kathryn. "Breeding Techniques and Court Influence: Charting a 'Decline' of the Spanish Horse in the Early Modern Period." *Court Historian* 24, no. 3 (2019): 221–34.

Renton, Kathryn. "Defining 'Race' in the Spanish Horse: The Breeding Program of King Philip II." In Guest and Mattfield, *Horse Breeds and Human Society*, 13–26.

Renton, Kathryn. "The Knight with No Horse: Defining Nobility in Late Medieval and Early Modern Castile." *Sixteenth Century Journal* 50, no. 1 (2020): 109–28.

Renton, Kathryn. "A Social and Environmental History of the Horse in Spain and Spanish America, 1492–1600." Ph.D. dissertation, University of California, Los Angeles, 2019.

Renzi, Silvia De. "Family Resemblance in the Old Regime." In Hopwood, Flemming, and Kassell, *Reproduction*, 241–52.

Renzi, Silvia De. "Storia naturale ed erudizione nella prima età moderna: Johannes Faber (1574–1629) medico linceo." Università degli Studi di Bari, Tesi di Dottorato in Storia Scienza, 1992–93.

Restall, Matthew. "The Renaissance World from the West: Spanish America and the "Real" Renaissance." In *A Companion to the Worlds of the Renaissance*, edited by Guido Ruggiero, 70–88. Oxford: Blackwell Publishing, 2007.

Restall, Matthew. *Seven Myths of Spanish Conquest*. Oxford: Oxford University Press, 2004.

Restall, Matthew. *When Moctezuma Met Cortés: The True Story of the Meeting That Changed History*. New York: HarperCollins, 2018.

Rey Bueno, Mar, and María Esther Alegre Pérez. "Renovación de la terapéutica real: los destiladores de su majestad, maestros simplicistas y médicos herbolarios de Felipe II." *Asclepio* 53, no. 1 (2001): 27–56.

Ricci, Saverio. *Il sommo inquisitore, Giulio Antonio Santori tra autobiografía e storia (1532–1602)*. Roma: Salerno, 2002.

Ritvo, Harriet. *The Animal Estate: The English and Other Creatures in the Victorian Age.* Cambridge, MA: Harvard University Press, 1987.

Ritvo, Harriet. *Noble Cows and Hybrid Zebras: Essays on Animals and History*. Charlottesville: University of Virginia Press, 2010.

Ritvo, Harriet. *The Platypus and the Mermaid, and Other Figments of the Classifying Imagination*. Cambridge, MA: Harvard University Press, 1997.

Ritvo, Harriet. "Race, Breed, and the Myths of Origin: Chillingham Cattle as Ancient Britons." *Representations* 39 (Summer 1992): 1–22.

Romano, Giacinto. *Cronaca del soggiorno di Carlo V in Italia (dal 26 luglio 1529 al 25 aprile 1530)* [da Luigi Gonzaga]. Milano: Urico Hoepli, 1892.

Ropa, Anastasija, and Timothy Dawson, eds. *The Horse in Premodern European Culture.* Studies in Medieval and Early Modern Culture. Boston: Walter de Gruyter, 2019.

Rosenberg, Gabriel N. *The 4-H Harvest: Sexuality and the State in Rural America*. Philadelphia: University of Pennsylvania Press, 2015.

Rossi, Paolo. *Francesco Bacone: Dalla magia alla scienza*. Turin: Einaudi, 1957.

Rubial García, Antonio. "Icons of Devotion." In *Local Religion in Colonial Mexico*, edited by Martin Nesvig, 37–61. Albuquerque: University of New Mexico, 2006.

Ruggiero, Guido. *Binding Passions: Tales of Magic, Marriage, and Power*. Oxford: Oxford University Press, 1993.

Rusnock, Andrea. "Biopolitics and the Invention of Population." In Hopwood, Flemming, and Kassell, *Reproduction*, 333–45.

Russell, Nerissa. *Social Zooarchaeology: Humans and Animals in Prehistory.* Cambridge: Cambridge University Press, 2012.

Russell, Nicholas. *Like Engend'ring Like: Heredity and Animal Breeding in Early Modern England.* Cambridge: Cambridge University Press, 1986.

Russo, Alessandra. "The Curator's Eyes: Sebastiano Biavati, Custodian of a Heterogeneous Artistic World." In *The Significance of Small Things: Essays in Honour of Diana Fane,* edited by Luisa Elena Alcalá and Ken Moser, 150–58. Madrid: Ediciones El Viso, 2018.

Sahagún, Fray Bernardino de. *General History of the Things of New Spain: The Florentine Codex.* 12 books. Translated and edited by C. E. Dibble and A. J. O. Anderson. Santa Fe, NM: The School of American Research/University of Utah, 2012.

Sahagún, Fray Bernardino de. *Primeros Memoriales: Paleography of the Nahuatl Text and English Translation.* Translated by Thelma Sullivan. 2 books. Norman: University of Oklahoma Press, 1997.

Sale, Kirkpatrick. *The Conquest of Paradise: Christopher Columbus and the Columbian Legacy.* New York: Alfred A. Knopf, 1990.

Salomon, Frank. "Inkas through Texts: The Primary Sources." In Shimada, *Inka Empire,* 23–38.

Sánchez, Antonio, and Henrique Leitão, eds. "Artisanal Culture in Early Modern Iberian and Atlantic Worlds." Special issue, *Centaurus* 60, no. 3 (2018): 135–40.

Sandstrom, Alan. *Corn Is Our Blood: Culture and Ethnic Identity in a Contemporary Aztec Indian Village.* Norman: University of Oklahoma Press, 1991.

Sapolsky, Robert. *Behave: The Biology of Humans at Our Best and Worst.* London: Vintage, 2018.

Schiebinger, Londa. "Why Mammals Are Called Mammals: Gender Politics in Eighteenth-Century Natural History." *American Historical Review* 98, no. 2 (April 1993): 382–411.

Schiesari, Juliana. *Beasts and Beauties: Animals, Gender, and Domestication in the Italian Renaissance.* Toronto: University of Toronto Press, 2010.

Schiesari, Juliana. *Polymorphous Domesticities: Pets, Bodies, and Desire in Four Modern Writers.* Berkeley: University of California Press, 2012.

Schmitt, Charles B. *Aristotle and the Renaissance.* Cambridge, MA: Harvard University Press for Oberlin College, 1983.

Schwartz, Stuart. *All Can Be Saved: Religious Tolerance and Salvation in the Iberian Atlantic World.* New Haven: Yale University Press, 2008.

Selwyn, Jennifer D. *A Paradise Inhabited by Devils: The Jesuits' Civilizing Mission in Early Modern Naples.* Burlington, VT: Ashgate, 2004.

Serjeantson, Richard. "Francis Bacon and the 'Interpretation of Nature' in the Late Renaissance." *Isis* 104, no. 4 (December 2014): 681–705.

Serrai, Alfredo, Maria Cochetti, and Marco Menato. *Conrad Gesner.* Roma: Bulzoni, 1991.

Sforzino, Francesco. *Tre libri de gli uccelli, con un trattato de' cani da caccia.* Vicenza: Il Megietti, 1568.

Shakespeare, William. *Othello.* The Annotated Shakespeare. New Haven: Yale University Press, 2005.

Shepherd, Gregory J. *José de Acosta's "De Procuranda Indorum salute": A Call for Evangelical Reforms in Colonial Peru.* Currents in Comparative Romance Languages and Literatures. New York: Peter Lang, 2014.

Shimada, Izumi, ed. *The Inka Empire: A Multidisciplinary Approach*. Austin: University of Texas Press, 2015.

Shimada, Melody, and Izumi Shimada. "Prehistoric Llama Breeding and Herding on the North Coast of Peru." *American Antiquity* 50, no. 1 (1985): 3–26.

Shopov, Aleksandar. "Between the Pen and the Fields: Books on Farming, Changing Land Regimes, and Urban Agriculture in the Ottoman and Eastern Mediterranean, ca. 1500–1700." Ph.D. thesis, Harvard University, 2016.

Shopov, Aleksandar. "Grafting in Sixteenth-Century Mamluk and Ottoman Agriculture and Literature." In *History and Society During the Mamluk Period (1250–1517)*, edited by Bethany J. Walker and Abdelkader Al Ghouz, 381–406. Göttingen: Bonne University Press, 2021.

Shryock, Andrew, Thomas R. Trautmann, and Clive Gamble. "Imagining the Human in Deep Time." In *Deep History: The Architecture of Past and Present*, by Andrew Shryock and Daniel Lord Small, 21–52. Berkeley: University of California Press, 2011.

Signorini, Rodolfo. "A Dog Named Rubino." *Journal of the Warburg and Courtauld Institutes* 41 (1978): 317–20.

Sigwart, Julia D., Mark D. Sutton, and K. D. Bennett. "How Big Is a Genus? Towards a Nomothetic Systematics." *Zoological Journal of the Linnean Society* 183, no. 2 (2018): 237–52.

Silver, Larry. "World of Wonders: Exotic Animals in European Imagery, 1515–1650." In *Animals and Early Modern Identity*, edited by Pia F. Cuneo, 304–5. London: Routledge, 2017.

Smith, Justin. *Embodiment: A History*. New York: Oxford University Press, 2017.

Smith, Justin. "Introduction." In Justin Smith, *Problem of Animal Generation*, 1–21.

Smith, Justin. *Nature, Human Nature, and Human Difference: Race in Early Modern Philosophy*. Princeton: Princeton University Press, 2015.

Smith, Justin. *The Philosopher: A History in Six Types*. Princeton: Princeton University Press, 2016.

Smith, Justin, ed. *The Problem of Animal Generation in Early Modern Philosophy*. New York: Cambridge University Press, 2006.

Smith, Pamela. *The Body of the Artisan: Art and Experience in the Scientific Revolution*. Chicago: University of Chicago Press, 2006.

Solazzo, Caroline. "Characterizing Historical Textiles and Clothing with Proteomics." In "Estudos sobre têxteis históricos/Studies in Historical Textiles," edited by A. Serrano, M. J. Ferreira, and E. C. de Groot, 1–18. Special issue, *Conservar Património* 31 (May 2019): 1–18.

Somolino d'Ardois, Germán. *Vida y obra de Francisco Hernández*. México: Universidad Nacional Autónoma de México, 1960.

Sousa, Lisa. *The Woman Who Turned into a Jaguar, and Other Narratives of Native Women in Archives of Colonial Mexico*. Stanford: Stanford University Press, 2017.

Spary, Emma. "Political, Natural and Bodily Economies." In *Cultures of Natural History*, edited by N. Jardine, J. A. Secord, and E. C. Spary, 178–96. Cambridge: Cambridge University Press, 1996.

Spitzer, Leo. "Ratio > Race," *American Journal of Philology* 62 (1941): 129–43.

Stephens, Thomas M. *Dictionary of Latin American Racial and Ethnic Terminology*. Gainesville: University Press of Florida, 1999.

Stern, Alexandra Minna. *Eugenic Nation: Faults and Frontiers of Better Breeding in Modern America*. Oakland: University of California Press, 2015.

Stradanus, Joannes (Jan van der Straet). *Equile Ioannis Austriaci*. Antwerp: Phillipe Galle, ca. 1578. Also known as *Speculum equorum* (Mirror of Horses).

Subialka, Michael. "Transforming Plato: 'La città del sole,' the 'Republic,' and Socrates as natural philosopher." *Bruniana & Campanelliana* 17, no. 2 (2011): 417–33.

Tarrant, Neil. "Giambattista della Porta and the Roman Inquisition: Censorship and the Definition of Nature's Limits in Sixteenth-Century Italy." *British Journal for the History of Science* 46, no. 4 (December 2013): 601–25.

Terraciano, Kevin. "Introduction: An Encyclopedia of Nahua Culture." In Peterson and Terraciano, *The Florentine Codex*, 1–20.

Terraciano, Kevin. "Reading Between the Lines of Book Twelve." In Peterson and Terraciano, *The Florentine Codex*, 45–62.

Terrall, Mary. *Catching Nature in the Act: Reáumur & the Practice of Natural History in the Eighteenth Century*. Chicago: University of Chicago Press, 2014.

Thatcher, Oliver J., ed. *The Library of Original Sources*. Vol. 5: *9th to 16th Centuries*. Milwaukee: University Research Extension, 1907.

Thomson, Rosemarie Garland. "The Beauty and the Freak." *Disability, Art, and Culture (Part II)* 37, no. 3 (Summer 1998): 459–74.

Tobey, Elizabeth. "Legacy of Grisone." In Edwards, Enenkel, and Graham, *The Horse as Cultural Icon*, 143–74.

Tobey, Elizabeth. "The Palio Horse in Italy." In *The Culture of the Horse: Status, Discipline, and Identity in the Early Modern World*, edited by Karen Raber and Treva J. Tucker, 63–90. New York: Palgrave Macmillan, 2005.

Tobey, Elizabeth. "The Palio in Italian Renaissance Art, Thought, and Culture." Ph.D. dissertation, University of Maryland, 2005.

Tonni, Andrea. "The Renaissance Studs of the Gonzagas of Mantua." In Edwards, Enenkel, and Graham, *The Horse as Cultural Icon*, 261–78.

Topic, Theresa Lange, Thomas H. McGreevy, and John R. Topic. "A Comment on the Breeding and Herding of Llamas and Alpacas on the North Coast of Peru." *American Antiquity* 52, no. 4 (1987): 832–35.

Torquemada, Juan de. *Segunda parte de los veinte i un libros rituales i monarchia indiana*. Madrid: Nicolas Rodrigo Franco, 1723.

Townsend, Camilla. *Annals of Native America: How the Nahuas of Colonial Mexico Kept Their History Alive*. Oxford: Oxford University Press, 2017.

Townsend, Camilla. *Fifth Sun: A New History of the Aztecs*. Oxford: Oxford University Press, 2019.

Trexler, Richard C. *Sex and Conquest: Gendered Violence, Political Order and the European Conquest of the Americas*. Ithaca: Cornell University Press, 1995.

Urton, Gary. "The State of Strings: Khipu Administration in the Inka Empire." In Shimada, *Inka Empire*, 149–64.

Valadez, R., B. Rodríguez, L. Manzanilla, and S. Tejeda. "Dog-Wolf Hybrid Biotype Reconstruction from the Archaeological City of Teotihuacan in Prehispanic Central Mexico." In *Dogs and People in Social, Working, Economic or Symbolic Interaction*, edited by L. M. Snyder and E. A. Moore, 309–29. Oxford: Oxbow Books, 2002.

Valente, Michaela. "Della Porta e l'Inquisizione: Nuovi documenti dell'Archivio del Sant'Uffizio." *Bruniana & Campanelliana* 5, no. 2 (1999): 415–34.

Valesey, Christopher. "Managing the Herd: Nahuas, Animals, and Colonialism in Sixteenth-Century New Spain." Ph.D. dissertation, Pennsylvania State University, 2019.

Vallières, Claudine. "Camelid Pastoralism at Ancient Tiwanaku: Urban Provisioning in the Highlands of Bolivia." In Capriles and Tripcevich, *Archaeology of Andean Pastoralism*, 67–86.

Van Damme, Ilja. "From a 'Knowledgeable' Salesman towards a 'Recognizable Product'?: Questioning Branding Strategies before Industrialization (Antwerp, Seventeenth to Nineteenth Centuries)." In *Concepts of Value in European Material Culture, 1500–1900*, edited by Dries Lyna and Bert De Munck, 75–102. Farnham, Surrey: Ashgate, 2015.

Van der Lugt, Maaike. "La peau noire dans la science médiévale." *Micrologus* 13 (2005): 444–47.

van der Sman, Gert Jan. "Print Publishing in Venice in the Second Half of the Sixteenth Century." *Print Quarterly* 17, no. 3 (September 2000): 235–47.

van Deusen, Nancy E. *Global Indios: The Indigenous Struggle for Justice in Sixteenth-Century Spain*. Durham: Duke University Press, 2015.

van Deusen, Nancy E. "Seeing Indios in Sixteenth-Century Castile." *William and Mary Quarterly* 69, no. 2 (2012): 205–34.

Varey, Simon, ed. *The Mexican Treasury: The Writings of Dr. Francisco Hernández*. Translated by Rafael Chabrá, Cynthia L. Chamberlin, and Simon Varey. Stanford: Stanford University Press, 2002.

Varey, Simon. *Sustaining Literature: Essays on Literature, History, and Culture, 1500–1800: Commemorating the Life and Work of Simon Varey*. Lewisburg: Bucknell University Press, 2007.

Varner, John Grier. *Dogs of the Conquest*. Norman: University of Oklahoma Press, 1983.

Vasari, Giorgio. "Giulio Romano." In *Le vite de' più eccellenti architetti, pittori, et scultori italiani, da Cimabue insino a' tempi nostri*. Firenze: Giunti, 1568.

Vasari, Giorgio. *The Lives of the Artists*. Translated by Julia Conaway Bondanella and Peter Bondanella. Oxford: Oxford University Press, 1998.

Vegetti, Mario, and Paolo Pissavino. *I Decembrio e la tradizione della Repubblica di Platone tra medioevo e umanesimo*. Napoli: Bibliopolis, 2005.

Velasco-Villa, Andres. "The History of Rabies in the Western Hemisphere." *Antiviral Research* 146 (October 2017): 221–32.

Vilar, Miguel. "Genographic Project DNA Results Reveal Details of Puerto Rican History." *National Geographic: Voices*. Posted July 25, 2014. http://blog.nationalgeographic.org2014/07/25/genographic-project-dna-results-reveal-details-of-puerto-rican-history/.

Vilar, M. G., C. Melendez, A. B. Sanders et al. "Genetic Diversity in Puerto Rico and Its Implications for the Peopling of the Island and the West Indies." *American Journal of Physical Anthropology* 155 (2014): 352–68.

Vinson, Ben III. *Before Mestizaje: The Frontiers of Race and Caste in Colonial Mexico*. New York: Cambridge University Press, 2017.

Virgil. *Eclogues, Georgics, Aeneid*. Translated by H. R. Fairclough. Loeb Classical Library, vols. 63 and 64. Cambridge, MA: Harvard University Press, 1916.

Wallen, Martin. *Whose Dog Are You? The Technology of Dog Breeds and the Aesthetics of Modern Human-Canine Relations*. East Lansing: Michigan State University Press, 2017.

Wang, Guo-Dong et al. "Out of Southern East Asia: The Natural History of Domestic Dogs across the World." *Cell Research* 26 (2016): 21–33.

Warsh, Molly A. *American Baroque: Pearls and the Nature of Empire, 1492–1700.* Omohundro Institute of Early American History and Culture. Chapel Hill: University of North Carolina Press, 2018.

Wheeler, Jane C. "Evolution and Present Situation of the South American Camelidae." *Biological Journal of the Linnaean Society* 54, no. 3 (1995): 271–95.

Wheeler, Jane C. "Llamas and Alpacas: Pre-conquest Breeds and Post-conquest Hybrids." *Journal of Archaeological Science* 22, no. 6 (November 1995): 833–40.

Wheeler, Jane C., A. J. F. Russel, and H. F. Stanley. "A Measure of Loss: Prehispanic Llama and Alpaca Breeds." *Archivos de Zootecnia* 41, no. 154 (1992): 467–75.

Whittaker, Gordon. *Deciphering Aztec Hieroglyphs: A Guide to Nahuatl Writing.* London: Thames & Hudson, 2021.

Wiesner-Hanks, Merry E. *The Marvelous Hairy Girls: The Gonzales Sisters and Their Worlds.* New Haven: Yale University Press, 2009.

Wiesner-Hanks, Merry E. "The Wild and Hairy Gonzales Family." *Apparence(s): Histoire et culture du paraître* 5 (2014). https://doi.org/10.4000/apparences.1283.

Wilkins, John. *Species: A History of the Idea.* Berkeley: University of California Press, 2009.

Wilmot, Sarah. "Breeding Farm Animals and Humans." In Hopwood, Flemming, and Kassell, *Reproduction*, 397–414.

Wilson, Duncan A., and Masaki Tomonaga. "Exploring Attentional Bias towards Threatening Faces in Chimpanzees Using the Dot Probe Task," *PLoS One* 13, no. 11 (2018): e0207378.

Woods, Rebecca. *The Herds Shot Round the World: Native Breeds and the British Empire, 1800–1900.* Chapel Hill: University of North Carolina Press, 2017.

Yates, Frances A. *John Florio: The Life of an Italian in Shakespeare's England.* Cambridge: Cambridge University Press, 1934.

Ziegler, Joseph. "Physiognomy, Science, and Proto-racism, 1200–1500." In Eliav-Feldon, Isaac, and Ziegler, *Origins of Racism in the West*, 181–99.

Zilsel, Edgar. *The Social Origins of Modern Science.* Boston Studies in the Philosophy of Science. Dordrecht: Springer Netherlands, 2003.

Zirkle, Conway. "Animals Impregnated by the Wind." *Isis* 25, no. 1 (May 1936): 95–130.

Index